Figuring Violence

Figuring Violence

Affective Investments in Perpetual War

Rebecca A. Adelman

FORDHAM UNIVERSITY PRESS

New York 2019

Fordham University Press also publishes its books in a variety of electronic formats. Some content that appears in print may not be available in electronic books.

Visit us online at www.fordhampress.com.

Library of Congress Cataloging-in-Publication Data available online at https://catalog.loc.gov.

Printed in the United States of America

21 20 19 5 4 3 2 1

First edition

for my dad

CONTENTS

ON THE COVER IMAGE: "VERTIGO AT GUANTANAMO"

Just a few inches above my name on the cover of this book there is another that might be more or less familiar, depending on the hypothetical reader's area of expertise: Ammar Al-Baluchi.

A Pakistani national, Al-Baluchi was captured by the Pakistani Intelligence Bureau on April 29, 2003. Roughly two weeks later, on May 15, he was turned over to the CIA and dropped into its system of renditions and secret prisons.[1] He has been detained at Guantánamo Bay since September 4, 2006. The U.S. government classifies Al-Baluchi as a "high-value detainee" (HVD) and holds him at Camp 7, the highest-security area of Guantánamo. His defense team has argued against the logic of the HVD categorization in general and has also avowed that Al-Baluchi's imputed status is an obstacle to him receiving a fair trial in a military commission.[2] Joint Task Force Guantanamo has deemed him "a HIGH risk, as he is likely to pose a threat to the US, its interests, and its allies," a "MEDIUM threat from a detention perspective," and of "HIGH intelligence value."[3]

He is also an artist. Because he is designated as an HVD, however, Al-Baluchi has only intermittent access to art supplies (detainees to whom less "value" is attributed have readier access to these materials). This book's cover image, "Vertigo at Guantanamo," is one of the few pieces he has been able to produce in these circumstances. Initially, Al-Baluchi created this image for his lawyers in an attempt to make visible one symptom of a traumatic brain injury (TBI) that he sustained during interrogation.

As I write this, however, I struggle with how to balance attention to Al-Baluchi's artistry with recognition of the suffering that inspired this artwork. I am aware that neither is my story to tell, and yet I feel I must say something.

Many others have spoken of Al-Baluchi before me. He is named, for example, in *The 9/11 Commission Report*, which identified him as the nephew of Khalid Sheikh Mohammed and a "financial and travel facilitator for [the] 9/11 plot."[4] He was fictionalized as the character "Ammar" in *Zero Dark Thirty* and tortured on-screen throughout the film's opening sequence. In February 2018, the UN Working Group on Arbitrary Detention issued a report that declared his detention illegal and called for his immediate release.[5] For my part, I have never seen or spoken to Al-Baluchi; besides the information I gather about him online, my only access to him is through his artwork and what his lawyer is able to tell me about him.

I first encountered "Vertigo at Guantanamo" on display at *Ode to the Sea: Art from Guantánamo*, an exceedingly controversial exhibition of visual art by detainees. I read through the maelstrom of news coverage before I went to see the show in January 2018, when it was up at the John Jay College of Criminal Justice. Consequently, I expected something very different than what I actually found: a thoughtfully curated and unassuming display tucked into a hallway of administrative offices. The curators—Erin Thompson, Paige Laino, and Charles Shields—managed to showcase the work and highlight the circumstances of its creation without fetishizing either. While critics of *Ode to the Sea* insisted that it lionized terrorists and defiled the memory of September 11, I experienced it as a quiet commentary on indefinite detention that did not ask for clemency, only due process and the renunciation of torture.

Most of the art in *Ode to the Sea* is figurative (landscapes, still lifes, model ships), and some of it is vaguely surrealist, but "Vertigo" was unique in the collection for its abstractness. For me, this was part of its aesthetic draw. And I loved the colors and the way that it showed, insistently, the process of its creation in the pencil lines and eraser smudges. Conceptually, the premise of "Vertigo" echoes a central preoccupation of *Figuring Violence*. Al-Baluchi made the work to render visible to outsiders an otherwise inaccessible facet of his interiority. The revelation is compelling but partial, essentially indecipherable. Even if we know it is meant to depict the feeling of vertigo, even if we know what vertigo feels like, we still cannot know, precisely, Al-Baluchi's individual experience of vertigo and how that sensation imprints on his subjectivity.

I wanted this painting to be the cover of my book, and this desire is not unproblematic. After all, the cover of a book is arguably its most commercial element, and I do not wish to commodify his art or spectacularize his experience of being tortured. I do not wish to sell books, get professional recognition, or earn a promotion by putting his injury on display. At the same time, featuring his art on the cover of *Figuring Violence* felt to me like a final and necessary piece of the project falling into place.

Although I talk at some length about creative production by detainees in Chapter 5, "Vertigo" is the only piece of a detainee's visual art that appears anywhere in the book. Indeed, I deliberately chose *not* to reproduce any of the detainee visual art that I discuss, largely because I did not have a mechanism for obtaining permission from the artists. Presumably, I could have made the case that such reproductions would have been covered by the protections of academic fair use. Moreover, I almost certainly would have been insulated from any consequences of overstepping on this, because the artists would be exceedingly unlikely to find out and would have had virtually no mechanism for redress if they did. But these rationalizations seemed too flimsy, too quick to replicate the very power dynamics that I seek to trouble here.

Obtaining Al-Baluchi's permission to use his art on the cover was a multi-step process. I first contacted Erin Thompson to inquire about the feasibility of my idea. She made an email introduction to Alka Pradhan, his lawyer. I provided Pradhan with a copy of an article I had published on creative production at Guantánamo, as well as a summary of *Figuring Violence* and a draft of the introduction. A few days later, Pradhan obtained his permission to use the image and asked that we work together on this writing about it.

Because any access to Al-Baluchi is mediated, in one way or another, I cannot say with any confidence that I know him at all. Yet because I have seen some of his artwork, and "Vertigo" in particular, I have the sense that I know something about him. And, to the extent that "Vertigo" depicts a part of his embodied experience of the world, I have the sense that this knowledge is intimate. But I have no way of verifying its completeness or accuracy.

Epistemological dilemmas like this, which activate a tension between knowing and not-knowing, are central to *Figuring Violence*. The affective and imaginative practices that I analyze are essentially attempts to deny, defuse, or paper over them. With "Vertigo," Al-Baluchi invites us to contemplate his interiority while revealing it only elliptically. As a documentary record of torture, the painting is provocative but inscrutable, testifying to a violence that most of us cannot know. But this unknowability is not a problem to be solved or a failure of the work. Instead, it persists in "Vertigo" as a reminder of what we can neither imagine nor afford to ignore.

Figuring Violence

INTRODUCTION

Fabricated Connections, Deeply Felt

"Stand here and look as long as you like. They can't see you." "Here" was a designated spot on a perfectly clean concrete floor before a large pane of one-way glass slotted into a painted cinder-block wall. "They" were a group of detainees finishing up their art class, a dozen or so men in jumpsuits, setting aside their drawing projects and tidying up their supplies in preparation for midday prayer. The promise that I could see them surreptitiously and at my leisure was guaranteed by the rather rudimentary technology of the facility: a corridor outside a recreation area for compliant detainees at Guantánamo Bay in the fall of 2012. And this opportunity was my reward for making it through the petitioning process that I had initiated more than eight months before.

I don't remember exactly what inspired me to try to visit Guantánamo in the first place. I was intensely interested in the legal, political, and cultural aspects of indefinite detention and curious about the place of the visual within this peculiar system, and I got a notion to try to see it for myself. I talked to a few colleagues, idly, about my idea; no one was optimistic about my chances. By this time, the military and government had already begun to loosen, or rearrange, their controls over the visuality of the facility, but the skepticism I encountered reflected the pervasive impression that Guantánamo was, and would remain, invisible. Nevertheless, I started contacting the media relations personnel there, introducing myself as an academic interested in media representations of Guantánamo, which seemed the most straightforward description I could offer. Explaining that I thought it would be valuable for me to see what journalists saw, I asked for a press tour of the facility. Because I, too, envisioned Guantánamo as a black hole, I did not expect to receive any reply at all. But I did, and a tepidly encouraging one at that, in the form of a polite request for more information. This initial success was short-lived. My request would be routed to one office or another. Someone there would write, respectfully asking for more details about me, my work, and my interest in visiting the facility. I would provide them and await a reply. Shortly thereafter, someone would reconsider, or the command would change, and my petition would get bumped back down the chain. Then I learned that I could not have a press tour because I was not a credentialed member of the press. Eventually, I just stopped trying. But then, after weeks of silence, I received an email from the Office of the Secretary of Defense, offering me a place on a trip the following week, if I was interested. I was. And so I joined a distinguished retinue of law professors, judges, and Department of Defense employees for a tightly scheduled day trip, during which we were treated very graciously.

Of course, the whole thing was heavily stage-managed; everyone who worked at Guantánamo was well aware that they had a problem with optics. Given this, the invitation to look at detainees served multiple functions: to vouch for the new transparency of the operation, to show that the detainees are reasonably content and well-treated, and—like the meals and briefings and group photographs and trip to the commissary for souvenirs—to keep the visitors entertained. Superficially, my detainee (non-)encounter was a surveilling one. But not only. I could not help but see the men before me, though I felt uneasy about looking at them. I felt uneasy about looking at them, but I looked anyway. At the same time, it seemed likely that, even if they could not know exactly when some outsider was watching them, they would be aware of the possibility of such an audience. I wondered if their apparent indifference to the large reflective surface on their wall was studied and deliberate, rather than genuine.

What I could not see in this encounter was as important, or even more important, than what I could. We had been told that the detainees were taking their art class, but I could not see what they were sketching. Practically, this was a happenstance of my awkward vantage point. But it was also a metaphor for the ways that militarized visual systems occlude the subjectivities of the people they identify as enemies, even as they promise an unobstructed view. Rather than a truly illuminating look at the detainees, this experience offered me instead the shadowy feeling of having seen them and a place from which to credibly imagine that I had acquired meaningful information about them in the process. This dynamic is central to *Figuring Violence.*

Despite the invitation, no one stayed long before the viewing window. We had other things to see, and the scene itself, deviating from more familiar and spectacular depictions of detention, was not especially interesting. I have spent much longer at Guantánamo in thought and memory than I did in person. This book marks something of a return, an extended reconsideration of the promise that the guards made to me there. Materially enacted on both surfaces of the one-way mirror, that promise was offered and fulfilled by the various structures that apportion visibility and power, coordinating the limitations and capacities of our sight as they linked us all, in disparate ways, to the state.

Figuring Violence is a book about imagination and affect in wartime and the beings around whom they converge. In the process of this convergence, these beings are abstracted into figures who appear not as political subjects but instead as receptacles for affective investment. Activated by the sense that these beings are suffering, the forces of affect and imagination position them as repositories for apprehension, affection, admiration, gratitude, pity, and anger. Yet the most troubling paradox of this figuration is that it affords affective intensity and imaginative visibility at the cost of eclipsing the actual beings upon whom those figures are patterned. In this book, I seek to document the harms engendered by this commingling of affect and imagination. I argue that the occlusion of the actual beings in question promotes the development of a shallow ethics in response to their suffering, the erasure of their political subjectivities, and, ultimately, propagation of the very militarism that begets their victimization.

Affect, Imagination, and Militarism in the Contemporary American Context

Doubtless, the early years of the Global War on Terror were marked by the primacy of affects like fear and insecurity, which functioned handily in the service of the state's drive toward intensive securitization and aggressive foreign policy. Those initial wartime affects were tolerably unpleasant at best and unbearable at worst. And they were not sustainable.[1]

Figuring Violence catalogs the affects that define the latter stages of this war.[2] This period, arguably inaugurated by the transition from the George W. Bush presidency to that of Barack Obama, is marked by phenomena like the receding of the popular memory of September 11; the open-endedness of U.S. military involvement in Afghanistan and Iraq; the proliferation of new enemies and new kinds of threats; and the continued bifurcation of the military and civilian populations alongside the widespread, if superficial, veneration of military personnel.[3] Ultimately, I suggest that the affective successors to the feelings that defined the early stages of the war are more variegated and often enjoyable, as they offer the ideological and emotional gratifications of profound sympathy or righteous indignation. They are also more compatible with the ambitions of a state embroiling itself in a perpetual and essentially unwinnable war.

The "figuring" that gives this book its title delivers these gratifications. I use the term "figuring" to describe a process of imaginative construction that dwells on the suffering of its objects and entertains fantasies of its amelioration. The *Oxford English Dictionary* tells me that the verb "figure" can be employed in roughly two dozen ways. Many of these are to do with representation and include "to picture in the mind; to imagine," as well as "to be an image, symbol, or type of." These types of operations form the core of the figuring I analyze here. Relatedly, there is "to embellish or ornament with a design or pattern," which to me is evocative of the ways that figuring refashions its objects. A different sense of "to figure" is mathematical: "to reckon, calculate, understand, ascertain" or "to estimate or calculate; hence, to work out, make out." The figuring with which I am concerned includes a version of this, primarily in its efforts to parse, measure, and quantify how much particular figures are suffering. The *OED* concludes its entry with a late addition, based on the incredulous colloquialism "go figure." This phrase is "used esp. as an invitation to consider something the speaker or writer considers bewildering, inexplicable, or ridiculous." This usage, too, is apposite in that figuring in response to militarized violence is occasioned by an encounter with the unknowable or the unthinkable. It is an effort to override the uncertainty and risk they engender.

Crucially, the actual voices of the beings on which these figures are patterned are absent, muted, or extensively mediated. Figuring happens at a distance—whether physical or epistemological or both—and so involves not only speaking, but feeling on behalf of its subjects. The figures in this book have in common their status as objects of intense emotional and discursive investment and as political subjects that are partially or fully unknowable: children both civilian and military, military spouses, veterans with post-traumatic stress disorder (PTSD) and traumatic brain injury (TBI), detained enemy combatants, and military dogs.[4] Affectively, these figures anchor contemporary American militarism, which I define as the complex of feelings, beliefs, and perceptions that make war in general—and our current wars in particular—seem necessary, if not inevitable, and ultimately beneficial.

Unlike fear or aggression or insecurity, which seem obviously compatible with militarization, apprehension, affection, admiration, gratitude, pity, and sympathetic anger often seem far removed from, or even irreconcilable with, violence. They are easily misrecognized as what Cynthia Enloe calls "demilitarized emotion."[5] But this is exactly why they are so pernicious. They follow channels carved by power that masquerades as sentiment. And they help to fabricate connections by which outsiders can imagine those others who are made vulnerable by war. But these processes leave very little room for the appearance of the actual beings in question. They reduce them either to their suffering or their apparent transcendence of it while stripping away the political significance of both phenomena.[6] This process is a variation on the long-established sentimental rituals of liberal modernity that compel sufferers to narrate their pain for the benefit of more privileged interlocutors.[7] Figuring retains the same power dynamics as these practices but circumvents the voices of its suffering objects. Figuring is not so much an aversion to knowledge of the actual suffering of these beings; it is a presumption that such suffering can be known, understood, and eventually ameliorated with certain affective investments.

These imaginaries are deeply felt but essentially superficial. Acknowledging that "even those who live in the most dire circumstances possess a complex and oftentimes contradictory humanity and subjectivity that is never adequately glimpsed by viewing them as victims or, on the other hand, as superhuman agents," Avery Gordon wonders why critics often "withhold from the very people they are most concerned with the right to complex personhood."[8] I do not purport, in my analysis, to reclaim that personhood; I cannot. Instead, I seek to provide a record of its disappearance under affective and imaginative practices that frequently go unquestioned. The foundation of my intervention here is the development of an affective history for each figure, a genealogy of the feelings that orbit them today. While these feelings often appear as automatic or untutored expressions of empathy, I seek to denaturalize them by documenting their origins in various political, social, and cultural systems. Against our cultural tendency to imagine that feelings, especially seemingly virtuous ones like admiration or gratitude, float free of power, politics, and history, I analyze the ways that these affects manifest their origins and operate in tandem with ideological formations.[9] The format of these histories varies from chapter to chapter, figure to figure. Mary Favret observes that "affect often eludes the usual models for organizing time such as linearity, punctuality, and periodicity," and so the histories of some affects have gaps or switchbacks or resonate and overlap with those of others in unexpected but significant ways.[10]

Histories of affect are also histories of power because they record patterns of interaction and recognition, negotiations over who qualifies as a subject.[11] As Ben Anderson contends, forms of power "work in conjunction with the force of affect, intensifying, multiplying and saturating the material-affective processes through which bodies come into and out of formation."[12] Here, I am expressly concerned with the question of who gets recognized as a political subject, under what circumstances, and at what cost. Political subjects are beings with the desire (and sometimes, but not always, the capacity) to think and act in ways consequential to their communities.[13] Political subjectivity includes, but is not reducible to, legal citizenship.[14] Following Marita Sturken's description of the nation as an "entity that is experienced at the level of emotion and intimacy," I suggest that political subjectivity is essential to that experience.[15] Only those who are recognized as legitimate political subjects have access to the emotional life and intimacies of the nation, and participation in those rituals verifies the legitimacy of political subjects.[16] Figuring proceeds under the guise of attending to another being's value, sentience, and suffering, but ultimately engenders denial or negation of their political subjectivity. The imaginative and affective processes of figuring cannot sustain true recognition of the other's political subjectivity, and any hint of that subjectivity is immediately followed by a commensurate withdrawal of sympathetic affective investment. In other words, figuring demands a trade-off in which another person can be recognized either as a political subject *or* as a being whose suffering is significant and worthy of response, but never as both.

Rather than recognizing their political subjectivity, figuring addresses its objects in a sentimental register. The figuring gaze regards the agency of its objects warily, discounting it, denying it, or identifying it as a threat. In her philosophical meditation on the nature of subjectivity, Kelly Oliver describes it as an "openness to the other," a formulation that marks its essential relationality.[17] Those who do the work of figuring, as by venting their anger, professing their gratitude, or expressing their admiration, cultivate their own subjectivities through these displays of openness.[18] But this openness is conditional, lopsided, and coercive. The other (the civilian or military child, the military spouse, the veteran with TBI, the detained enemy combatant, the dog) cannot participate in this exchange except through the proxy of another person's idealized vision of them. Although the work of figuring proceeds on the conceit that the other can be known or accurately and fully imagined, it is actually predicated on the inability of the other to respond to these overtures.

Most theoretical and critical descriptions of affect emphasize its intersubjective and relational character, yet the figures I focus on here have in common their inaccessibility, their unavailability for direct interaction.[19] This is not to say that the apprehension, affection, admiration, gratitude, pity, and anger I describe here are somehow disingenuous or fake.[20] For the people who experience them, I'd wager that these affects are very real indeed. But they are also fabricated, in the sense that they are derived from imagined connections with their objects, their intensity camouflaging the enormous, often unbridgeable distances separating them from the objects of their sentiment. Sara Ahmed writes that "emotions are not simply 'within' or 'without' but . . . create the very effect of the surfaces or boundaries of bodies and worlds."[21] My analysis here demonstrates that when they entwine with imagination, they can also create a sensation of boundarylessness, the feeling of having an emotional connection with a distant stranger. This sense of connection, however, is more of a ricochet of our own emotions than a true window onto theirs.

Phenomena like sensations, feelings, and emotions are related, but not identical to, affect.[22] For my purposes, the crucial distinction between emotion and affect is that affect requires some kind of interactivity, exchange, or partnership. In Kathleen Stewart's terms, "The affective subject is a collection of trajectories and circuits."[23] Affects are oriented outward. If someone has a crush, for example, there is a difference between the emotions she might feel when daydreaming privately about her beloved and the affects she experiences when she directly interacts with the object of her ardor in some way. The operative affects and emotions in these two cases are connected, but they are not the same. Affects require partners to activate, refine, or intensify them.[24] Affect is an exchange; an affective investment, then, arises from a desire to specify its parties and set its terms. The work of figuring conscripts its objects into these partnerships, but unilaterally and in radically limited ways. The affective investment of figuring is a presumption to know how the figure in question thinks or feels; the dividend is the sense of sharing in that feeling and responding appropriately to it.

By attending to the six affective positions I focus on in this book, I do not mean to suggest that other affects are irrelevant or that every citizen will feel the same thing about every figure. All people inhabit what Barbara Rosenwein describes as "emotional communities," fashioned by smaller-scale institutions like families and jobs as well as larger ones like the nation-state. These emotional communities collectively "define and assess [feelings] as valuable or harmful to them; the evaluations that they make about others' emotions the nature of the affective bonds between people that they recognize; and the modes of emotional expression that they expect, encourage, tolerate, and deplore."[25] Insofar as groups have preferred and distinctive emotional practices, we can analyze them as coherent phenomena. In the case of national emotional communities, militarism tends to fortify these practices as well as the stakes of getting them right. "Affect," John Protevi writes, "is inherently political: bodies are part of an ecosocial matrix of other bodies, affecting them and being affected by them; affect is part of the basic constitution of bodies politic."[26] These bodies politic are marked by salient affective patterns. They organize themselves through what Protevi calls "entrainment," defined as the "assumption of a common frequency, that is, the falling into step of previously independent systems."[27] The common frequency, however, can accommodate many variations, and in my analysis, I mark the complex manifestations of the affective investments in each figure even as I track the larger patterns that they follow.[28] When I use the pronouns "we" and "us," it is to locate myself among the participants in the phenomena I describe.

Historically, sentimentality has been a powerful organizing force in these affective collectivities, particularly because it aligns so well with discourses of American benevolence and exceptionalism. Sentimentality enables emotional reconciliation for phenomena like disenfranchisement, inequality, and violence. Sentimentality, Lauren Berlant argues, "operates when relatively privileged national subjects are exposed to the suffering of their new intimate Others, so that to be virtuous requires feeling the pain of flawed or denied citizenship as their own pain."[29] Sentimentality entails a temporary identification with those suffering others, and the citizen's capacity for this kind of imaginative work then becomes proof of their goodness and noncomplicity with the systems that mete out harm to others. This basic structure of sentiment scaffolds all of the affective investments I analyze in *Figuring Violence*.

In sentimental cultures, emotion operates as a convincing alibi. Crucially, sentimental identification does not necessarily entail any actual contact with its objects and may indeed function more smoothly without it. It requires only a plausible vision of its object. Identification, or at least the feeling of identifying with someone else, can be satisfying whether it is accurate or not. Yet most theorizations of affect accept those feelings of connection and interactivity on their own terms. In those models, it is enough to believe that I am being affected by another. Consequently, the assertion that some kind of affective exchange has happened, which necessarily implies an intersubjective connection, requires a report from only one party.[30] These unilateral declarations of intersubjectivity are the emotional equivalent of speaking for others and elide their subjectivity at the very moment of proposing to recognize it. This matter of intersubjectivity becomes even more confounded when the relationship in question is mediated. The term "mediation," as I employ it, encompasses various means of transmission and representation that enable audiences to experience and to imagine distant people, places, and phenomena. That distance can be literal or geographical (wherein mediation provides access to things that would otherwise be out of reach) and epistemological (whereby mediation provides the illusion of understanding that which would otherwise be unknowable or incomprehensible). Like the one-way mirror in that Guantánamo hallway, mediation facilitates certain kinds of connections while precluding others. Here I am particularly concerned with the mediation of suffering. Whose suffering gets relayed? How? By whom? To what audiences? For what purposes? And with what consequences?

Scholars in visual culture continue to debate whether representations of suffering draw audiences closer to the afflicted parties or introduce additional distance between them, as by spectacularizing suffering or inducing weariness in the form of "compassion fatigue."[31] Describing the emotional functions that mediated depictions of suffering perform, Lilie Chouliaraki contends that "the media not only exposes audiences to the spectacles of distant suffering but also, in so doing, simultaneously expose them to specific *dispositions to feel, think, and act* toward each instance of suffering."[32] Sometimes media representations, even of profound suffering, leave us feeling nothing or contradictory somethings.[33] Wendy Kozol, analyzing visual representations of the suffering wrought by distant wars, reveals their potential to generate ambivalent spectatorships. Attending to ambivalence, she writes, enables the examination of "instabilities, frictions, and contradictions within representation, as well as those that arise intertextually within discursive and material contexts that shape both production and circulation."[34] Even as my work is indebted to this scholarship, my present inquiry diverges a bit from these lines of analysis. I focus less on the emotional content of media representations or how particular representations might make audiences feel; instead, I begin from a place of curiosity about what forms, if any, of intersubjectivity are possible in mediated encounters. Do such representations actually transmit the feelings of the suffering beings that they depict? Or do they instead serve to anchor a popular imagination about how those beings are, or *ought to be*, feeling? And what is lost in the process of misrecognizing the latter dynamic as the former?

Rather than accept claims of being affected at face value, I argue that, in the case of the figures I analyze here, it is imagination rather than actual connection that sustains these circuits of feeling. All of the emotional upwellings that I track, from the worry over children's classroom exposure to visual evidence of wartime atrocity to the outrage over the killing of a puppy in Iraq, purport to be responding to the feelings of the figures in question but proceed with virtually no input from them. They are fueled, instead, by a militarized imagination that provides a gratifying fiction of intimacy or communion.

We can speak of both individuals and groups, like nation-states, possessing the capacity of imagination. Imagination is always at least partly volitional, even if our imagination can sometimes run "wild" or "away with" us. Consequently, I prefer the framework of "imagination" over that of "fantasy," with its psychoanalytic connotations of unregulated delusion or unarticulated desire. Imagining is a motivated act of imposition, the lamination of a preferable alternative onto a particular set of circumstances. It is an experiment not only of thought but also of feeling. Of course, imagination is key to any political practice and does not always beget the kinds of harms that figuring causes. At its core, figuring is problematic because it involves relatively privileged actors imagining the feelings of others who are far more imperiled and vulnerable.

Imagination is a key vector along which state actors consolidate power.[35] As Jacqueline Rose reveals, "Over and above its monopoly of legitimate violence, the modern state's authority passes straight off the edge of the graspable, immediately knowable world."[36] And against the tendency to dismiss imagination as a "mere historical epiphenomenon," Raymond Geuss intimates that imagination may actually constitute "political reality."[37] Part of my project, then, is to specify who exactly is imagining whom and what material realities they construct in the process. This work is a turn on Benedict Anderson's canonical theorization about the role of imagination in the life of the nation-state. Anderson argues that imagination is integral to the modern experience of citizenship, as it enables citizens to understand themselves as linked to all others of a common nationality, despite the fact that they will meet only the tiniest fraction of them in a lifetime.[38] This binding imagination conceives of fellow citizens in generic terms as a people with a shared history, similar experience of the present, and a common vision for the future. This, Anderson writes, is part of what inspires citizens to join militaries to fight, kill, and die on behalf of strangers. The imagination I describe here, however, purports to know certain types of beings intimately, while refusing to acknowledge their complex subjectivities. And, insofar as it is predicated on a split between civilian and military populations, it helps sustain the division of labor that insulates the vast majority of the country from direct contact with war itself.

Imagination at the levels of both citizens and nation-states is intensely and generatively compatible with militarization, particularly in its contemporary counterterror and counterinsurgency forms.[39] According to Joseph Masco, during the Cold War, the United States forged a pervasive "national security affect." Following September 11, this affect was activated by "U.S. officials [for] a conceptual project that mobilizes affects (fear, terror, anger) via imaginary processes (worry, precarity, threat) to constitute an unlimited space and time horizon for military state action."[40] He describes imagination as one of the many infrastructures along which militarization proceeds and contends that "the goal of a national security system is to produce a citizen-subject who responds to officially designated signs of danger automatically, instinctively activating logics and actions learned over time through drills and media indoctrination."[41] In a compatible analysis, Louise Amoore notes that in the aftermath of September 11, U.S. officials blamed their inability to prevent the attacks on a "failure of imagination." This, in turn, "has led to the imagination of 'all possible links' to identify 'potential terrorists' and to 'build a complete picture of a person'" who might pose a threat.[42]

In the prosecution of its War on Terror, the state has engaged citizens' imagination in a range of ways. Part of this work, as Donald Pease notes, has unfolded in the cultivation of various "state fantasies" that render citizens amenable to limitless wars and willing to relinquish civil liberties to support them.[43] Eva Cherniavsky describes the era of the War on Terror as one distinguished by a "fundamental derealization of political life."[44] She argues that the Bush administration scarcely attempted to make a rational or ideological justification for the invasion of Iraq, "proffering instead an array of disarticulated catchphrases and soundbites that functioned more on the order of feedback, a kind of constant, staticky interference."[45] This derealization of politics, she says, operates not through an effort to coerce citizens to a particular way of understanding the world, but rather by framing the world as not intellectually or rationally understandable, and so giving free rein to imagination instead. These dynamics form the backdrop for the phenomena I am concerned with here.

Importantly, these processes are not entirely governed by the state; even as the state ultimately benefits from the figuring processes I describe, it does not always, or even often, direct them.[46] Indeed, militarism, the theory that inspires the practice of militarization, is more than just a feeling that trickles down from a hawkish government.[47] While states oversee the functioning of their militaries, making decisions about policy, recruitment, and deployment, phenomena of militarization do not proceed under the state's exclusive authority. In short, militarization is the material transformation of an object in the name of preparation for war. This process, as Enloe describes it, is "far more subtle" than the act of joining the military.[48] Militarization requires material, something to act upon, imprint, and change. And anything, anything, anything can be militarized: people, consumer goods, relationships, hopes, wishes, fears, and feelings.[49] Militarization renders patriotic love of country aggressive.[50] And the contemporary American version of militarization offers citizens, and especially civilians, a range of enjoyments and compensations as part of this transformation. Indeed, as Andrew Bacevich argues, most Americans suffer no ill effects from contemporary militarism.[51] And when the state does not have recourse to measures like a draft to force citizens to participate in militarization, the work of maintaining support for militarism falls to other entities that must make it tolerable because it is not mandatory.[52]

Affect and imagination are readily conscripted into this mission. They are the engines of the figuring processes that I describe here and set the patterns by which they proceed, facilitating and justifying them. The affects of apprehension, affection, admiration, gratitude, pity, and anger are nearly ubiquitous in contemporary American public cultures of militarization. Again, I am not suggesting that every citizen will feel precisely the same way about these figures or that every citizen would experience or express those feelings identically. Yet it is possible to generalize about how these figures are constructed by certain core affects. Notably, many of these practices become established without any sort of official mandate. Of course, politicians and public figures may seek to cultivate certain emotions or discourage others, but the results are not guaranteed.[53] While it is heartening to realize that there are emotional paths available other than those preferred by the state, this delocalization also makes it difficult to pinpoint the origins (and identify the beneficiaries) of these emotional processes. Writing about the spate of anti-Muslim hate crimes in the aftermath of September 11, Andrew A. G. Ross classifies them as "patterned, but also uncoordinated," a subtle but significant distinction that speaks to how affects and sentiments, propelled by certain histories, can move into alignment.[54] Anker describes the circulation of melodrama in similar terms, arguing that the "use of melodrama is not forced or coordinated across media outlets or political parties; its popularity across two centuries of cultural media make[s] it readily available to multiple sites of power and address for depicting political life."[55] In some of the cases I analyze, like the presidential proclamations I consider in Chapter 2, Americans receive clear and specific emotional direction from the head of state, but few of the other examples are so explicit. Consequently, as I wrote these chapters, I found myself relying much more on passive-voice constructions than my writerly sensibilities would normally allow. I have revised as many of them as possible, for reasons both stylistic and analytical, because passive voice obscures cause, effect, and responsibility. But I retained these constructions in places where there simply was no single agent to be found or where there are too many agents to be captured in a single noun.

Despite all of this, I am not suggesting that imagination and affect ought to be excluded from political life, as if such a thing were possible. Imagination is a key aspect of political subjectivity, and collective imagining is necessary for all political projects, including resistance to militarization.[56] Rather, my critique of imaginative and affective figuring focuses on the obligations that these processes foist onto their objects. To identify a particular being as a promising repository for feelings is to become attached to it. As Berlant argues, "All attachments are optimistic. When we talk about an object of desire, we are really talking about a cluster of promises we want someone or something to make to us and make possible for us."[57] The abstracted civilian and military children, military spouses, injured veterans, detainees, and war dogs provide outsiders with affective oases in the broader landscape of militarism and an illusion of reckoning with its consequences.

A New Methodology for the Study of Militarized Affects

Geographically, this book begins in Guantánamo, but practically, it began where it ends, with my analysis of dogs in wartime and that one dead puppy in particular. The analysis that became part of Chapter 6 was the first piece I wrote; the second was the portion of Chapter 1 on scandals about atrocity photos in the classroom. I originally conceptualized these as discrete and stand-alone projects. But my research revealed deep historical entwinements between sentimental care for domestic animals and concern for children, and as I studied these two upwellings of feeling, I discovered affective parallels between them as well, as both were centered around unknowable subjects. So I began looking for other targets that drew this kind of intense and militarized sentiment, beings who are simultaneously hypervisible in public culture and popular imagination and somehow inaccessible as political subjects. This led me to the figures that would become the foci for the remaining chapters of this book. Superficially, civilian American children traumatized by the sight of the Abu Ghraib photos and one mongrel puppy killed on a vicious lark by two U.S. Marines in Iraq appear to have very little in common with one another or with military children and spouses, veterans with PTSD and TBI, and Guantánamo detainees. But they are all deeply attractive to affective investment. And my analysis of the affective work that happens around these populations showcases their connections to one another and their alignments on the broader cultural landscape of contemporary American militarism.[58]

Even as it provides important insights into emotional and interpersonal phenomena that defy straightforward explanation, the field of affect theory tends toward speculation and abstraction.[59] These can be important ways of knowing and explicating complex and dynamic emotional and social processes. However, I am also concerned that affect theory's turn away from the material risks occluding the high stakes of these inquiries or obscuring of the real, often embodied, consequences of the phenomena they seek to describe and examine. In this book I seek to remedy that limitation and to devise and model a new way of studying affect as it materializes in practices of representation and through public culture. In particular, I demonstrate how we might trace the militarization of affect by identifying specific figures that become points of dense and vibrant affective investment and mapping the networks of images, documents, and objects that construct and connect them. As an effort to name and understand these often ineffable phenomena, *Figuring Violence* is necessarily and constitutively interdisciplinary, drawing inspiration from fields like American studies, affect theory, visual culture studies, media studies, political theory, and gender studies. Taken together, these modes of inquiry assemble into a prismatic view on questions of militarization and recognition. Moreover, they provide a foundation from which to develop my own inquiry about how specific, politicized, affect-laden fantasies travel through various media and settle, heavily, on a range of bodies.[60]

Although my analysis is primarily concerned with domestic rather than foreign politics (though the separation is never so tidy), I am also indebted to recent work on the role of emotions in international relations. These include the studies by Ross, who calls for an expanded emotional lexicon in international relations beyond the usual paradigms of fear and hatred, as well as Christine Sylvester's advocacy for a more systematic study of emotions in war.[61] One of the most compelling examples of how to approach the question of emotion in political life is Elisabeth R. Anker's *Orgies of Feeling*, as she analyzes how "melodramatic discourse solicits affective states of astonishment, sorrow, and pathos through the scenes it shows of persecuted citizens."[62] Anker focuses on the evolution of a single affective style, melodrama, over time. Alternatively, I frame my analysis around the militarized constellation of apprehension, affection, admiration, gratitude, pity, and anger, unified by their role in the construction of the figures I describe. And while melodramatic political discourse is largely self-involved (Americans feeling bad for themselves or other Americans), the process of figuring, while also solipsistic, is oriented primarily toward others, a notional experience of recognition.

Recognition alone is an inadequate remedy for structural disenfranchisement of or disinvestment in marginalized bodies. Berlant offers a succinct critique of the limits of recognition in general—namely, that it can "provide a means for making minor structural adjustments seem like major events, because the theater of emotions is emotionally intense."[63] For the figures I study, recognition itself has been militarized, which compounds the risks that Berlant notes. In *Frames of War*, Judith Butler queries the relationships between representation, recognition, affect, and the valuation of various lives and deaths in war, arguing that beings we do not regard as human become easy targets of violence and, relatedly, that their deaths elicit little in the way of emotional upset.[64] Complicating this framework, I want to attend to the specific mechanisms by which certain figures are simultaneously invested with emotional meaning and excluded from the field of recognized political subjects.[65] Importantly, Oliver argues for a move "beyond recognition," because philosophical models of recognition rely on zero-sum designations of subject and object and a presumption of distance between them. But while Oliver suggests that we are closer than we think, that the space between us is full of connective possibility, my argument here is that we are further from these bodies than affective and imaginative structures of figuring lead us to believe.[66]

The notion of "affect" engenders thorny questions about agency. With their emphasis on affect as presocial or somehow extradiscursive, most prevailing theoretical depictions of affect situate it as something beyond the control of those who are affected: a rush of feeling that exceeds any effort to constrain it. Surely, anyone who has ever been, or felt, overwhelmed could intuitively confirm that description. But Rei Terada introduces a provocative complication with her argument that "like intentions, emotions entail beliefs and apply to objects. To this extent, they are less sensations that happen to one than thoughts that one pursues."[67] This suggests that emotions demand active nurturing—whether through resources, action, or imagination—and so too do the affects I analyze here. In her reflection on militarization, Enloe reminds us that it is rarely "simple" or "easy," but instead requires continual maintenance and decision making to support it.[68] So it is, too, with the militarization of affects. Part of the work of this book, then, is to identify the institutions and practices that enable them to flourish.

Key texts in the fields of affect theory and affect studies provide blueprints for tracking the political lives of various emotions. Sianne Ngai's *Ugly Feelings* offers a critical taxonomy of the political potentials of unpleasant and seemingly useless feelings like irritation or boredom. My work in this book offers something of the obverse of that, identifying the problematics embedded in feelings, like gratitude or admiration, often deemed politically or socially productive. Ngai reads ugly feelings as "situations of suspended agency" and argues that many of them do not have distinct objects.[69] Alternatively, the affective investments I consider here require a suspension of the agency of the other and have very specific objects in the form of the figures at which they are directed. Berlant's *Cruel Optimism* exposes the destructive potential of holding tightly and hopefully to political, economic, and relational ideals. But whereas Berlant focuses primarily on the damaging effects of such optimism for the optimists themselves, I attend to the perils of such imagining for those absent others who become its objects and show how its costs are distributed outward onto bodies that are already vulnerable. Like Protevi's *Political Affect*, these analyses of affect illuminate key facets of its structure and significance, but none of them engage substantively with questions of war or militarism. Here, I endeavor to fill that silence.

Figuring Violence aspires to develop a critique of contemporary American militarism that looks beyond the state. For example, while the state, via its military, orchestrated my view of the detainees that afternoon at Guantánamo, it was not the only participant to benefit from the encounter. Nor did it set limits on our participation ("look as long as you like"), leaving us instead to craft our own enjoyments. State-sanctioned affective and imaginative projects are relatively easy to identify; states are rarely subtle or coy about their wishes. Here, however, I focus on affective and imaginative practices that emanate from other sources and operate to buttress militarism precisely by expressing sympathies for the casualties of the violence it engenders.

One of the challenges inherent in the study of affect is the difficulty of locating it; consequently, much of the scholarship on affect tends toward abstraction. While this is an important way of knowing about phenomena that are sometimes ineffable, this book models a different approach to the study of militarized affect, one that is concerned with the concrete operations of feeling in wartime. Beginning from the supposition that affect is always materialized, in one way or another, whether on the individual level of a sensation in the body or at the scale of national ritual, my research tracks its most tangible forms.[70] This book surveys the vast cultural landscape of contemporary American militarism as it has taken shape over the last decade and a half. To choose my objects of inquiry, I looked for sites where I could clearly identify the operation of these affects. In that way, the writing of the book began in much the same way that the project itself did when I was in that Guantánamo hallway: with watching. As I began assembling my archive, I wanted to uncover the resonances among various objects and figures and remain open to serendipitous connections, as affects do not always move predictably.[71] While some elements of this archive might be unexpected or counterintuitive, all of the examples illuminate different facets of the affects orbiting these figures.

A core commonality among these affects is that they are all pleasant; even anger offers the solace of righteousness. They are also generally regarded as socially useful or constructive. As Berlant trenchantly observes, "In the liberal society that sanctions individuality as sovereign, we like our positive emotions to feel well intentioned and we like our good intentions to constitute the meaning of our acts."[72] Given the circumstances of terror in which the war began, my selection of affects might be surprising. But I argue that with the gradual subsidence of the memory of September 11, fear and insecurity are no longer the dominant modes through which most Americans experience the war. As this conflict spills past its fifteenth year and onto new territories, always moving toward a horizon that recedes as almost as quickly as it appears, it generates a different and newly variegated affective environment. In part, this shift reflects what Masco describes as the "domestication" of fear and terror over time, a softening that renders them more familiar and palatable.[73] It is also a function of the division between civilian and military in

American life. In Bacevich's estimation, the 99 percent of Americans who do not serve in the military look upon those who do with "cordial indifference," content to offer amorphous support to a war that proceeds without any expectation that they change their behavior, pay higher taxes, or risk their own lives in it.[74] War, for us civilians, comes at virtually no cost. Instead, by investing affectively in the figures I describe here, Americans reinvent compassion itself as a sacrifice and then reap the intense pleasures of making it.[75]

The affects and imaginaries I describe here persist, in short, because they *feel* good and offer different gratifications to a range of stakeholders, both pro- and antiwar. Psychologists speak of the "valences" attached to particular emotions: our perceptions of them as positive or negative, as when we feel good, or bad, about the fact that we feel a certain way.[76] The affects I analyze here persist because they feed off a general consensus that they *are* good: justified, ethical, and productive. We might describe this as their political valence. This political valence explains the centrality of these affects in contemporary antiwar discourse. Yet I argue that this antiwar affective traffic is especially damaging because that discourse often valorizes these affective excursions as forms of resistance; my critique, then, is aimed at unmasking their vexed legacies and outcomes. I seek to dislodge these affects from their privileged positions in American political life and popular culture and interrupt the work of figuring that animates them.

Early in each chapter, I document how the relevant affect(s) functions in the case of the specific figure. Of the six, *apprehension* makes the briefest appearance; it describes a generalized orientation toward the mediated wartime experiences of civilian children, an anxious wish to prevent these encounters, motivated by a certainty that they will damage an otherwise pure (and apolitical) child. I invoke *affection* to name a muted form of love that operates at a distance and resembles the filial care of adults for children.[77] Like affection, *admiration* is an articulation of desire, an urge to draw nearer to a person we find attractive in some way. Desire, as Berlant observes, seems to come from outside, activated by the presence of another person, but really comes from inside.[78] The military children and spouses (read "wives") who are the objects of this admiration anchor a desirable vision of the ideally heteronormative military family, which in turn sustains a vision of an ideally heteronormative nation-state. Both military children and spouses are frequent recipients of *gratitude* for their service and sacrifice, derivative versions of the more intense and direct forms lavished on military personnel, especially those who are injured. This urge, apparently ubiquitous and irrepressible, to express appreciation in the form of a "thank you" to the troops, inaugurates a vexed relationship of indebtedness that ultimately returns the burden of continued service to military personnel. All of these feelings are commingled with *pity*, sorrow triggered by a feeling of compassion for their struggles. Pity, however, operates in tandem with and only extends as far as the perception that its objects are innocent. This innocence affirms that they do not deserve the suffering with which they are afflicted and that they are helpless to contest it, the main criteria by which pity is allotted. This is easy enough to negotiate in the first three instances (in the case of the veterans, I argue, their injury offsets the unwieldy fact of their involvement in militarized violence), but becomes more complicated in the case of the detainees. In order to prime them as surfaces for pity, they are figured as apolitical and eager to forgive the United States its transgressions. By the last chapter, which begins from a short cell phone video of two marines killing a puppy by throwing it off a cliff, all that remains is *anger*. In the figuration of humans, anger operates anxiously, always as a vicarious thing that outsiders feel on behalf of their objects (especially in the case of the detainees) and an obstruction to figuring that must be filtered out. Here, however, it explodes beyond the constraints of sentiment, washing out everything else.

Unknowable, Unthinkable, Unimaginable: Figuring Violence on a Continuum

My analysis begins with the study of six affective investments in American children during wartime, and in each successive chapter, one affect drops away. Throughout, I explore how the affects surrounding a particular figure interact: sometimes harmonizing, sometimes standing in tension. Either way, they are bound together by a common militarized intensity, with a figure acting like a magnet to draw imagination and bind these affects together. Ahmed has written of the "stickiness" of emotions as they adhere to objects; my analysis reveals how affects stick to one another as well.[79] As we move from figure to figure, the affective palette grows narrower, so that fewer affects are operative in the construction and maintenance of each figure. The previous affects do not disappear from the field altogether, but rather recede and become gradually less salient.

In organizing the book, I arranged the figures from Chapters 1 to 6 along a gradient of decreasing knowability and comprehensibility, two different epistemological qualities. Knowability is essentially a matter of logistics. It is easier, typically, to learn what a military child is thinking than to speak to a detainee at Guantánamo, but while such communication is theoretically possible with detainees, there is no such option for communicating with dogs. Comprehensibility refers to the extent to which we can understand—or, more precisely, *believe* we can understand—the interiority of another being. It is relatively easy to imagine, with a sense of plausibility or certainty, what a military spouse might be thinking, feeling, or worrying about. The embodied experiences of PTSD and TBI, however, with their characteristic scrambling of cognitive systems for memory, relationality, and organization, are much harder to envision. Importantly, unknowability and incomprehensibility do not undermine the pleasures of imagining and feeling for these figures; in fact, they intensify those pleasures by making them seem frictionless and uncomplicated. Periodically, and especially in the case of the detainees, I invoke the idea of unthinkability as well. Unthinkable subjectivities grate against the imaginative conceits of the figuring that seeks to define them, complicating its gratifications. This continuum also corresponds to a decreasing number of affects surrounding the figures in question. The book provides not only an inventory of these figures and the emotions that comprise our fictive intimacies with them, but also a study of how affects circulate in wartime, carried along by deeply political imaginings.

I begin my analysis with the figure of the civilian child. Unavoidably, militarism imprints the political subjectivities of children, and its harms settle unevenly across different demographics. In Chapter 1, I attend to how those harms are distributed, modulated, and imagined by a range of institutions, both public and private, that seek to regulate what civilian children see and feel during wartime. I begin with a brief overview of the Western construction of childhood as a cognitive and emotional developmental stage, but also an imaginary. Children are unfinished political subjects, surely, but I contend that they are not quite so incomplete as we might assume. Highlighting the problems with understanding children as somehow pre- or apolitical, I foreground the reality of children's militarization. Focusing on a cluster of educational scandals around children's classroom encounters with the sights of wartime atrocity, I demonstrate that all of these controversies were catalyzed, problematically, by their imagined vulnerability and entitlement to safety, which often preclude recognition of other beings who might suffer during wartime.

In the next chapter, my focus moves from civilian to military children, tracing the looping discourses of duty and sensitivity through which they are imagined. Military children are often figured as being more emotionally mature and complex than their civilian counterparts. Two websites designed for them, *Sesame Street for Military Families* by the Sesame Workshop and the Department of Defense's *Military Kids Connect*, seek to help them catalog and superintend their feelings, and in Chapter 2, I analyze the differences in their approaches to this task. All the examples in Chapters 1 and 2 are linked by an abiding and apparently loving concern with the management of children's sight, senses, and sensibilities. But not all children are afforded the same protection, and this chapter is haunted throughout by the adolescent shadow of Omar Khadr, the child soldier who would become the first person to be tried at Guantánamo. I invoke Khadr here not merely as a foil, but to illustrate the stark differences in how particular childhood sensibilities are privileged above others. Highlighting the obverse of imagining certain young people as fragile and precious, I also explore the trade-off revealed in the treatment of Khadr. Khadr was widely recognized as a dangerous political subject and consequently banished from the realm of sympathy. In this way, Khadr dramatizes the exchange that figuring demands: sympathy swapped for recognition of one's political subjectivity.

Like the children I consider in Chapters 1 and 2, military spouses (predominantly wives) find themselves at the center of clusters of institutions, and Chapter 3 focuses on the figure of the military spouse. They are identified as being in need of protection, even as they are also positioned, admiringly, as the embodiment of a noble wartime ideal. Military spouses are hypervisible, but narrowly. Their exposure increases in proximity to sacrifice, their own or their partner's, but only through the lenses of patriotic romance or the emerging trope of the victim of a traumatized soldier's domestic violence. To contextualize this figuring, I first provide an account of pervasive affective investments in and expectations for military spouses that informs my subsequent discussion of women's militarization, both in general and in the specific case of the Global War on Terror. I then turn to an analysis of presidential appreciations of military spouses for their service and sacrifice. Operational Security (OPSEC) guidelines for military families, on the other hand, reveal the underside of official regard for military spouses, identifying them as vital but weak links in national security. Alternatively, the American Widow Project, a network organized and maintained by military widows, departs from these official discourses, recognizing widows' sacrifices but also embracing a vision of widowhood that is independent and pleasure-seeking. The work of the AWP intimates that within the affective strictures established by prevailing systems for recognizing military spouses, their full visibility is possible only in the radical absence of their husbands. I conclude with a consideration of emerging research on military spouse PTSD, a clinical discourse that accounts for the women's victimization but not their subject-positions.

Chapter 4 is situated at the fraught intersection of trauma and gratitude and analyzes how they interact on the figure of the veteran with PTSD and/or TBI. After providing a brief history of PTSD and TBI, with a particular attention to the linkages between World War I and the War on Terror, I turn to a theoretical account of how pity, gratitude, and anger have shaped discourses about injured and disabled veterans. I then reflect on the ethical and political intricacies of the compulsory appreciation embedded in the imperative to "support the troops." This informs my subsequent analysis of various nonprofit organizations that work to institutionalize this gratitude and their approaches to the matters of veteran affect and responsiveness. From there, I turn to an analysis of the clinical and exacting standards by which the Department of Defense awards Purple Hearts for TBI but refuses them for PTSD. These metrics stand in stark contrast to the bestselling *Thank You for Your Service*, a nonfiction account of members of an infantry battalion returned damaged from Iraq that is the next object of my analysis. That book, whose title reminds us that traumatized veterans are often recipients of the most public and profuse thanks, leverages the narrative of the victimized spouse to explore its central thesis about the irreparability of the emotional, cognitive, and physiological injuries endemic to this war. While *Thank You for Your Service* attempts to make these signature injuries intelligible by emphasizing their relational consequences, a Department of Defense–sponsored brain tissue repository has identified distinctive patterns of scarring on the brains of people with TBI, which promises to lend a different kind of visibility to the injury. This work depends on families who have lost military personnel to "give a gift" by donating their brain tissue. Here, the injured veteran is radically deconstructed in an effort to make this kind of trauma recognizable and intelligible.

On the broader affective spectrum that I trace in *Figuring Violence*, veterans with PTSD and/or TBI are a hinge, figures who span the book's axes of unknowability and unthinkability. The men detained at places like Guantánamo Bay embody political subjectivities that are, at least while they are in detention, unknowable and often unthinkable to outsiders. In Chapter 5, I analyze efforts to figure detained enemy combatants as more sympathetic. These anti-detention advocacy projects abstract the detainees, rendering them blankly compatible with the prevailing values of liberal American modernity. I begin by tracing the dynamics of pity and anger operative in these constructions and their contingency on the denial of any possibility that detainees might be angry. I then provide a brief history of anti-Guantánamo activism as it has evolved over the last decade and a half.

This informs my subsequent study of a collection of municipal actions in which communities voted to welcome select "cleared" detainees into their neighborhoods. While these initiatives proceeded without any input from the detainees, other projects take a more collaborative approach, including the *Witness to Guantánamo* documentary project. This archive-in-progress includes a collection of interviews with former detainees that document the minutiae of detention within an editorial apparatus that ultimately reinforces American exceptionalism and minimizes detainee emotions that might seem unsavory. The new availability of detainees' artistic and cultural productions intensifies the illusion of access to their interiority. Subsequently, reflecting on *Poems from Guantánamo: The Detainees Speak*, Mohamedou Ould Slahi's *Guantánamo Diary*, and detainees' visual art, I consider what kinds of knowledge are produced and foreclosed by their circulation.

Compared to the more affectively and imaginatively laborious process of figuring the detainees as objects of sympathy, the response to the 2008 video of two marines throwing a puppy off a cliff required no such machination. I begin Chapter 6 with a detailed description of the video and its visual and narrative elements. From there, I outline the discursive crisis it provoked by opening a fissure between two prevailing forms of sympathy in contemporary American culture: a humane care for companion animals like dogs and an imperative to support U.S. troops during wartime. While these emotional investments are often compatible (as in the popular fascination with the military working dogs involved in the killing of Osama bin Laden), in this video, they could not be reconciled. As an alternative to the untenable possibility of sympathizing with the humans pictured in the video, sympathy with the puppy seemed to be the predominant reaction to it, but not an entirely satisfying one. I consider the politics of this emotional move, situating recognition of the animal as both necessary and impossible. Among the most remarkable, but apparently unnoticed, dimensions of the video is the dog's accomplishment, as it somehow surmounted mediation to inspire interest and care among spectators who are often inured to the distant suffering of others, particularly during wartime. I conclude the chapter with a reflection on what kind of partial and imperfect justice might be possible in cases of animal cruelty. Given the affective residues that such instances leave behind, in the book's Conclusion, I consider how we might make ethical sense of these remnants. I set forth the possibility of a reaction to suffering that refuses the familiar comforts of sympathy. Indeed, my analyses suggest that meaningful responses to suffering might require detachment from sympathy altogether.[80]

Affect, in Stewart's description, is "fractally complex," and the primary sources I introduce for each figure illuminate different dimensions of the affects directed toward them.[81] Affects are variegated, but they are also durable; Ross argues for the importance of watching for "instances where seemingly dissipating responses are actually morphing into some alternate form."[82] Thus, in accordance with Berlant's description of the "affective event" as "an effect in process, not a thing delivered in its genre as such," I approach the phenomena of apprehension, affection, admiration, gratitude, pity, and anger kaleidoscopically, tracing different scenes as they come into view.[83] In the case of the figures I analyze here, these networks are marked by variety and sprawl. Every chapter ended up longer than I intended or expected. And every time I mentioned one of my case study figures to someone, he or she invariably suggested additional sites of inquiry—like self-help books for military spouses or the PTSD narratives surrounding the spate of husbands who murdered their wives at Fort Bragg in 2002 or the 2016 movie *Billy Lynn's Long Halftime Walk*—that I simply did not have room for. My map of this constellation of objects, therefore, is detailed but not exhaustive. Almost daily, I run across something that could introduce another dimension or embellishment or wrinkle to this analysis. Yet the overall affective composition of American militarism seems to remain the same, and the six figures I concern myself with here remain its anchors, hypervisible and invisible at once.

My work on this book began, somewhat inadvertently, because of my preoccupation with the video of the marines throwing the puppy off a cliff. Although I tend toward squeamishness in general, I find myself largely inured to the effects of graphic imagery; I am not disinterested in the suffering of others—quite the opposite—but tend to respond to it intellectually rather than emotionally. But this video unsettled me, and I could not think my way out of that reaction. As I began drafting the argument, I tried, and failed, to write around it, because there persisted an unthinkability at its core. As I reflected on that, I marveled too at all that I could not know about the detainees, despite having seen them up so close. And so I began to wonder if there might be intellectual work to be done in this space. My difficulty with the video of the puppy reminded me that despite my intellectualizing, I am nonetheless a product of the same national affective and imaginative histories that generate the figurations I describe here, and I wanted to retain that self-reflexivity.[84] Accordingly, I refer to my own habits of imagining these figures periodically throughout the chapters, to demonstrate the pervasiveness of these militarized feelings and the various ways that they might speak through us.[85] Against the tendency to conceptualize affects as automatic or spontaneous, and hence apolitical and ahistorical, I work throughout the book to contextualize them and mark the very specific circumstances in which they arise and the means by which they circulate. My ultimate aspiration for this book is that it might help refine anti-militarist critique by exposing the limitations of its reliance on affective intensities while revealing how and why they have become so deeply entrenched.

CHAPTER 1

Envisioning Civilian Childhood

"Can you draw a picture of Freedom? What does this mean?" Near the end of *We Shall Never Forget 9/11: The Kids' Book of Freedom / A Graphic Coloring Novel on the Events of September 11, 2001*, there is a page that is blank but for those questions and the thin black line that borders it as an anticipatory frame.[1] The preceding text-heavy pages narrate the events of September 11, the official response, the beginnings of the War on Terror, the motivations of the attackers ("Children, the truth is, these terrorist acts were done by freedom-hating radical Islamic Muslim extremists. These crazy people hate the American way of life because we are FREE and our society is FREE"), and an overview of the Bill of Rights.[2] Children can color the smoke drifting from the World Trade Center, George W. Bush's tie, the posters that families hung for their missing loved ones after the attacks, and the veil covering the face of Osama bin Laden's wife—behind whom he is hiding—during the raid on his compound. They can complete a freedom-themed word search or "write a story about what freedom means" and then send it "to someone in China, Russia, or the Middle East."[3] The "Can you draw a picture of Freedom?" page is unique, both in the narrow context of the coloring book and in the larger sweep of children's culture

A page from the *We Shall Never Forget 9/11* coloring book. Although the book is problematic in many ways, this page is a rare invitation for children to experiment with the exercise of political subjectivity.

of the Global War on Terror, in that it invites American children into an unregulated, creative, and expressly political act. This page allows them to conjure a vision of their nation, affording them room to enact and explore their political subjectivities, with all the messiness and risk those acts entail.

Doubtless, there are many reasons to be critical of *We Shall Never Forget 9/11*: its jingoism and hawkishness, its drastic simplification of complex histories, its overdetermined symbology, its enticements into American exceptionalism. The blank page appears near the end of a text that is didactic and prescriptive, replete with guidance about how freedom ought to be understood, and so the invitation to draw is far from unstructured. Still, there is something in its open-endedness that I find oddly refreshing, given the nervous stringency with which children's encounters with militarization are otherwise managed.

To contextualize anxieties around childhood encounters with militarized violence, I begin with a brief history of the construct of "childhood" in America, from the Colonial period to the present. This history demonstrates that predominant beliefs about childhood innocence are not simply natural responses to their vulnerability. Instead, it reveals these beliefs to be historically variable, socially constructed, and unevenly applied to different types of children. This history provides the background for my discussion of the six affects—apprehension, affection, admiration, gratitude, pity, and anger—operative in the figuring of civilian children. These affects are activated most pronouncedly when the vision of the innocent, apolitical child is threatened, insulted, or troubled. And so, in this chapter, I focus on a collection of scandals related to the exposure of civilian children to images of militarized violence. These scandals are fueled by the six affects I describe, and all of them unfold in classrooms, which are frequently the first point of entry into the public sphere for civilian children. In every case, a range of stakeholders—including parents, administrators, and journalists—rush to the defense of the children's senses, especially their sight, to insist that they deserve a view of the world unclouded by violence.

Constructing Childhood Innocence

As legal and economic changes excluded children from the marketplace in Western nation-states during the late nineteenth and early twentieth centuries, the child, as Patricia Pace notes, "gained value as an object of love, sacred in and of herself." Children, no longer valued in economic terms, became sentimentally "priceless." Henceforth, their worth would be calculated in "moral, altruistic terms" instead, which prized their seeming purity, vulnerability, and emotional authenticity.[4] This view of childhood has been generally endorsed by governments as well, and their interventions on behalf of minors tend to prioritize two modes of protection: for the fragile bodies of young children and the impressionable minds of older ones.[5] As the state, prompted by humanitarian reformers, began excluding children from the labor market and, by extension, the public sphere, it relegated children to the category of "objects of the state rather than actors within it" as a measure of its benevolence.[6] This paradigm emerges most clearly in the case of militarization; modern states exempt minors from military service, even as they make decisions that will directly affect the lives of children whose parents enlist. And at the level of public culture, various institutions collaborate to shield civilian children from all but the most benign encounters with militarization.

Yet among marginalized groups, children are unique because they will outgrow at least one element of their disenfranchisement as they age. If disenfranchised groups like women, people of color, people with disabilities, and those who identify as LGBTQ need structural changes to redress their disempowerment, theoretically children need only wait until they come of age to attain full citizenship. James Schmidt describes children's condition as one of organized "exclusion before inclusion" and identifies this as the "central paradox" of children's political circumstances. In the liberal model of childhood, minors are always "citizens in waiting" who need to be educated before joining the polity.[7] Practically, this education involves lessons in history and civics, but I argue that it also aspires to cultivate sensibilities and sentiments and to foreground the affective as a key way of engaging the nation-state and geopolitics. For civilian children, "freedom" is often cast as freedom from the cognitive and physical burdens of militarization. The logic here is circular, holding that children should be shielded from militarization because they are innocent and that children are innocent because they are detached from militarization. The seemingly descriptive quality of this innocence becomes prescriptive, which results in a denial of children's political subjectivity as well as a refusal to acknowledge the various ways that militarization seeps into nearly every facet of daily life. As a political subjectivity, childhood is partial and in flux, but not entirely incomplete, a status that necessitates a rethinking of conventional understandings of their responsibility, capacity, and agency.

With the invention of the printing press and concomitant expectation of literacy, education became the defining experience of Western childhood. Schooling occupies a central place in liberal democracies, training students not only for gainful employment but also for adult citizenship. Sara Ahmed argues that it has also served, since the eighteenth century, as the mechanism by which children learn how to be absorbed into societies and their structures of authority. Predominantly, she contends, this happens through discipline of the child's will. Following the work of Jean-Jacques Rousseau, she writes that through education, "the child is made to will according to the will of those in authority without ever being conscious of the circumstances of this making. This is how will becomes central to the formation of not only moral character but also social harmony: the child becomes willing in a way that agrees with how the child is willed to will, without becoming conscious of this agreement."[8] In concert with acting on the child's will, curtailing or redirecting it as necessary, I suggest that the education system also acts on the child's senses and sensitivities, training them on what counts as a real affront and whose feelings or injuries matter, and how. This is key to the process of socializing them for participation in public culture and political life.

Victorian-era middle-class parents sought to cultivate in their children the capacity for what Peter N. Stearns describes as positive and "fervent emotional expression," and expectations about the responsiveness of children persist today.[9] Sentimental culture and the notion of childhood innocence functioned reciprocally to secure the venerable status of the child, who would function both as a privileged object of sentiment (by virtue of his or her vulnerability) and as an idealized vehicle for it (by virtue of his or her innocence). This was the essence of the Romantic understanding of childhood: "inherently virtuous, pure, angelic, and innocent." The perceived corollaries of these attributes—children as "immature, ignorant, weak and vulnerable"—suggested to concerned adults that children needed and deserved protection.[10] Firmly established by the latter part of the nineteenth century, these idealizations persist well into the contemporary moment, as late-twentieth century claims for children's rights subsequently complicated this model, but did not displace it.[11] Indeed, as Schmidt notes, most prevailing ideas of "'children's rights' imagine young people as non-citizens, non-participants in the state," and so reaffirm their innocence and hence their exclusion.[12]

Overall, the affective experience of childhood in the West changed quite drastically during the late nineteenth and early twentieth centuries, particularly among the middle classes. Stearns attributes this to three major factors. First, the decrease in infant mortality freed parents to become more attached to their children while making the sickness, suffering, and death of children less commonplace and hence seem more lamentable and extraordinary. This also meant that women gave birth to fewer children, and the decrease in birth rate and family size meant that parents could invest more time and attention in the children they had. Finally, the transition from an industrialized economy to one based more on managerial and service skills brought with it the need for more advanced education, including training in the "emotional restraint" necessary for success.[13] In short: despite its masquerading as ahistorical and transcendent, the model of childhood familiar today is not so much a natural artifact as the product of intersecting coincidences of medicine, demography, and economics. This history shapes the way civilian children are imagined and so inflects the forms of apprehension, affection, admiration, gratitude, pity, and anger directed toward the figure of the civilian American child encountering the War on Terror.

Relative to the other figures I consider in this book, civilian children are the least threatening and the most transparently knowable, or at least can be most readily imagined as such. Identifying with children is relatively easy; after all, every adult was once a younger person who aged progressively in a process that felt organic and unwilled, apolitical to the extent that it appeared natural. The idea of the vulnerable child underpins a variety of discourses ranging from pop psychology to global human rights paradigms.[14] These discourses reinforce and are reinforced by children's legally and politically marginal status, making them especially appealing repositories for affective investment, through differences among children in race, class, and ability determine the allocation of that affect.[15] Robin Bernstein notes, for example, that the discursive and material protections afforded to white children from the nineteenth century onward were never extended to African American children.[16] However, the civilian child figured in the instances I consider here is devoid of all markers except national affiliation. This does not mean that gender, race, class, and ability are insignificant, just that they are overshadowed in these figuring practices by a discourse of expressly American childhood innocence.

Affective investments in civilian children hinge upon their putative innocence, read also as a synonym for their vulnerability. Children are ideal objects of the melodramatic attachments that Elisabeth R. Anker describes, which "confe[r] virtue upon innocent people who unjustly suffer from dominating power."[17] Discourses of childhood innocence operate descriptively and prescriptively at once. And the preoccupation with children's victimization by terrorists, or militarization more generally, obscures the systemic facts of their domination by people close to them and structural disempowerment by a range of institutions.[18] In the cases I analyze, children seem to suffer most at the hands of their teachers, represented either as carelessly incompetent or dangerously politicized, who redouble the harms done by terrorists or other villains. Melodramatic political styles operate in situations of "overwhelming vulnerability,"[19] and in the context of a war framed as an existential threat to the nation, civilian children seem especially imperiled and especially worthy of protection.

Figuring the Civilian Child in the War on Terror

All the operative affects feeding into the figure of the civilian child are underwritten by apprehension. This apprehension takes the shape of worry over children's mediated encounters with graphic truths of terrorism, war, and militarized violence. While the news stories about these incidents include reports that some children were discomfited, most of the harm that adults objected to was hypothetical or future, and none of the controversial incidents I consider in this chapter involved children being put in any physical danger. The perils are cognitive, emotional, and sensory as they encounter information about the attacks of September 11, the torture at Abu Ghraib, the beheading of Nicholas Berg, and ISIS recruitment of young people, all phenomena that were themselves heavily mediated. I do not mean to minimize the possibility that images of these events might be distressing; my interest here is in the broader cultural response to that potential distress. The apprehension that characterized this response is rooted in a century and a half of worry about the emotional dangers of communication technologies, made more acute by the ubiquity and instantaneity afforded by contemporary forms of mass media.[20] This history, to the extent that it cues us to particular kinds of concern over civilian children—and systematic unconcern with others—needs to be examined.

I acknowledge that the critiques I offer might read suspiciously, or even coldly, as inadequately attentive to the children's sensitivities. I do not mean to downplay the upset that these experiences might have caused for actual children. My focus, rather, is on the discourses activated in these controversies, the ways that they discount the political subjectivity of children while tacitly endorsing the continuation of the forms of militarization that beget this kind of violence. Moreover, this concern over civilian children's mediated encounters with warfare obscures the more immediate dangers faced by military children and, of course, the ongoing threat of violence that haunts children who grow up in the countries where these wars unfold. It is worth noting, also, that these figured children incite apprehension only to the extent that they appear nonthreatening. Sunaina Marr Maira writes, for example, about the state's targeting of Muslim youth after September 11 through various forms of profiling, infiltration, and surveillance.[21] Those young people inspired a different kind of apprehension, rooted in fear that they might be, or become, radicalized against the United States. The children involved in the classroom scandals, by contrast, inspired apprehension when it appeared that their apolitical natures might be jeopardized or that they might learn the wrong kinds of lessons about American militarism.

Apprehension is deeply compatible with affection, articulated in filial forms of intimacy, protection, and care. The public sphere that most civilian children inhabit is designed to operate *in loco parentis*, an arrangement that naturalizes public affection for them as a bond like the love of a parent for their offspring. But love is always political, perhaps most especially when it claims otherwise.[22] This kind of protective affection for the civilian child is historically conditioned, directed less at specific children and more at the whole socially constructed category of "the child." The attributed innocence that inspires apprehension over and affection for the civilian child also sparks admiration. Lauren Berlant's analysis of "infantile citizenship" is instructive here. Infantile citizens, as Berlant describes them, are impressionable and ingenuous: earnestly wishing to belong to the nation-state, certain of its abiding goodness, and bearing no responsibilities for its failings or violences.[23] With an essential naïveté, the infantile citizen is appealing, lovable, admirable because it is blameless and largely helpless; Berlant writes that these young model citizens appear to embody "a consciousness of the nation with no imagination of agency."[24] The repeated identification of children as ideal American citizens in turn secures a vision of America itself as innocent.[25] Maira notes that "young people symbolize the unknown future or possible direction of the nation and become the site of projection of adult hopes and fears about their society."[26] Accordingly, veneration for infantile citizens is tinged with appreciation for what they might be capable of, how their earnest patriotism might be channeled and rewarded. For its capacity to anchor this kind of identification, the figured civilian child inspires a form of gratitude, offering a vision of citizenship that is pure and only lightly politicized, vouching for how good the nation-state as a whole could be if it were only untainted by war or other geopolitical concerns.

Civilian children garner pity for their imagined delicate sensibilities, which appear all the more imperiled during wartime. In response to this, only adults are allowed to be angry, provoked to righteous and justifiable ire by the things that rob children of their innocence. This anger manifests as outrage at teachers and school administrators who permit or require civilian children to encounter information about militarized violence. Adults deploy this anger on behalf of children, who—ideally—would not be capable of or inclined toward it. Childish displays of anger threaten to expose childhood innocence as a fiction, and angry children are much harder to pity; consider the frustration that parents feel when a child's sniffling plaint turns into a screaming tantrum. Indeed, the eradication of the child's propensity for anger, along with regulating their other base impulses, has been a central task of Western parenting for hundreds of years.

During the Colonial period, prevailing Calvinist doctrines identified children, particularly babies, as the embodiment of original sin, which compelled parents to chase the devil out of them by force. This "doctrine of infant depravity," Bernstein notes, remained ascendant through the late eighteenth century. At that point, a "competing doctrine" emerged that posited children variously as "innocent," "sinless," "absent of sexual feelings," and oblivious to worldly concerns. In this model, even the very young could be "holy angels leading adults to heaven." Concomitantly, the Scientific Revolution challenged the notion of original sin and promoted what Stearns characterizes as an "intellectual redefinition of childhood" in terms more physiological and developmental than theological, but also "explicitly emotional."[27] By the mid-nineteenth century, the notion of essential childhood innocence was firmly sedimented as a cornerstone of sentimental culture and approaches to childrearing.[28]

While experts began urging parents away from the use of fear and anger in childrearing in the middle of the nineteenth century, they also promoted stringent standards for childhood emotions. These included the expectations that children be obedient and demonstrative in their affections toward family members. Parents, in turn, were urged to enforce these new mandates with guilt, rather than rage of their own.[29] In keeping with idealized visions of the home as a space free from violence and coercion, early twentieth-century literature on child development evinced a new concern about children's own anger (rebranded as "aggression") as a sign of pathology or maladjustment.[30] In their encounters with contemporary militarization, civilian children are given limited license to express their anger, provided that it does not devolve into a critique of the state or its policies. Particularly in the aftermath of September 11, children were permitted, perhaps even expected, to be patriotically upset by the attacks, responsive at the level of emotion but not politics.

The children's literature industry was one of many enterprises that aspired to train children whose reactions were not automatic or ideal. Surveying the vast array of books that began appearing shortly after the attacks, Kenneth Kidd notes that many of them were sensationalized and graphic. At the levels of both aesthetics and plot, he claims, these books endeavored to make the young reader's literary experience "traumatic itself—as if . . . children can't otherwise comprehend atrocity."[31] Kidd found that this literature, which began coalescing almost immediately after the attacks, uniformly described September 11 as "the ultimate and easily knowable affront to self and nation." Moreover, he notes, they did this work by "appropriating the vulnerable / dead child as the representative American."[32] In these representations and across public culture more broadly, civilian children became the embodiments of an inherent natural goodness and sensitivity. In this literature, fictionalized children encountering the September 11 attacks served as ideal models of childish response to terrorism. In most of the news accounts detailing the scandals I analyze, this ideal is secured by the silence of the children concerned. The involved children are rarely, if ever, given the opportunity to voice their experiences, but they all share an imputed status of victim—whether of their teachers, of the media, of terrorists, or of all three working in frightening concert. In these narratives, the actual children concerned are conspicuously absent, appearing only through the traces of the work they produced in the classroom and represented primarily by their adult proxies, eager to recount the grievous harms that have been done.[33]

"That's Something Kids Should Get in Trouble for Drawing": Children's Art and September 11

In the Library of Congress's American Folklife Center (AFC), there is an American flag with eighteen stripes and no stars.[34] Despite this oversight at the level of detail, it was made with obvious consideration and thoughtfulness, each stripe cut and placed just so. It is preserved with a similar care, now stored in a box lined with tissue, available for viewing only by advance request and elevated to the status of archival artifact. This flag is part of a collection of sixteen similar works held by the AFC, all of which were created by third-graders in Knoxville, Tennessee shortly after September 11. Originally commissioned by a local writer's guild for display in a public memorial space, these pieces capture third-graders' recollections of the attacks. Drawn on construction paper in combinations of crayon, ink, pencil, marker, and colored pencil and with varying degrees of maturity and skill, these pictures convey their creators' understandings of the attacks and their emotional resonances. Although they are now stored with an eye toward preservation, the drawings themselves are in relatively poor condition. They are laminated, but in addition to the staple holes where they were hung, they all show signs of sun or water damage.

Were it not for the curatorial largesse of the Library of Congress, these drawings likely would have met the same end that most children's artwork

This drawing, by Knoxville third-grader Hannah Beach, is part of a collection of sixteen pieces of children's art about September 11, held by the American Folklife Center at the Library of Congress. By their inclusion in the archive, these impressions of the attacks are endowed with historical significance. Credit: September 11, 2001, Documentary Project collection (AFC 2001/015), American Folklife Center, Library of Congress.

does—thrown out or more lovingly dispatched into indefinite storage. After they had served their purpose in Knoxville, someone involved with the memorial project offered the collection to the Folklife Center for their archive of personal recollections of September 11. The center agreed to house it because the drawings seemed so expressly personal, despite the fact that they did not fit the criteria for the kinds of narratives the AFC originally wanted.[35] Children's writing, drawing, and play can be a unique and valuable source of information about the historical moments from which they emerge, but such artifacts, often deemed insignificant, rarely make it into the archives.[36] By contrast, these pieces were essentially hung on the national refrigerator. The archiving of these artworks is, on the one hand, an important acknowledgment of the nascent political subjectivity of children and their capacity to register events of broader historical significance. On the other hand, the act instantiates a particular norm about how children ought to respond to such events. It privileges children's sight and visual sensibilities, provided that they demonstrate the right affective calibration toward the event, the correct degree of receptivity. The ideal childhood political subjectivity is capable of being traumatized, but naïve enough to experience the trauma only as trauma, not as politics.[37] Reflecting on the children's literature of September 11, Kidd writes, "Picture books about 9/11 insist upon a traumatized reader, but they also redefine trauma as the stuff of pop-psychology, emphasizing—and delimiting—choice, pleasure, and action."[38] This is a trauma that is intensely individualized but rarely connected to larger communities or phenomena.[39]

None of these drawings reveal any new information, yet their inclusion in this archive suggests that they are historically meaningful. They do not record the events of September 11 themselves, but rather childish impressions of them. Many of the images are profoundly anthropomorphic; for example, the Pentagon steels itself for attack, the Statue of Liberty cries, as do the towers of the World Trade Center, which in one drawing reach out for one another.[40] The contours, rather than the contents, of the drawings matter here. Apparently documenting the corruption of childhood—and perhaps also national—innocence by the attacks, they retroactively verify the presence of that innocence in the first place. Of course, not every child's experience of September 11 is afforded the same kind of iconicity; for example, Maira recounts that many Muslim youth felt silenced after the attacks and notes that they are rarely, if ever, allowed access to public forums to share how the War on Terror affected their lives.[41]

By contrast, the young artists whose work is held at the AFC became privileged representatives of their generation. Their artwork resonated with the broader national sentiment, and they seemed to get it right, at the level of intention if not always of detail; they knew that it was time for an American flag, even if they did not know exactly how to reconstruct it. Strikingly, some of the drawings include playful details like smiling suns and singing birds that seem out of sync with the overall gravity of the subject. More than childish incomprehension, these details suggest a prehistory of affective pedagogy, an untrained approach to trauma, how it looked to children before they were taught how it was supposed to feel. But these poignant little errors also seem to verify the status of the drawings as authentic and spontaneous reflections of how the children interpreted the attacks: apolitically and through a merciful screen of ignorance. By contrast, apprehension is activated at any indication that children are producing September 11 artwork with a different lens.

Superficially, it was a nonevent, a tempest sized perfectly for the teapot of small-market local news. In September 2012, reports surfaced that a fourth-grade teacher in El Paso, Texas "forced" students to draw pictures commemorating the September 11 attacks, focusing on their most graphic elements. Parents complained that their children had been frightened by the assignment, a few local and national news outlets seized briefly on the story, and the school district vowed to investigate.[42] I was unable to find any further reports about how the story ended, but any resolution would have been largely superfluous to the work of figuring these children.

A fourth-grade teacher in El Paso, Texas, gave students a drawing assignment about September 11 and provoked outrage from parents. This freeze-frame is from an ABC News story about the incident.

Both ABC News and Fox News frame their stories about the assignment in terms of parental "outrage" or "furious"-ness. All of the anchors who report the story do so with mild incredulity in their voices as they hit key details like the age of the children (fourth grade) or the specificity of the instructions that the teacher allegedly provided. In a Fox brief on the incident, one of the anchors warns (adult) viewers that they might find the drawings "disturbing." The lengthiest segment, from a local Fox affiliate in Texas, includes an interview with Ivie Gremillion, a parent who describes the distress of her daughter and a classmate. Gremillion insisted that children under other circumstances would and "should get in trouble for drawing" this kind of thing: "people being murdered" and "committing suicide." She went on to express a doubled concern that the children were being frightened *and* that the teacher had couched the assignment in racist terms. Her daughter was especially worried, she said, because her husband was expecting deployment to Afghanistan before the end of the year. Gremillion emerged as the main character in this drama, but the news stories—with their references to groups of irate parents—suggest that all of the students in the classroom were discomfited by the assignment and that her child's trauma was simply more acute. Gremillion's proposed remedy is that a "counselor" go back to the students and "re-explain" to them what had happened. Of course, I do not doubt that Gremillion is concerned about her daughter's well-being, and she does offer a multi-layered critique of the assignment, which gets buried by the more sensational elements of the reporting. My interest here is in why and how an incident like this becomes national news.

Beginning in the nineteenth century, children's art acquired a purpose. Middle-class children of the Victorian era had ready access to painting books, precursors to coloring books. Patterned on great works of art but cheaply produced, these books were intended as affordable means to cultivate taste and discernment in young people.[43] Coloring books replaced painting books by the middle of the twentieth century; Crayola introduced their resplendent boxes of sixty-four crayons in the 1960s, when "suburban parents were desperate to keep their kids occupied."[44] While these leisure activities were relatively unstructured, institutionalized art education began with practical motivations during the Industrial Revolution. Schools began offering art classes because of a marketplace demand for artisans and craftspeople who, in Donna Darling Kelly's terms, "might provide a cheap, but skilled, labor force for the burgeoning urban factories." With the subsequent ascendance of psychoanalysis, children's art took on a new significance as a "mirror" of the child's psyche. Ultimately, Kelly suggests, this gave way to a "window" view of children's art making, which considers its aesthetics and communicative intent and understands the image as the "child's reality, and the act of representation [as] the goal, not the truth behind the goal."[45] Yet while this generous approach to children's art might offer more recognition of their creative agency, I suggest that it also raises the stakes of their creating and intensifies the scrutiny their artwork elicits while prioritizing the sensitivity and perceptiveness of the child artist above other concerns. Indeed, the news stories about this assignment all included some form of close reading of the drawings, reflecting the persistence of this approach to children's art.

In its public statement, the school district emphasized that it "regrets the insensitivity" of the assignment. But insensitivity to what, or to whom? It is significant that Gremillion calls for a mental-health professional—rather than a different kind of expert—to teach the students about September 11, a suggestion that the news stories tacitly endorse by intimating that the children made their drawings under duress and that this process was traumatic. In general, however, drawing is a common way that children reconcile confusing or frightening events, and art therapy was a key element in the recovery of many children who experienced the events of September 11.[46] Art therapists and child psychologists encourage honest, age-appropriate, and reassuring conversations with children who encounter disturbing news stories in order to help them make sense of the events. Thus, in 2002, the New York University Child Study Center and the Museum of the City of New York released a glossy book entitled *The Day Our World Changed: Children's Art of 9/11*, a project that was lauded by community members and government officials alike.[47] Crucially, those drawings are virtually indistinguishable from the ones held at the AFC and from those produced in El Paso. The controversy in the latter case, then, was the explicitness of the instructions that the teacher allegedly gave and presumably the fact that this kind of information was disseminated in a classroom at all, while a teacher required that the students engage with it, thus violating their otherwise apolitical sensibilities.

Relative to the educational scandals I consider in the next section, this one was localized and short-lived, but it nonetheless reveals something about conventional understandings and underestimations of children's political subjectivities and their capacities to reckon with the outside world. Although the news reports intimate that the teacher gave students explicit instructions on what to draw, the small sample of drawings shown in the report evince the uniqueness of each child's interpretation; some drew the towers, others the planes or hijackers, while one added dialogue to people jumping out of the towers, saying "I love you" as they fell. This artistic license hints at a capacity to invent, imagine, and renarrate the events of the day. The scrutiny of the drawings for any sign of trauma and refusal to read them as inventive ways of processing historical facts obscure their content in an ostensible attempt to parse it. Simultaneously, it reifies an understanding of sensitivity as vulnerability, as opposed to a means by which children might come to know the world.[48]

Prior to the nineteenth century, childhood innocence was a theological matter; subsequently, it was reframed as a moral and epistemological one.[49] Bernstein writes that as this construct developed, "to be innocent was to be innocent *of* something, to achieve obliviousness. This obliviousness was not merely an absence of knowledge, but an active state of repelling knowledge."[50] In the case of the classroom drawings from El Paso, the children did not have the option to repel the knowledge: hence the controversy. Yet the conviction that children ought to be sheltered from the realities of death and mortality has a fairly short history. As late as the 1870s, the expert and lay consensus held that children should be trained how to grieve and participate in the elaborate mourning rituals of the Victorian era. Subsequently, within the span of fifty years, the prevailing wisdom shifted in the opposite direction, with experts advising parents to keep children away from the knowledge, sight, and ritual engagement with death and dying.[51] Over time, the perceived innocence of children shaded into arguments about their vulnerability, so that the attributes eventually became functionally synonymous.[52] Once interwoven, innocence and vulnerability acquired moral valences whereby helplessness was coded as virtue, and this elision keeps the affects in orbit around the figure of the child besieged by contemporary American militarism. Children who respond to this threat appropriately earn the unquestioned and public defense of their parents. They get their art preserved in a national repository, a testament to the rightness of their vision and the preciousness of their sight. More generally, affection for the helpless child starts to seem automatic, unquestionable. With helplessness as the embodiment of goodness, admiration follows shortly. If goodness comes under attack by discomfiting knowledge, pity follows. And this bastion of goodness starts to look extra precious, unusual in a world apparently defined by danger and cruelty, and morphs into a reason to be grateful. For all of this, ultimately, the response is anger.

"These Staff Members Exercised Poor Judgment": Classroom Encounters with Atrocity Images

Eight years before this, around the time the El Paso fourth-graders were born, in the spring of 2004, there was apparently an epidemic of visual misjudgments by American high school teachers. A Massachusetts current events teacher was briefly removed from his classroom after assigning his students to view and write about the Abu Ghraib photos; he was subsequently reinstated after the American Civil Liberties Union (ACLU) intervened on his behalf. In Texas, two teachers were suspended for the remainder of the school year after showing their classes the Internet video of the beheading of Berg. Similar screenings were reported in high schools across the country. All such screenings caused controversy in their communities, and the news reports followed a common narrative pattern: teachers expose their students to this troublesome content, parents become concerned and complain, and school administrators reprove or censure the teachers.

Unlike the apparently isolated incident of the September 11 drawing assignment, these stories were more numerous, more widely reported, and more thoroughly investigated. On the surface, this allocation of attention seems odd. After all, the children involved were much older and hence presumably better equipped to deal with this kind of information. And watching is generally considered to be a more passive act than drawing. But here, the discourse of childhood innocence takes on a new urgency for adolescents. In these cases, legalistic rhetorics of rights were deployed instead of the commonsense, emotional appeals to ideas about the vulnerability of younger children that informed the response to the drawing assignment. I argue that this reaction was especially intense because the experience of seeing others suffering, rather than the witnessing of collective trauma by which all Americans could claim victimization, placed the fiction of the apolitical civilian child in even greater jeopardy.

Ultimately, the debate over educators' judgment and students' maturity distracts from the rights of the people pictured in the images and obfuscates the real significance of these visual events and the kinds of spectators and citizens they train students to be. While most observers were preoccupied by the notion that students would not be able to cope with the sight of such violence and suffering, I suggest that the more urgent questions are whether they would be able to apprehend the actual significance of what they are seeing and to empathize with the people at whom they are looking. The issue is not so much that students might be traumatized by these sights or even that they would become inured to this sort of violence. Rather, my concern is that there is no mechanism in these displays for promoting students to think through their relationships to them or to attend to the intricacies and obligations inherent in their roles as spectators and citizens.

The first Abu Ghraib photographs appeared before the American public on April 28, 2004, in a story on *60 Minutes II.* Almost immediately, they became iconic and ubiquitous, and for a time, it seemed that these scenes of American military personnel presiding over tableaus of debasement and sexualized torture might prove fatally damaging to the Bush administration or its war effort. But once the official apologies were blandly issued, the few token punishments were meted out, and sufficient time passed, Abu Ghraib receded from public view, memory, and concern.[53] In the period before the attention dissipated, Abu Ghraib provoked a range of questions about the facts of what had happened and how, about the legal status of the prisoners and the abuse they had suffered, about the culpability of their torturers and the people who commanded them. Images of atrocity are uniquely, and sometimes contradictorily, generative; Elizabeth Dauphinée observes that because its meaning is not inherently stable, "the image of the body in pain animates and makes possible a whole host of political activities, from torture to military intervention to anti-war activities to critical social science scholarship."[54] The pluripotentiality of the Abu Ghraib photos was vividly dramatized when they reappeared in classrooms full of American children.

Because the story of Berg's beheading broke at the peak of the debate about Abu Ghraib, the two visual events fused into an intertext, each inflecting the interpretation of the other. The American public had been familiarized with the new genre of the beheading video through the story of Daniel Pearl in February 2002, but Berg's death was far more visible. Whereas news organizations treaded carefully around the grisly details of Pearl's beheading, the Berg video was widely available on the Internet before any news organizations reported it, and scores of people were already viewing the video online while traditional news outlets dithered about what to do with it. Berg's beheading became the second-most-popular search on the Internet in May 2004, after *American Idol.*[55] If many spectators seemed unsure about the true or full significance of the Abu Ghraib photographs, there was very little equivocation in American media about how vividly the beheading of Berg testified to the depravity of the enemy.

Both sets of images were targets of intense regulation as various actors—ranging from the U.S. government to news agencies to individual high school teachers—tried to control public access to them, their circulation, and the debates about their meaning. In the high schools where teachers showed the photos from Abu Ghraib and the video of Nicholas Berg, everyone seemed to agree in advance that the images were overwhelmingly powerful in myriad ways and that the students were defenseless against them. Participating in this fiction about the omnipotence of the images requires ceding human agency to these documents of atrocity and obscures a dense and tangled network of relationships in the process. The conceptualization of teachers as mere transmitters of the images and students as passive consumers of them promotes a helpless vision of citizenship in which Americans are assailed by grisly images at which they can do nothing but look, compounding their initial victimization by the terrorists and consequently redoubling their innocence.

None of the news stories offered any indication that students were discomfited by what they saw in these lessons. Brian Newark, who taught in Bellingham, Massachusetts, assigned his students to log onto either CNN or MSNBC to look at the Abu Ghraib archive. He claims to have allowed students to choose whether or not to participate and insisted that he gave them the option of an alternative assignment and that none of the students exercised it. Still, a parent complained, and the high school responded by barring him from teaching his apparently very popular current-events class in the fall. With the support of the ACLU, he sued on the grounds that being thus prohibited was a violation of his civil liberties; he won and was reinstated to his classroom in a decision deemed a victory for free speech.[56] This ruling was made, ostensibly, to protect Newark's rights to show these images as a form of protected speech or, perhaps indirectly, on the basis of a claim that the students had a right to see them. This paradigm cannot account for the rights of the people pictured in the images. Excluding them from the realm of people who might be affected by these displays ensures that the only possible victims here are American, whether the students or Newark himself.[57] And in the process, the photos and the events that they depict become significant only to the extent that they impact Americans, and American children more specifically.

This Abu Ghraib controversy was apparently isolated, but there were widespread accounts of teachers showing the Berg beheading. A Google search turned up instances in California, North Carolina, Nebraska, Pennsylvania, Alabama, Washington, South Dakota, Oklahoma, Ohio, and Canada. The case of Michelle White and Andy Gebert, both teachers in Justin, Texas, seems to have garnered the most attention. The special assistant to the superintendent for their school district opined that "these staff members exercised poor judgment" by choosing to expose their students to this imagery.[58] Suspended for the remainder of the school year, they wrote apology letters to parents. One report indicated that that they got the video from their students and that the teachers themselves were sent to counseling. For what, precisely, is unclear, but it is significant that this incident was addressed as a mental health issue rather than a political or ethical one, as when Gremillion requested that a counselor come in to talk to the students about September 11.[59] All of these controversies, which attribute a childish innocence even to adolescents, have in common a narrow focus on students' feelings and teachers' actions to the exclusion of the rights of the people at whom they are looking.

Worrying excessively about harm done to students overemphasizes their vulnerability, a spurious concern given their status as American citizens, which places them in a privileged position relative to the detainees at Abu Ghraib and most of the suffering others at whom they will look. Emphasis on student vulnerability rearticulates a post–September 11 fantasy of generalized American vulnerability *qua* innocence, recentering American suffering—which is always framed agonistically, exclusively, and superlatively—at a visual moment when it was threatened with eclipse. As Donald E. Pease argues, "The state's representation of a vulnerable civilian population in need of the protection of the state was fashioned in a relation of opposition to the captured Taliban and Iraqis who were subjected to the power of the state yet lacked the protection of their rights or liberties."[60] The relative privilege of American spectators, even pre-adult American spectators, over the detainees at Abu Ghraib makes it easy to objectify those detainees and hence dismiss their suffering; on the other hand, the highly visible death of Berg makes it easy to revert to the notion that all Americans are at risk.

These adolescent spectators inhabit a liminal place on the cusp of adulthood and the full citizenship linked to reaching the age of majority; their proximity to legal adulthood, though based on a somewhat arbitrary number, is a not-insignificant variable. Historically, the figure of the juvenile has posed a challenge to the ideal of the perfect child, highlighting the inevitability of outgrowing innocence and the ineluctable onset of full political awareness and subjectivity. Maira writes that adolescents "are assumed to fall between the cracks of 'innocent' childhood and 'stable' adulthood, dangerously outside of normative social structures and always teetering on the brink of revolt."[61] Their technical status as children has legal implications and raises questions about what kinds of reasoning and discernment can be expected of these students, what kinds of political subjectivities they can be said to have.[62] But citizenship is a matter of both rights and obligations, and there is not broad cultural consensus on what sort of either can be attributed to adolescents. Adolescents have partially outgrown the physical helplessness that characterizes early childhood, and so their agency is lopsided or uneven, but it is agency nonetheless. Even if they are not yet old enough to vote or serve in the military, they have emerging ideas about the world and thus are emerging political subjects defined by much more than vulnerability.

For educators seeking a more structured approach to teaching about the Abu Ghraib photos, PBS offers a lesson plan for grades nine through twelve as a companion to a *Frontline* documentary called *The Torture Question*. Insofar as PBS operates under the aegis of the government, this curriculum represents a semiofficial approach to the images.[63] In this module, students watch the documentary and subsequently participate in a range of complementary activities, like reading the Geneva Conventions, deliberating about what counts as torture, and engaging in thought experiments about the sort of counterterrorism policies they would enact. The multipart curriculum is thoughtful and layered, but it does not include any structured invitation for students to reflect on their own status as spectators; insofar as the curriculum is concerned with questions of American and global citizenship, this omission suggests that there is no link between these different roles, when in fact they are intimately connected. There is a warning to teachers that the subject matter might be difficult and a statement that the photographs are "unquestionably repulsive," but no advice on how to talk about what that means. There are factual questions about how the photographs became public, a reference to "humiliation and ridicule" of prisoners, but no suggestion that students need to be introspective about looking or that teachers need to be self-reflexive about showing. Certainly, some teachers may be doing this work on their own, but the oversight in this prominent resource is telling.

Compared to the act of displaying the Abu Ghraib photos in a classroom, recirculating the Berg video raises a different set of affective, ethical, and political questions. CBS News ran a story about classroom showings of the beheading but wagered that "the reality is none of the students needed a teacher to see the video."[64] However, the likelihood that students would have already seen the video does not mean that showing it to them again is ethically uncomplicated or neutral. We might question the wisdom of exposing young people to the sight of such gruesome violence, particularly because the video likely would have been otherwise inscrutable to most of these students, as the killers read their statement in unsubtitled Arabic. But this is not the most important issue, as concern about the students' sensibilities risked crowding out consideration of Berg's privacy, which apparently did not figure in the public debate about the classroom screenings. The result is a vision of citizenship that is terrorized but also solipsistic, in which Americans (whether Berg himself in the video or spectators of it) are wantonly victimized and under no obligation to be self-reflexive about the cost of exercising their putative rights.

One version of the story of Berg in the classroom alleges that a teacher in Texas showed the beheading video during a pizza party.[65] Ever since I ran across this account, I have tried to imagine how that might have happened, by what logic the teacher might have come to believe this was a good idea. Was the intention to make the festivities somehow topical or educational? Was it a Hail Mary attempt to make the students pay attention? How did the students react? Were they disinterested, insouciant, hostile, troubled, or bored? What did they say to one another as they were scattering for their next class, or to their parents in response to the question of what they did in school that day? The thought of the combination of the sensory and social landscapes of the high school classroom pizza party with those of the beheading video makes me feel a little queasy. It is obviously gratuitous, self-evidently unjustifiable, but it is the very gratuitousness of the act that secures our moral outrage. And this outrage is problematic because it reinforces an ideal of American citizenship as innocent and virtuous, embodied most perfectly in the figure of the civilian child, entirely untouched by militarized violence.

When spectators are children or students and the objects of their looking especially graphic, the young people might seem even more like victims, particularly on questions of militarization. J. Marshall Beier writes that in the discourse about children's participation in wartime national cultures, "complex subject positions collapsed into juridical categories and the rhetorically necessary construction of victimhood insisted upon children's separation from politics and the denial of their agency."[66] In the case of high school students—just a few years away from being old enough to enlist themselves—viewing images like the ones in question here, these conceits are unsustainable. This is not to suggest that we should be unconcerned about students; rather, I argue that worrying over their vulnerability is the wrong place from which to begin their education. The preoccupation with students' fragility is rooted in a deeply flawed assumption of equivalence between their spectatorial suffering and the bodily, often mortal suffering to which they bear witness. Prioritizing their susceptibility to visual harm also establishes a host of agonistic relationships and defines their citizenship as a locus of injury. The Americanness of the perpetrators in the Abu Ghraib photos imbues them with the potential to be even more politically traumatic, as the teenage students will encounter this brutality when they are on the verge of becoming full citizens.

"Age," Maira writes, "is a category that is deployed by the state, and that is used to understand what constitutes 'good' citizenship or mature selfhood," a way for the state to distribute its attention, resources, and power.[67] It is also a gradient along which the state determines accountability. Preemptively, then, the blame for this potential undoing is foisted onto the teachers who showed the images and onto the images themselves. Moreover, the apprehension around these classroom happenings is predicated on the assumption that students' responses to the sights of torture or beheading would be negative or unpleasant. Many young people regularly, habitually consume media that "disembody" war and violence and so encourage them to see these phenomena as spectacularly intense.[68] Worrying that students might be overwhelmed by the documentary evidence of violence ignores the possibility that they might feel nothing at the sight of it and denies the possibility that they might feel a kind of pleasure. It presumes that they are already compassionate, empathetic to a fault.

Most troublingly, this configuration places students in an antagonistic relationship to the images, which risks devolving into a lack of concern with or even enmity for the people pictured within them. In a context where teachers are reprimanded for showing such pictures, images are refigured as weapons with the capacity to unsettle or unnerve students and interrupt their progression toward uncomplicatedly patriotic adult citizenship. The images are recast as obstacles, the people pictured within them rendered incidental or perhaps the cause of all the unpleasantness; they are never recognized as victims or subjects. In fact, the greatest risk inherent in showing these images is reinforcing the idea that the most important thing about them is how they make students feel, with the implied corollary that they should never have to feel anything unpleasant.

"Rip It Up": Children Imagining ISIS

More recently, ISIS began generating atrocity videos apace, and such artifacts remain apparently tempting visual aids for classroom instruction. Once again, teachers are making questionable visual choices, and once again, the news is taking note. The example of ISIS further complicates the entanglements of childhood and militarization in light of the group's success at recruiting Western young people to support and occasionally fight for their cause. Many of the ISIS-related assignments are similar to their Al Qaeda–inspired predecessors, and the news stories about them follow the same pattern I described previously, wherein the denouement is typically the action or, less commonly, the inaction of school or district administrators. For example, an English teacher in the Cincinnati area screened the video of ISIS operatives burning a Jordanian Air Force pilot alive in a cage; one female student reported that the teacher told students who did not want to watch to put their heads down on their desks, but that hearing without seeing was "worse" than seeing and hearing at once.[69] And a sixth-grade teacher in Alvin, Texas, in a lesson on reading comprehension, assigned students to read an MTV News story about ISIS and identify information that was confusing, while analyzing the writer's use of evidence. Parents objected because the story contained a reference to beheading. The district opted not to reprimand the teacher, citing the likelihood that students would already know about ISIS from peers and social media and asserting that the classroom is a place where they can gather more information and practice analyzing it.[70] According to my research, this instance is an outlier in that the school opted not to censure the teacher in any way: no public disavowals, alternative assignments, mandatory counseling, official reprimands, or dismissals.

Perhaps the most interesting incident is the one from Salem, Utah, where a first-year teacher assigned her students to create a propaganda poster for ISIS, ostensibly to illuminate how the group recruits.[71] Unsurprisingly, parents were incensed. And, again unsurprisingly, the district concurred with their reactions. According to the local news, the school's administration canceled the assignment and then "'sat down'" with the teacher. But some ambitious students had already turned it in; the school responded by shredding the posters they had submitted. One mother was shocked to discover that her daughter, excused from turning in her assignment, had failed to dispose of it, and ordered her to "'rip it up.'"[72] A story in the *Washington Post* includes a photo of the assignment sheet, front and back. One side contains a full page of text explaining why "young Muslims join ISIS."[73] The other instructs students to draw a poster that is "neat, colored, professional, etc.," and includes a fine-print parenthetical telling students to request an alternative assignment if they are uncomfortable with this one. But most of the page is an empty rectangle, waiting to be filled with the product of the students' imaginings of what might appeal to a would-be member of ISIS.

Here again: the dizzying potential of the blank page for the inquisitive and artistic young person. And an invitation to imagine a haunting pair of political subjectivities: that of the ISIS recruiter and that of the young person who might be persuaded by what they produce. Of all the stories about children's classroom's encounters with the War on Terror, this one is singular because of where the parents' anger is directed. Beyond the corrective sit-down, the reaction to the teacher seemed rather muted, particularly in comparison to the destruction—whether by mechanical shredding or mandatory ripping-up—of what the students produced in response, when their creation could not be preempted altogether. The anger usually reserved for the educator that leads them astray was instead redirected, albeit obliquely, at the students themselves and their temporary refusal to enact a subjectivity that is innocent and apolitical.

Civilian children who deviate from this apolitical ideal expose the fiction of childhood innocence and hence that of national innocence at the same time. The affective investment in the civilian child is ultimately contingent on their eschewal of politics. And this figuring prescribes how children should have felt during their encounters with these images and subsequently imagines that they actually did, which intensifies the perceived need to protect children from the truths of militarization. Yet at any inkling that they are not politically innocent, the affective investment begins to splinter: suddenly the apprehension is useless and too late, the affection, admiration, pity, and gratitude begin to look misplaced, and the anger—parental and disapproving—begins to turn on the children themselves. Of course, this is uncomfortable, grating as it does against the affective and political pleasures of figuring these children in this way. The figure remains intact as long as there is someone else, whether a teacher or a terrorist, to blame for any corruption. Ultimately, this results in a privileging of the sensitivities of civilian children above those of almost any other being, including their military counterparts, who are the focus of my next chapter.

CHAPTER 2

Affective Pedagogies for Military Children

The Department of Defense wants military kids to breathe from their diaphragms. To facilitate this, the DoD's National Center for Telehealth and Teletechnology (T2) created an app called "Breathe 2 Relax." And while the app itself seems targeted more toward adults, the DoD also promotes it across its *Military Kids Connect* website, where deep breathing is a suggested remedy for nearly any kind of distress that a military kid might experience.[1] "Breathe 2 Relax" allows users to select a soothing background image (like a rainforest, a beach, or a science-fiction-y "cosmos") and customize their background music (the options are all tinny, New Age) as support for a guided breathing exercise. Once they tap "*Breathe*" and use a slider bar to rate their stress, a robotic female voice begins talking them through the process. At the end of the designated number of inhale-exhale cycles, along with an offer to continue breathing deeply and encouragement to "take time" before returning awareness to the immediate surroundings, the user can again rate her perceived stress. Reviews on the App Store are overwhelmingly positive, though the app does warn users to return to their normal breathing rhythms if they "experience any unpleasant sensations . . . have difficulty breathing" or feel "dizzy, nervous, or out of control."

No doubt, deep breathing helps. But even as it offers some steadiness or momentary respite, it also arguably fosters acceptance of one's situation. Breathing deeply literally binds one more tightly to her environment, whereby inspiration establishes a deeper bodily connection with it, a circuit.[2] In his explanation of how individuals locate themselves affectively, bodily, and politically in their worlds, John Protevi writes:

> Individual bodies politic are cognitive agents that actively make sense of situations: they constitute significations by establishing value for themselves, and they adopt an orientation or direction of action. This cognition is co-constituted with affective openness to that situation; affect is concretely the imbrication of the social and the somatic, as our bodies change in relation to the changing situations in which they find themselves.[3]

In the case of "Breathe 2 Relax," the situational openness created by breathing—a dilation of the passages by which the outside world is internalized, which also makes that world more bearable—instantiates a more peaceable coexistence with militarization. The paradox, of course, is that the military is both the cause of the stress and the purveyor of the means for ameliorating it, but it falls to the breather herself to resolve this tension and "relax" into it.

This dilemma recurs across media for military children, whose interiorities are figured differently than those of their civilian counterparts. While civilian children are to be insulated from any potentially damaging encounter with the fact of militarization, the task for military children is to adjust to the ways that their parent's enlistment, and militarized violence more generally, impact their daily lives. As evidenced in the previous chapter, concerns about the consequences of war for civilian children generally take the shape of worries about emotional surplus: too much exposure to too much unsettling information, which threatens to erode the protective sensory and epistemological (hence, political) boundaries around childhood. On the other hand, for those children that experience war more directly, the effects of war are generally constraining: decreased security, less stability, and fewer resources available for their care, both financial and emotional.

The management of the wartime experience of civilian children generally falls to very few institutions, primarily the family and the educational system, with a bit of guidance from the news media. Alternatively, military children encounter a surfeit of them. In practical terms, Patricia Lester and Lt. Col. Eric Flake write, "Military children are embedded in an array of systems—family, school, health care, spiritual, and local and national communities—all of which may affect how they experience and negotiate their parents' deployments."[4] But this also means that an array of institutions—both state and non-state—oversee, regulate, and mediate children's experience of deployment. In this chapter, I explore how these entities figure military children and negotiate the conceptual oxymoron embedded in the term "military child" itself. Focusing on Internet resources for military children, I analyze how they represent the emotional experiences of military childhood back to that very audience. Although many charitable organizations provide services for military children, the websites I discuss here are perhaps the most prominent sources of digital content designed expressly for them.[5]

My discussion of American military children begins with a necessary detour through Afghanistan and then Canada in the story of Omar Khadr, a stark illustration of the politics of recognizing the militarized political subjectivities of young people. I then frame my affective history of the figure of the military child as a genealogy of the so-called military brat. After an overview of the various ways that the military has interfaced with children, I explore how these histories inform the investments of affection, admiration, gratitude, pity, and anger circulating around military children today. With this background in place, I turn to the core of my analysis: a comparison of two websites aimed at military children, *Sesame Street for Military Families (SSMF)* and *Military Kids Connect (MKC)*. Military homes on Sesame Street are characterized by warmth, intimacy, and intense focus on children's needs; nondeployed parents, almost always mothers, are expected and happy to make the necessary investments of energy and resources to secure this ideal. By contrast, *Military Kids Connect* presumes a military household marked by varying degrees of stress, constraint, and dysfunction. It endeavors to empower military kids to assume more responsibility and manage hardship independently, on the presumption that neither deployed nor at-home parents (usually mothers, who often appear as taxed and fragile) will be available to offer much care or support. Although the overall visions of *SSMF* and *MKC* differ, they share preoccupations with the emotional dynamics of the military household and the threat that home front instability poses to the efficacy of deployed personnel. In disparate ways, both websites acknowledge and deny the impact of militarization on children. Consequently, even as they offer potentially important resources for military children and families, the websites also instrumentalize their emotional well-being and transform coping into a child's patriotic obligation.

"Honest Omar": Political Subjectivity and the Retraction of Sympathy

In July 2002, when Khadr was just a few months short of turning sixteen, Sergeant First Class Christopher Speer, a U.S. Army medic, was killed. Ostensibly, the lethal weapon was a grenade thrown during a firefight in a compound in Afghanistan where the Khadr family was staying. In 2010, Khadr pled guilty to war crimes leading to Speer's death. A Canadian citizen taken to Afghanistan by his father, Khadr maintains his innocence and avows that he confessed with the aim of expediting his release from Guantánamo Bay. He claims that, during his years in American detention, he was humiliated and subjected to physical harm that included stress positions and sleep deprivation.[6] In 2012, Khadr was transferred to a Canadian prison, and in a contested decision, a Canadian judge presided over his release on bail.[7] While he was detained, Khadr was the youngest man held at Guantánamo and the first to be tried under the Combatant Status Review Tribunal system, the quasi-legal mechanism devised specifically for processing detainees.

Presumably for reasons of national interest, the Canadian press paid much more sustained attention to Khadr's story than American news outlets did. In her overview of the coverage of Khadr's story, Sherene Razack argues that Khadr was always faulted for his lineage, either in the implication that he had fanaticism in his blood or that he was a child soldier in need of repair.[8] In either instance, Razack argues, Khadr got "caught in a humanitarian current that demands innocent Third World Children who are in need of rescue by an enlightened West."[9] Khadr, for his part, was almost universally recognized as a political subject and, more precisely, a militarized political subject. Coupled with his status as an enemy, this recognition placed Khadr beyond the pale of compassion that is typically extended to children. This retraction of sympathy is licensed by the notion that Khadr was a willing combatant, rather than an unwilling child soldier forced into violence by his family or a wrong-place–wrong-time bystander. The attribution of agency to Khadr also precludes any meaningful recognition of his vulnerability or entertaining of the notion that he deserves adult protection. Even if the acknowledgment of Khadr's political subjectivity is specious, it nonetheless dictated the terms of his treatment during detention.

In February 2003, a team of Canadian and American military personnel interrogated Khadr for four days, and their sessions were recorded in seven hours of video. The United States gave the videos to the Canadian Security Intelligence Service, which in turn released them to Khadr's defense attorneys after Canada's Supreme Court ordered them to do so on July 15, 2008.[10] Heavily abridged versions—eight to ten minutes long—began circulating almost immediately thereafter.[11] The editing achieves a jarring effect. The cuts are quick and rough, and the video moves, often in a matter of seconds, from Khadr chatting placidly, even amiably, with the interrogators to sobbing and lifting his orange uniform shirt to reveal injuries sustained during detention. Khadr's words are sometimes muffled, whether because he is crying or because of poor audio quality, and interpretations differ on what he is actually saying at key moments.

While we cannot always understand Khadr, we can always see him; by contrast, the interrogators' faces are blacked out, but their voices are always intelligible. Nowhere in the excerpt do they get angry or even raise their voices beyond exasperation, as when they counter Khadr's claims that he has been mistreated or injured.[12] Because of a shrapnel injury sustained during the firefight, Khadr is blind in one eye. At one point during his interrogation, he tries to persuade the military personnel that his body is failing. Slumped, with his forehead in his left hand, he says that his injuries are not healing, that he cannot move his arm. "I lost my eyes," he says. "I lost my feet, everything." From off-screen, a man counters, loudly, "No, you still have your eyes, and your feet are still at the end of your legs, y'know." Khadr does not look up or reply. The man continues: "Look, I wanna take a few minutes, I want you to get yourself together, y'know, relax a bit, have a bite to eat, and we'll start again. . . ." Referencing their gifts of hamburgers and chocolate, the good medical care at the facility, and the temperate Cuban climate, Khadr's interrogators aver that he is being treated well. In exchange, they want cooperation. They ask for "honest Omar" and tell him that he can, and must, help himself by providing the information that they request. Implying that Khadr is willfully dissimulating, the interrogators attribute a subjectivity to him while urging him to incriminate himself.

The interrogators likewise acknowledge, and regularly attempt to leverage, Khadr's connections to his family, his wishes to go home.[13] The edited video of his interrogations shows some of these exchanges. At one point, Khadr murmurs that he wants to call his grandparents. His interrogator replies, with a little sigh, "Well, I can't arrange that." In another clip, Khadr mumbles, "I'm thinking of the day I'm gonna get out of here."

The interrogator asks what he'll do on that day, and Khadr replies that he'll go home, and in response to the subsequent question about where home is, Khadr clarifies that he means Canada. And then there is a gap in the audio. When the interrogator becomes audible again, he is asking if Khadr wants "a chocolate bar or something." Khadr declines and (apparently) says again that he wants to go back to Canada. The interrogator asks, rhetorically, "You wanna go back to Canada?" And answers his own question, "Well, there's not anything I can do about that." Seemingly apropos of nothing, he continues, "I wanna stay in Cuba with you. Can you help me with that?" An indecipherable response. "No?" the interrogator asks again. Seemingly perplexed, he goes on, "Weather's nice, no snow. . . ." The next we hear from him in the edited video, he's asking Khadr, "What other interesting things do you want to tell me about?" Had these references to family and home been made by another child, they would signify in radically different ways, appealing to the sentimental histories I referenced in the previous chapter. Spoken by Khadr, now thoroughly marked as a recalcitrant political subject, however, they are readily dismissed with non sequiturs and evasions.

In 2010, Khadr pled guilty to five war crimes. Although he claims to have no memory of the firefight in which Sgt. Speer was killed and that he is innocent of his murder, he wagered that a guilty plea was the only way he would be allowed to avoid indefinite detention, possibly for the rest of his life, and return home to Canada.[14] Apparently, he was right. Khadr is one of the few individual children who appeared identifiably in the War on Terror. His example illuminates the lopsided distribution of care and concern in this affective landscape and the resulting divergence in how various types of children are treated. In Razack's estimation, this creates an absurd theater of compassion in which citizens of the same Western nation-states that condone torture also regard the "spanking of children" as a "grievous harm."[15] Norms about innocence, culpability, and punishment do not adhere equally to all children's bodies. The diversion of responsibility away from children who are not Omar Khadr also opens them to a range of deep affective investments, for which Khadr was rendered ineligible from the outset.

Genealogies of the Military Brat

Most discourses of children's rights—like those marshaled against the teachers that exposed them to wartime atrocity in the cases I described in the previous chapter—amplify impressions of their vulnerability and malleability. The background noise of innocence muddies thinking about relationships between childhood and militarization by positioning the terms as essentially incompatible, particularly in the Western context where ideals about childhood are firmly established.[16] Khadr thus serves as a foil for American children, who are figured as apolitical and distant from war. Yet his adolescent shadow haunts all of this: as the threat of what children might become or as a sign of the savagery against which the United States is ostensibly fighting—either way, he will never draw the same affective investment. His best hope was for clemency.

The precedent for distancing children from combat was established during the earliest wars between nation-states in the sixteenth and seventeenth centuries, when "limited" wars were generally fought by professional armies on faraway battlefields.[17] But as the nature of warfare changed, so too did children's proximity to it. During the Revolutionary War, the population of the colonies was relatively young; the legal age for enlistment was sixteen, and some soldiers went much younger than that.[18] Conscription during the Civil War left many families without fathers in the home, and their long—and in some cases permanent—absences led to their reduced significance after the war. This war begat ambiguous consequences for the militarization of children, which settled unevenly on families of different classes. On the one hand, it inspired parents, especially among the middle class, to preserve the ideal of childhood and shield their progeny from the violent realities of the outside world.[19] But at the same time, educational publications like textbooks about the war explicitly aimed to politicize young learners.[20] Concurrently, reliance on child labor in industrial settings bifurcated the young into groups deemed "useful" (laborers in factories and on farms, who contributed actively to the maintenance of their families) and those worthy of being "protected" (the relatively insulated children of the middle and upper classes). Over time, this ideal of protected childhood spread to the working classes as well.[21]

As the "total" war paradigm became ascendant in the early twentieth century, the separation between the battlefield and the home front began to erode. World Wars I and II curtailed Western parents' abilities to protect their children, both physically and psychically. Worldwide, 10 percent and 45 percent of casualties in World War I and II were civilians, respectively.[22] In both Europe and the United States, children's educational and entertainment media became key loci of their militarization and acculturation to the facts of "total war."[23] World War II politicized American youth culture and entertainment. Moreover, as parents mobilized for war, children were regularly left unshielded from its realities and were expected to deal with them on their own, without benefit of parental nurture. Even young children were expected to participate materially in national defense, and the historical record suggests that they did and even believed themselves to be valuable contributors to and protectors of the nation.[24] For example, in her study of youth culture and the militarization of agriculture during World War II, Lisa L. Ossian describes how these young "soldiers of the soil" immersed themselves in the work of farming, often forgoing both school and sleep for the good of the nation.[25]

During the second half of the twentieth century, the relationship between young people and militarization changed multiple times. The Cold War entailed a different set of demands on children.[26] Cold War children were expected to be aware of the pervasive threats of both Communism and nuclear annihilation and hence prepared to protect themselves and defend American values. The widespread emphasis on not just obedience to authority but conformity to a rather narrow set of ideal behaviors intensified scrutiny of children, their actions, and their social networks.[27] This pattern, whereby militarization tends to tighten strictures on children's behaviors and establish more constraints on their autonomy, persists today. Vietnam, of course, catalyzed the politicization of American youth, particularly adolescents and teenagers, around antiwar and anti-draft activism. The ratification of the twenty-sixth amendment in 1971 decreased the voting age to eighteen, lowering the threshold for a key index of full citizenship.[28] Then, in 1973, the United States shifted to an all-volunteer military. Removing the threat of conscription defused some of the charge of teenage anti-militarist activism and stripped the lethal significance from the age of majority. For most American young people today, turning eighteen entails the acquisition of rights (like voting and buying tobacco) rather than the assumption of responsibilities (like military service).

The transition to an all-volunteer force also drew families into closer proximity to military spaces and activities.[29] This entailed unprecedented logistical challenges for military bureaucracies, established new contexts for military children's development and socialization, and begat the cultural motif of the "military brat." Anneke Meyer suggests that the notion of childhood innocence has a performative force, as the protective systems it begets "can produce the kind of personal lack of social experience and social vulnerability that the discourse of innocence portrays as innate to children."[30] This dynamic is even more complicated in the case of military children, whose identities and statuses are more overdetermined than those of their civilian counterparts. The trope of the "military brat" is ambivalent. Its emphasis on the exceptional pluckiness of military-connected children is both approving and fretful, so that military children can be regarded either as uncommonly poised and mature or too independent and worldly for their own good. Jennifer Sinor writes, "As military dependents and as children, these individuals are doubly disenfranchised, seldom seen as individuals and almost never as complicated subjects with conflicting desires and needs."[31] She argues that there are two, mutually exclusive "subject positions" available to them: "victim or warrior—which both represent them in static, one-dimensional ways."[32] The category of innocent victim denies their agency and capability to reason about their circumstances, while the idea of the heroic warrior empties out any tensions or contradictions in their nascent politics.[33] My consideration of resources designed to help military children "cope" queries the affective pedagogies orbiting these positions.[34]

The focus on the "military brat" can also obscure the commonplace ways that civilian children encounter militarism. Cindy Clark notes that two of the three major celebrations of American militarism (Memorial Day and the Fourth of July) correspond with the summer vacation from school and, hence, the ultimate experience of childhood freedom, and wagers that this coincidence amounts to a subtle militarist pedagogy. These holidays are also often observed with rituals that entail strong affective demands to show gratitude for American troops, and her study indicates that children participate eagerly.[35] While the militarization of American childhood hides behind the innocent fun of parades and fireworks, Western observers tend to find the militarization of children in the East or Global South much easier to imagine, detect, and criticize. The discourse of the "child soldier," as Katrina Lee-Koo incisively notes, hinges on the iconic image of the "lone, unsmiling, armed African boy."[36] While polities in the Global North tend to conceptualize the militarization of young people—as through organizations like the ROTC—as "empowering," they often criticize analogous phenomena in the Global South as "disempowering" and markers of uncivilizedness or barbarity.[37] She argues that these representational strategies, like the image of the child soldier with no visible ties to his family, persist because they cohere with a paternalistic model of colonialism.[38] Even as these conceits fuel Western indignation about the harms that militarization visits on children worldwide, the example of Khadr demonstrates that this indignation is selective and limited in its application.

In the United States and the West more generally, early modern interactions between the child and the state centered around labor. Once the Romantic ideal of childhood, which I described in the previous chapter, began to take hold in the nineteenth century, it inflected the way that the state regulated children's behavior and understood children as citizens. The prevailing model was one of childhood "incapacity." Restrictions on child labor were intended to acknowledge and protect this incapacity, but also perpetuated it.[39] Importantly, however, the state's approach to childhood incapacity fractures around the figure of the military child. The state expects and depends on military children to be capable and composed and to forgo, of necessity, the emotional and practical privileges afforded to their civilian counterparts. The atmosphere of compulsory veneration of "the troops" demands that the state exhibit concern for their families and that the armed forces acknowledge explicitly that family well-being is essential to their overall readiness. After all, servicemembers cannot afford to be distracted by family problems, and the military cannot afford a mass exodus of personnel necessitated by unhappy spouses and children.

Military leadership is aware that the duration of the War on Terror has led to what Lester and Flake characterize as "extraordinary demands" on military families.[40] Nonetheless, it cannot concede too freely that military children are in need of additional support or otherwise imperiled by the rhythms and stresses of deployment and military life. The burden of reconciling this tension thus devolves to military children themselves, whom the military positions as responsible for their own well-being and even the well-being of their families. This is a turn on the pervasive understanding, which became ascendant in the late nineteenth century, that the emotional status of the child is key to the overall functioning of the family and the household. Those discourses were intended to protect emotionally delicate children and reorient the attention and resources of the household accordingly. In the case of the military child, however, the focus is a disciplinary one and raises the stakes of the child's comportment to a national or even international level. Accordingly, official discourses situate military childhood as simultaneously average and exceptional, a logic imprinted on the resources provided for military children. Sinor notes the damaging effects of the "work undertaken by both institutions and individuals to reproduce as ordinary the extraordinariness of war," arguing that this obscures the "ordinary trauma" that military children experience, which is rendered "unexceptional" by its repetition and ongoingness and the military's insistence that nothing troubling is happening.[41]

When divorced from other markers of identity, the conceptual category of "the child" is transcendent, applicable to all humans under the age of majority. But in practice, not all children receive equal protection, an allotment that varies predictably by the child's gender, race, class, and sexuality. "Not all innocences are alike," Anne Higonnet observes, "or equal."[42] While cultural studies of childhood now attend regularly to the role of gender, race, class, and sexuality in determining how a child will be treated and perceived, the differential of civilian or military status has been less fully theorized. Yet I argue that these categories are often determinant in the attribution of innocence or perceptions of a child's vulnerability.

My thinking on this is indebted to Robin Bernstein's work on the racialization of childhood pain in the nineteenth century. Black and white childhoods, she argues, were represented differently, divided primarily along perceptions of susceptibility to pain. Broadly, she argues, the emphasis was on the "tenderness" of white children and the unfeelingness of black ones, an attribute most clearly expressed in the stereotype of the pickaninny.[43] In cultural productions for white audiences, both child and adult, pickaninnies either appeared "quak[ing] in exaggerated fear" before threats that white audiences knew to be harmless or sustaining obviously serious injuries without experiencing pain.[44] Overall, according to Bernstein, black children were represented as "resistant if not immune to pain" but also quick to exaggerate symptoms to avoid responsibility.[45] This racialized understanding of childhood vulnerability and the concomitant idea that children of color might exaggerate their suffering appears operative in the treatment of Khadr as well.

This differential understanding of susceptibility to suffering parallels the divergent representations of civilian and military children, which construe military children as tougher than their civilian counterparts but also inevitably corrupted, at least a little, by their proximity to militarization. The institutions that depend upon the compliance of military children, like the military itself, must extend them certain considerations without assuming too much responsibility for their suffering or distress. On the whole, military children are figured as differently, or marginally less, precious than their civilian peers. Military children do enjoy some symbolic privilege by virtue of their association with the military, though this veneration of the military is often superficial, as I demonstrate in the next two chapters. But the emotional resources directed at them are rather thin and utilitarian compared to those afforded to civilian children.

The apprehension surrounding civilian children becomes moot in the case of military children and falls away, an ironic withdrawal given that military children face much more concrete perils than the hypothetical cognitive and emotional harms at stake in the classroom scandals. Absent this apprehension, military children start to look more capable and the intrusion of militarization into their childhood less lamentable. The figure of the military child is still the beneficiary of the affection directed at civilian children, this time colored by questions about whether their own parents are available to love them. Admiration, on the other hand, gets bequeathed differently on the figures of the civilian and military child. Always, that admiration is contingent on the attributed innocence of the child: the civilian child oblivious to and unsullied by politics, the military child forced by circumstance to sacrifice for the national good but demonstrating uncommon fortitude in meeting these demands. In other words, civilian children earn admiration through their innocence, while military children draw it through the sacrificial loss of their innocence to militarization. In the previous chapter, I invoked Lauren Berlant's theorization of the infantile citizen to explain the admiration that civilian children elicit. But that infantile citizen is expressly civilian; for military children, the vexed dynamics of adulation that Berlant describes become even more fraught. Military children are not permitted to be infantile. Instead, they are expected to display savvy and maturity in order to figure as ideals.

This admiration folds into gratitude as recompense for their juvenile service and sacrifice. A recent estimate counts 4,000,000 "military-connected children" in the United States.[46] Gratitude on such a scale can only be abstract and general and so manifests more at the level of feelings than of actions or resources. Various actors offer their gratitude in response to the staggering demands placed upon these young people but translate it only inconsistently into material support. This gratitude is often tinged with pity, usually tied to the notion that they are deprived of contact with the deployed parent, a perception that presumes all parent-child relationships are loving and healthy. However, in the case of the military child, their attributed innocence is decoupled from assumptions about their vulnerability, because acknowledgment of their vulnerability would amount to a criticism of the costs of militarization. And anger here must be wielded carefully, directed only at the external geopolitical forces that compel the United States to militarize, never at the country or its military directly.

All of this affection, admiration, gratitude, pity, and anger flow on the condition that military children feel appropriately about militarization and their experiences of it. Each affective investment entails a corollary expectation on the child it figures. To be worthy of affection, the child must have affection for her country and the parent deployed to protect it. To be worthy of admiration, the child must be appropriately admiring of military personnel and government. To be worthy of gratitude, the child must be self-sacrificing. To be worthy of pity, the child must be sad about terrorism, death, and the absence of his parent, but in this case, not too sad, lest he appear selfish or insufficiently patriotic. To incite anger on his behalf, the child must not be angry, or only rightfully: at the terrorists or the enemies, but never at her government or his parents. In other words, the military child is licensed to be angry, provided that it does not devolve into a critique of the state and that it lead only toward reintegration into military life.

In her study of juvenile literatures of the War on Terror, Laura Browder describes their imperatives for military audiences. Broadly, the literature confronts the consequences of a long war fought through extended and multiple deployments, during which "children have to think about themselves as soldiers and the entire family becomes militarized."[47] Uniformly, she says, authors display a commendable awareness that children will have strong feelings about their parents' deployment and obligations. Yet Browder also underscores that these authors identify good children as those who know how to "'be brave' and repress their feelings," and literature for older children offers guidance on how to "support the parent who is deployed as well as the one who is still at home."[48] The books often recast overseas deployments as an "extension of parenting," whereby the deployed parent protects nation and child alike while earning the money necessary to maintain the family, thus foreclosing any grounds on which the child might rightfully question the parent's absence.[49] The websites I consider here come freighted with similar goals, but undertake a much more direct and intensive form of affective work, emphasizing the tasks of naming, expressing, and managing feelings.

Friday Family Game Nights and Taco Tuesdays: Military Childhood on Sesame Street

Elmo is among the many conscripts to the mission of helping military children reconcile the contradictions inherent in their roles.[50] *Sesame Street for Military Families* (*SSMF*) is a partnership between the Sesame Workshop and the Defense Centers of Excellence (DCOE).[51] Aimed at the estimated 700,000 children under the age of five who have a parent in the military, *SSMF* provides a range of content aimed at both young children and their parents: information, advice, activities, sketches featuring *Sesame Street* characters, and mobile apps.[52] Organized around themes like Relocation, Homecomings, Self-Expression, Deployments, Injuries, and

Grief, the site emphasizes communication and togetherness as remedies for the myriad stresses of military life.[53]

The DCOE, as part of the Department of Defense (DoD), describes its mission as "to [improve] the lives of our nation's service members, veterans and their families by advancing excellence in psychological health and traumatic brain injury prevention and care." It undertakes this work through research and services, which it provides in both direct-care settings and remotely through its National Center for Telehealth and Teletechnology, which also runs *Military Kids Connect*. Sesame Workshop, the educational organization that runs *Sesame Street*, began its work with military children independently in 2006, with a multimedia campaign called "Talk, Listen, Connect: Helping Families During Military Deployment," and partnered with the DCOE a few years later to expand its reach.[54] The DCOE is *SSMF*'s most prominent partner, but it is not the only surprising bedfellow. Alongside the more predictable donors, like various philanthropic foundations, the Corporation for Public Broadcasting, and the USO, are a range of corporations. These include Walmart (noted for providing "major support"), American Greetings, and financial institutions like New York Life and BNY Mellon, as well as defense contractors like Lockheed Martin, BAE Systems, and Oshkosh Defense.[55] This long list reveals the great number of stakeholders invested in the work of getting it right in terms of military children's feelings.[56]

SSMF is part of a broader series called *Little Children, Big Challenges* that includes content for young people on the themes of divorce and parental incarceration. And in other parts of the world, adaptations of *Sesame Street* are acutely topical, as in the case of the HIV-positive character Kami in the South African version, *Takalani Sesame*. Yet even as *Sesame Street* content acknowledges the complexity of some of its young viewers' situations, it also endeavors to restore them to the innocence and playfulness that characterize more typical projections of childhood, optimistically emphasizing strategies for repair.

The theme of "Relocations" features prominently on *SSMF*. Relative to phenomena like deployments, injuries, and death, the matter of relocation is rather anodyne. Accordingly, the site encourages parents to "make moving a positive experience for your child," recommending that they treat the news of the move as a "special announcement," "make packing exciting," and keep doing fun things together as a family even as the members are settling into their new home. Alongside concessions that children might be "anxious or upset," the page recommends helping children talk about their feelings, preserve happy memories, and look forward to making new friends. While the relocating child is expected to go gamely along with various activities meant to smooth the moving process, this strategy also places tremendous burdens on the parent—likely to be a mother, presumably already contending with the stress of moving, and less able to call on a partner who is busy preparing for deployment for assistance—to facilitate them. The parent not only has to modulate her own feelings in front of the child, but take extra steps (like researching the route to the new home to point out exciting landmarks and preserving routines like "Friday Family Game Night" and "Taco Tuesday") to keep up the child's spirits. Practically, this promotes familial togetherness. But this prescriptive emotional content suggests that happy compliance is the default behavior of all children and reinscribes a gendered mandate of cheerfulness onto military wives and mothers (in the next chapter, I will discuss affective expectations for military wives at greater length). Cynthia Enloe notes that women are often tasked with reestablishing normalcy in the military home after a war has ended.[57] *SSMF* suggests that they must actually do this work at every stage of militarization.

Across *SSMF*, difficult emotions appear as phenomena that can be ameliorated and even prevented with proper planning, as in the section on Homecomings. Noting that the reality of "any highly anticipated event" will likely deviate from lofty expectations, "Homecomings" emphasizes the importance of preparing the child ahead of the parent's return and readjusting slowly.[58] The content also includes gentle reminders about the dangers of a child being too exuberant at a parent's homecoming. This lesson is adorably dramatized by Elmo in the role of a military brat, who greets his newly returned father with a long list of fun things he wants them to do together. Here, too, *SSMF* places its faith in patience and the simple passage of time. Elmo's father reassures him that they will have plenty of time to do all of it, and—after Elmo realizes that he "doesn't really know what he wants"—that it will just take time to figure out how the pieces of the family will fit back together. Indeed, one of the suggested activities on the site is a Family Puzzle for the child to decorate and experiment with by removing and reconnecting pieces. Although the acknowledgment that the transition of homecoming might be difficult is an important corrective to fairy-tale expectations of seamless reintegration, "Homecomings" depoliticizes this difficulty. The section emphasizes how much family members, and especially the child, will have grown and changed during deployment and attributes the difficulty to these processes,rather than locating it in either parent or his or her wartime experiences.

Sesame Street for Military Families promotes communication and the passage of time as remedies for the difficulties that military families encounter. Elmo's family provides an adorable model for how families should enact this kind of openness and patience.

Along with the passage of time, *SSMF* posits self-expression as a powerful remedy for the difficulties of militarization. In early 2016, the site summarized the emotional circumstances of military children as follows:

> Children of military families are unique in many ways. They face a number of challenges that are difficult even for most adults to handle. They may also have amazing opportunities, like living abroad, that most children their age only dream of. Help your child build resilience, maintain a positive outlook, and find ways to express how he feels about being part of a military family.[59]

The structure of the final sentence suggests that these three activities are synonymous or interconnected: that expression of feelings will help sustain a positive orientation toward life in a military family and perhaps to militarism more generally.

This optimistic depiction obfuscates a common dilemma for military children: the need for extra support before and during deployment, junctures at which both parents are likely to be less able to provide it. The military parent might be busy with predeployment preparations or be unreachable overseas, while the nondeployed parent will be making her or his own preparations for deployment and subsequently managing a household single-handedly, and so be short on time and energy.[60]

Conveniently, *SSMF* children are surrounded by friends; for example, Elmo's daddy reassures Elmo that he won't be lonely during their separation because he has so many friends. And they all live in neighborhoods equipped with ample resources and safe, fun places to play, a depiction that understates the social disruptions that military children experience because of frequent moves and presumes that they will find every community welcoming. However, these social networks are represented as a bonus for these military children, as the parents and other adults depicted in *SSMF* seem to have not only love in abundance, but also bottomless resources of care, patience, and enthusiasm. They are always eager to hear their children's feelings and always ready to respond constructively to them.

For a time, *SSMF* featured the video for an *Electric Company* song called "Let It Out."[61] Featuring military parents and children, "Let It Out," intimates that children possess a veritable "superpower" in their capacity to describe what they are feeling. Most of the video focuses on helping children expand their emotional vocabulary, offering more complex synonyms of common feeling words and linking these feelings to their corporeal manifestations:

> Say you're feelin' mad
> I mean, really really mad
> Like there's smoke pourin' out of your ears
> [Grrr.]
> And your fists are clenched
> And your face is scrunched
> And you feel like you could shout for years
> [Ahhhhh]
> You could say, "I'm mad,"
> "I'm angry"
> "Annoyed, frustrated, and furious"
> Or you could say, "I'm infuriated!"
> That also means "mad," if you're curious.

Every verse is punctuated by dancing, and the exhortation to "Let it out, let it out! Get it off your chest! Let it out, let it out! You got the words to express! Let it out, let it out! Don't keep it inside! Let it out, let it out! You got nothing to hide!" Overall, the video traces a movement of feelings from embodied sensation to verbal expression to shared communication, as it concludes with children passing handwritten signs with feeling-words to their military parents, who accept them smilingly.[62]

The Electric Company's "Let It Out" video provides military children with a nuanced vocabulary for expressing their feelings and promises that their parents will always be happy to hear them.

Indeed, the overall tone of *SSMF* is relentlessly upbeat. It validates all feelings as acceptable but privileges the process of expiating them and so returning everyone to normal, a militarized homeostasis. Even in the section on grief, it advises parents, "Allow your children to be children. This applies especially to older children, who may have new responsibilities but still need opportunities for fun and play."[63] This generally laudable effort to free children from the emotional burdens of grief simultaneously constructs childhood as a terrain where grief cannot adhere, where militarization never intrudes. Overall, *SSMF*'s ostensibly permissive attitude toward all kinds of feelings coexists with an evaluative framework that determines which feelings are ultimately good or bad.

Following the rise of a consumerist model of childhood, according to Peter N. Stearns, "American emotional patterns began to be distinguished by an unusually clear ability to distinguish between good and bad emotions on the basis of their pleasurability, and a reluctance to believe that any utility could be constructed from the bad."[64] In its section on Self-Expression, *SSMF* replicates this logic and encourages parents to "look for signs of stress" in their children, including "acting out, being extra clingy, being withdrawn or overly active, experiencing nightmares or sleeplessness, and losing developmental milestones such as bathroom skills."[65] This instruction is noteworthy because it deputizes parents as mental health experts who should recognize their child's "stress" as a somatic cluster of symptoms aligned with regression, incompatible with the forward-looking vision of *SSMF* and, by extension, healthy military families.

Cuteness is key to *SSMF*'s bright vision for military families, as embodied in its trademark Muppets. Beginning around the turn of the twentieth century, manufacturers began producing cuddly toys meant, in Stearns's words, to "receive the love and affection of the very young,"[66] thus re-orienting their most ardent affections at least partially away from their parents and families. Durable, soft, and comforting, these objects embody the essence of "cute." Sianne Ngai notes that cuteness as an aesthetic emerged during the reconfiguration of the domestic sphere as an active node for, rather than a haven from, consumerism.[67] The aesthetic property of cuteness typically adheres to beings and objects that are "amorphous or bloblike," with squishiness and cuteness increasing in direct relationship.[68] Furry, floppy, and loveable, these Muppets are, in short, the epitome of cute.

Ostensibly, cute products are meant to invite affectionate contact, purchased with the intent of being snuggled and squeezed. Ngai suggests, however, that cute objects provoke and channel a more complex and confusing form of "desire . . . not just to lovingly molest but also to aggressively protect them."[69] Ngai's formulation points to a complex politics operative here. These cute characters experience the anxieties of deployment, the challenges of reunion, and even the experience of having a parent injured or killed. In the process, they become the transmogrifying objects through which those feelings are reconfigured into palatability. To the extent that the cute creatures of *Sesame Street* court such stealth aggression, they may also serve to divert military children's hostility away from other plausible targets, like the state or its disruptive actions.

Such a detour is especially urgent in the case of the injury or death of a parent. The buttons that hyperlink to the *SSMF* sections on injuries and grief include little tags labeling them as "sensitive material" and advise parents to preview the content and then discuss it with their children, installing them as emotional gatekeepers. The "Injuries" section pertains to physical and mental traumas and focuses on stabilizing the child's reaction to what has happened to the parent and reframing it as an opportunity to grow as a family and establish a "new normal" together. For example, the video entitled "For Families: Coming Home," hosted by Queen Latifah, features stories from actual military families dealing with the return of a seriously injured parent.[70] With other information only sketchily filled in, the video leaves ample room for resignification. In one vignette, a father's traumatic brain injury (TBI) presents an opportunity for him to learn his ABCs and 123s all over again alongside his children, while a son marvels over the coolness of the mechanical hand his father receives after amputation.

However, the relentless optimism of *SSMF* stumbles on the matter of emotional changes precipitated by TBI, as if it cannot quite countenance the idea of a family where parents are not always warm, loving, and emotionally harmonized with their children. There is no illustrative Muppet for TBI or PTSD. By contrast, the prominent character in the "Injuries" section is Papi, the father of Elmo's friend Rosita. Papi is in a wheelchair, which means that he can enjoy modified versions of activities like dancing with his daughter and that his emotional capacities are unafflicted by his wartime experience. Acknowledging that "invisible injuries" like TBI might be the most confusing for children, *SSMF* includes a whole page of guidelines for dealing with those. These anticipate that the child will detect inexplicable affective changes in the injured parent—who might be "angry," "sad," "irritable," "emotionally unavailable," or "just seem 'out of it.'" In turn, *SSMF* encourages the non-injured parent who must explain this to the child to compare an invisible injury to a stomachache: a hurt that is real and deeply felt, but not visible to anyone else.

Yet this benign comparison falters against the live-action content, as one of the women featured in the video describes being terrified of her husband, saying that he could "scare the daylights out of me just by the look in his eyes," while her daughter reports that her father would get so angry at her that he'd have to "go away." These are not feelings that can be readily ventriloquized by Elmo, who disappears until the end of the "Coming Home" video and returns to query the psychiatrist who visits *Sesame Street*: "Petey Esdee? Who's that?" The psychiatrist smiles and explains that PTSD is not a person, but a name for "scary feelings," thereby depoliticizing those feelings and detaching them from their battlefield etiologies. Of course, I am not suggesting that young children should get explicit or detailed information about combat-related PTSD; rather, I highlight this example to show how *SSMF* struggles to manage the unwieldy emotional possibilities engendered by war.

By comparison, the site's content about death and grief reads much more confidently. This, perhaps, is a more navigable emotional terrain. This section features the Muppet Jessie, Elmo's cousin whose father was killed during deployment, though the circumstances of his death are only ever implied. Jessie is struggling to cope with her father's death; apparently the most worrisome symptom of her difficulty is that she is becoming noncommunicative. She doesn't want to talk about it, even hides from her family to avoid it, until Elmo's father, her own uncle Louie, shares his own feelings of sadness and anger. Jessie relents by confessing her newfound sense that life isn't fair, that she can't understand why her dad is gone. Louie directly confronts this existential question, affirming that he, too, is mystified. As in the section on Injuries, the live-action content is more explicit, including stories from children who lost parents in nonmilitary circumstances as well as a vignette about a husband, father, and marine helicopter pilot who committed suicide. On this count, *SSMF* urges gentle candor with children, not hiding the fact of suicide but instead explaining it as an "illness" of the brain.

No matter the trauma, however, the expression of feelings appears as the panacea for recementing the family, a certainty encapsulated in the *SSMF* mantra "Talk, Listen, Connect." *SSMF* suggests that parents not hide their own grief and encourages them to cry with their children if they feel like it. For her part, Jessie seems relieved by the experience of sharing her feelings, conceding, "I do feel a little better," while another video, "Big Feelings," assures viewers that "there are no feelings too big to talk about." In this furry affective universe, the greatest peril is not grief itself, but the failure to vocalize it.

Uniformly, *SSMF* urges its audiences to process their feelings. Both the discourse and practice of emotional "processing" originate in a pervasive cultural antipathy toward bad or unpleasant feelings and the sense that they can be cathartically expressed into nonexistence. In her crucial work on "ugly feelings," Ngai dwells on emotions that she describes as "noncathartic," feelings that are relatively mild but refuse to be expunged, and so aggravate by their persistence. Contrary to a pop-psychological notion of processing feelings, Ngai suggests that ugly feelings are difficult to process in part because they are difficult to localize. She continues, "The unsuitability of these weakly intentional feelings for forceful or unambiguous action is precisely what amplifies their power to diagnose situations, and situations marked by blocked or thwarted action in particular."[71] For military children, the weakness of these feelings, marked by their inability to effect structural change, is a property not so much of the feelings themselves but their residence in the bodies of children, who are tasked with identifying and expressing feelings precipitated by national and geopolitical events, phenomena by which the children are massively outsized.

The operative notion that the default, or natural, emotional position of the child is happiness took root in the early twentieth century, when experts began opining that "normal" children are "actively cheerful" and tasking parents to intervene as necessary to ensure this.[72] This model constructs a normal childhood as an idyllic one in which bad or sad things are unexpected and aberrant intrusions. Bad feelings are expressed and expelled, presumably, to enable the flooding back in of the good feelings that naturally belong there. Fencing childhood off into a purely positive emotional state prevents recognition of the political dimensions of emotions, especially those experienced by the young.

Paradoxically, *SSMF*'s emphasis on childhood expressivity also inhibits recognition of children as possible political subjects by continually declining to link their feelings to the root cause of their suffering: militarization. The child's responsibility in this dynamic is to learn how to name and vocalize his feelings and to make them visible, particularly through guided artistic activities. Among the many worksheets offered on the site are "Feelings and Faces" and "Feeling Flower." Both alert children to the likelihood that they will feel all sorts of things in response to relocation or a parent's deployment and reassure them that all of these feelings are "OK!" "Feelings and Faces" shows different *Sesame Street* characters making faces corresponding to various feelings (courageous, worried, sad, frustrated, and angry); before coloring them in, the worksheet asks the child to point to "how you are feeling right now, and say why." The "Feeling Flower" is less structured, requiring more emotional initiative and dexterity: once a parent has helped them cut out the petals, children completing the "Feeling Flower" are to write a feeling word on one side of each and illustrate it on the other, ultimately assembling them all into a flower, something lovely and essentially harmless.

Stress Detectives: The Militarization of Coping in Military Kids Connect

Compared to *Military Kids Connect* (*MKC*), *SSMF* is oblique; it foregrounds the sweet aesthetic of *Sesame Street* and the idealized vision of the family over more direct confrontations with the "military" aspect of its mission.[73] *SSMF* approaches the actual fact of militarization elliptically; it does not euphemize, but it makes explicit reference to military life only rarely. For example, the "Homecomings" section describes the deployed parent as "back," but does not specify from where. Across *SSMF*, the word "military" appears primarily as a modifier for "children" or "families" and very infrequently on its own. For its part, *MKC* seems to presume a parent-child relationship much more fraught and alienated than those of *SSMF*. Unlike

SSMF, which envisions parents and children co-viewing its content, *MKC* is directed almost exclusively toward children, housing its materials for parents under a separate tree of links.[74] *MKC* focuses intently on teaching kids how to manage in the absence of parental support, occasioned both by the literal distance of the deployed parent and the figurative, emotional distance of the one who remains at home. In this way, the military kids envisioned by *MKC* share a common predicament with Omar Khadr in their need to manage a world that seems to have forgotten them.

As a T2 initiative, *MKC* is designed to "enhance resiliency, provide support and facilitate family readiness for military youth dealing with the unique psychological challenges of military life" by offering "age-appropriate interventions."[75] A key feature of the site, according to its official description (though this doesn't seem to pan out in practice) is its provision of an "online military community . . . to improve peer-to-peer communication so that military youth's experiences can be normalized."[76] Unlike *SSMF*'s relentless promotion of parent-child communication, *MKC* envisions a more preoccupied parent and emphasizes military kids connecting with one another. It also reflects what Ann S. Masten describes as the U.S. "military's implicit or explicit belief that children's well-being influences the successful functioning of their service member parents," while outsourcing the responsibility for maintaining that well-being to children themselves.[77]

Overall, *MKC* allows for the possibility of an incipient political subjectivity on the part of military children and so must work to keep it consonant with broader military needs and objectives. As it does this, the site instrumentalizes the military child as an essential piece in a larger fighting machine. Whereas emotion in *SSMF* serves, via expression, to fortify bonds between families, it is portrayed in *MKC* as an inconvenience at best, and often a peril, a threat to be assessed and a problem to be solved. *MKC* aspires to equip military kids with a "unique set of skills to draw on in order to get through long and often difficult separations and situations." In turn, emotionally healthy military children relieve their servicemember parents of a mental and emotional burden.[78] Compared to *SSMF*, *MKC* seems less invested in validating the military child's feelings and more concerned about managing them so they don't become disruptive.[79]

T2 launched *MKC* on January 18, 2012. As of September 2015, the site had garnered 242,443 lifetime unique visitors, and T2 proudly lists the awards *MKC* has won in the process.[80] It contains a tremendous amount of content, which makes it difficult to summarize, but also seems rough around the edges: rather simplistic visual elements, a lack of aesthetic uniformity, writing that is awkward and littered with typos, and so on. Many of the elements are slow to load or display with errors. Its promise of "age-appropriate" content is belied by murky distinctions among the three target age groups of kids (six to eight years old), tweens (nine to twelve years old), and teens (thirteen to seventeen years old), as some of the content recurs across the sections for all three groups. Originally, the content for teens appeared alongside that for kids and tweens, but it is now housed on a page called "Military Teens Connect," which has a different background and layout. The most significant distinctions between these populations appear in *MKC*'s presentation of the kind of problems or dilemmas each age group might be expected to face, though the overall aspiration of "coping" remains the same for all children.

Unlike in *SSMF*, here emotions are a liability for the child to manage mostly alone. Given its concession that parents are unlikely to be much help in reducing stress, *MKC* offers solutions in the form of guidance from and connection to other military kids, but these are the areas where its seams become most glaring. The video testimonials from other military kids, along with the short, scripted pieces meant to dramatize the challenges of military life, have the didactic tone and low production values of after-school specials. The interactive features, likewise, seem to miss the mark. So touted by the official descriptions, these looked especially underutilized when I visited the site; nearly every post in the discussion boards had replies numbering in the single digits, and many had no replies at all. One user, when starting a thread looking for information about Fort Rucker, ended the post with, "Does anyone use this site? Probably not my mom made me."[81] These spaces of childhood sociality are intensely managed; every discussion board is moderated, and posts must be approved before they appear online. Some of the posts seem to be written by adults masquerading as young people—the grammar is too polished, the advice too sage, the optimism too relentless.[82]

MKC content seems to have four main ambitions: to help military children get oriented in the world; to keep them occupied; to help them cope with stressors and manage their reactions to them; and to connect them to one another. The site itself is constructed on a background that looks like a bulletin board, and users click on the age-specific button for their group, which leads to games and activities, content related to deployment, video stories from other military children, and resources to help with coping. Tweens and teens have content meant to help them connect directly with other military children and share their feelings. A cartoon character named "Stampy the Global Guide" giggles and bounces in the corner of the main page and appears periodically across the sites for kids and tweens. More goofy than cute, Stampy reframes the life of a military child in terms of adventure and exploration.

While Stampy apparently embodies what is best about military childhood, *MKC* encapsulates its difficulties through the discourse of "stress." *SSMF* offers military kids an elaborate emotional vocabulary, but *MKC* reduces a range of cognitive, emotional, and physiological symptoms to "stress," or being "stressed out." Here, *MKC* partakes in a broader cultural tendency to employ "stress" as a powerfully communicative shorthand for a range of unpleasant stimuli and their psychic and physiological consequences. The "stress" that we typically speak of today is the conceptual descendent of a range of deleterious conditions that might previously have been identified as "hysteria, passions, vapors, nerves, neurasthenia, worry, mental strain, and tension."[83] The idea of stress originates in engineering, a way of describing strains on buildings and machines, and this idea carries over into contemporary associations of stress with overwork.[84] Although stress is most readily associated with the economic, social, and political pressures of adult life, as early as the mid-nineteenth century, experts were beginning to acknowledge that children, too, could be stressed.[85] With the rhetoric of stress popularized by the mid-1980s, it was often positioned as the opposite of, or anathema to, happiness.[86] By the late twentieth century, it had become a catchall term meant to describe both a range of tensions (stressors) and the consequences, both psychic and physical, of encountering them.[87]

Superficially, the discourse of stress seems to favor the powerless, people who find themselves in overwhelming and usually negative situations that they cannot change but must learn how to tolerate. Concern for stress would seem, therefore, to be a way of advocating for the interests of vulnerable or disenfranchised populations. But the history of stress as a concept suggests something different. Before stress became a matter of scientific study, the psychological understanding of stress was fairly loose. Until it was medicalized, the term implied what Russell Viner categorizes as fairly generic "'troubles of life,'" understood as "part of expected human experience, each hardship being separately potentially disastrous, but together having little collective scientific identity or consequence."[88] However, the medical study of stress, championed by endocrinologist Hans Selye in the mid-twentieth century, changed this, labeling stress as something that was both scientifically verifiable and hazardous to the life and flourishing of the organism.

Importantly, Viner points out that some of Selye's earliest champions were in the military; his theories seemed to promise a "way of maximizing operational efficiency" and even a "potential weapon against Communism" during the Cold War.[89] Reflecting on this dubious history, Kristian Pollock wagers that the idea of stress as "pathogenic" might be disempowering, causing people to become risk-averse and accepting of the status quo. [90] Contemporary discourses about stress identify everything as potentially stressful, an orientation that can "legitimi[z]e existing social arrangements." She continues, "The potentiality for harm, or illness, may then be regarded as residing in the diffuse, impersonal 'condition of things,' about which nothing much can be done," while also "undermin[ing] the confidence which people have both in their health and their capacity to cope with ordinary problems of living."[91] In other words, conventional uses of "stress" might serve to bolster or even legitimate existing social, economic, or political arrangements by pathologizing the sorts of feelings that might otherwise serve to catalyze change or resistance.

Unlike *SSMF*, *MKC* explicitly names the unique challenges that comprise military childhood, but by reframing them as "stress," it strips them of their political context and presents them as ambient and, hence, unavoidable. *MKC* urges kids to take a forensic and administrative approach to their feelings, responding to them as one might to an intrusion. Feelings, in *MKC* parlance, often seem to emanate from somewhere outside of the child. For example, it explains a military kid's persistent tummyache thusly: "It's like your heart gets filled up and those feelings have to go somewhere—so they go to your stomach." It invites even its youngest users to play "Stress Detective" and "have fun learning the clues to stress in your own body." The activity features the silhouette of an androgynous child with yellow bullseyes over different body parts, including the head, the chest, the stomach, the hands, and the legs.[92] Clicking on a bullseye opens up a range of symptoms, illustrated with little cartoon drawings about the things that stress might do: put butterflies in your stomach (a small swarm), make your hands sweat (little droplets), or lead to persistent negative thoughts (a thought bubble with a thundercloud in it). Each of these comes with an explanatory narration, as well as a multiple-choice question about what behaviors might relieve the symptoms. Tweens and teens can take another step by generating their own "Stress Management Plans," to be printed out and prominently displayed in the home.[93] In its explanation for the origins of stress, *MKC* clarifies that these reactions originally served an evolutionary purpose, now outdated and operating as a detriment to health.

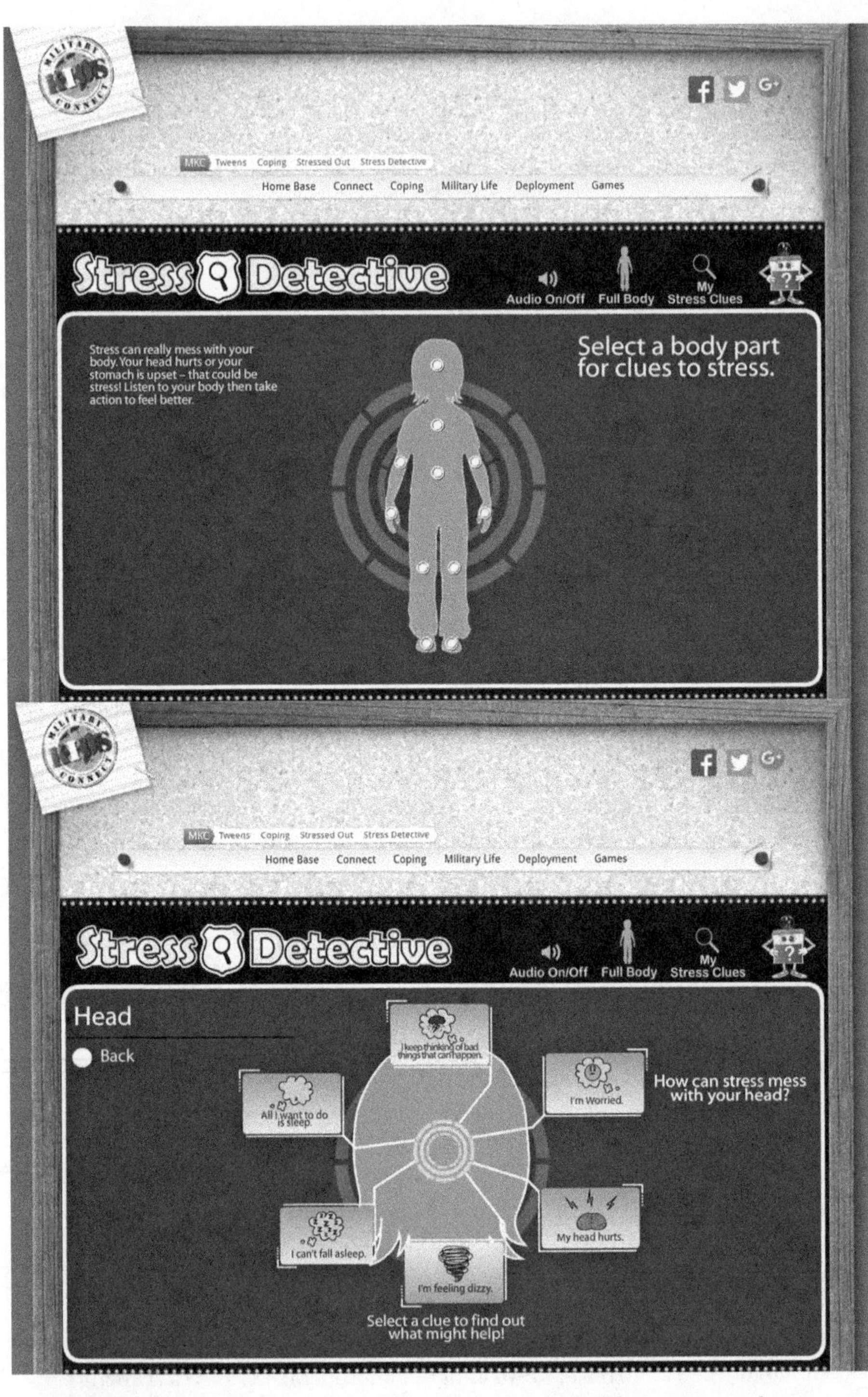

Military Kids Connect intimates to its young audience that they will have to manage the vicissitudes of military life with only minimal support from their parents. "Stress Detective" is an activity designed to help in this process.

In many ways, *MKC* replicates the logic of expediency that governs battlefield mental health interventions. Except in cases of acute duress (and perhaps not even then), most battlefield psychiatry proceeds according to the PIES formula: interventions based on proximity, immediacy, expectancy, and simplicity.[94] PIES-based treatments involve offering the servicemember brief respites from the frontline—without going too far away—that focus on restorative creature comforts like warm food, actual beds, showers, and quiet. *MKC* similarly reiterates the importance of such fundamentals for "stressed out" military kids and repeatedly reminds them to eat healthy foods, exercise, and get adequate rest. Battlefield psychiatry endeavors to reequip the warfighter to return to the fray as soon as possible; these techniques for military kids fold into *MKC*'s other goal of keeping children occupied. Toward that end, the site includes instructions for a range of crafts—most of which do not require parental assistance—that are both culturally neutral (like a baking soda and vinegar volcano) and not (like Arabic calligraphy).

Some of the content that *MKC* provides to keep kids busy is also designed to help them get oriented in the world. Presumably, this would help them to adjust better to overseas relocations and hence decrease the likelihood of troublesome behavior. For example, the site includes extensive sections of complicated recipes, including the local cuisine of common places where military personnel and families might get deployed, like Panama, Korea, and Afghanistan. The concern with increasing kids' cultural literacy about the places they or their parents might find themselves is reflective of the new emphasis on cultural sensitivity in U.S. military training.[95] For example, the "Where Are You Going?" feature is a militarized travelogue; it begins with a map of the world where thumbtacks indicate the location of various military installations. Users can "Click to explore," and Stampy provides a voiceover for a short video featuring information about local attractions, customs, climate, and basic vocabulary.

For kids who need to be entertained, *MKC* has plenty of games. Superficially, they are designed merely to give kids something to do. But unlike *SSMF*, which accentuates the importance of offering children playful respites from the exigencies of military life, *MKC* gameplay aspires to something weightier. "Word Recon" is a word-search where kids can hunt for words grouped according to themes like military acronyms, deployment locations, and "deployment feelings," which include "euphoria" and "pride" along with "guilt" and "resentment." Other games attune their senses to a more militarized bearing. "Operation Care Package," which requires using a cartoon cannon to launch a box of supplies around a number of obstacles, seems designed to refine both coordination and patience. Others simulate, implicitly, the visual habits necessitated by the amorphous landscapes of the War on Terror. In "The Amazing Wardrobe of Awesomeness," children select from an array of accessories to dress a mannequin in the correct traditional clothes for boys and girls in various countries; here, cultural awareness shades into learning how to recognize a local. All three age groups can play "What's Different?," which features two side-by-side photos of local scenes, like Afghan textile or produce markets, digitally altered to introduce subtle differences between them. Alerting players that something is "wrong" in the scenes, the game requires them to scour the photographs to identify minute changes from one to the next. It is very difficult. Often, even after I gave up and asked the game for a hint, I could not identify what the difference was. This may be another manifestation of the bad design that plagues the site, but the intense scrutiny that the game requires also mimics that of surveillance, a ludic introduction to the habits of attention necessary for a dangerous world.

Ultimately, all of this content tropes toward the virtues and practices of coping: keeping the body well, the mind active, and the child appropriately, constructively networked. Coping here is an expressly militarized process, and even the simple act of keeping busy, whether to pass the time or soothe a troubled mind, further acculturates children to militarism. Berlant describes the processes by which conditions like depression or obesity become legible as public health crises, arguing that "the disease becomes an epidemic and a problem when it interferes with reigning notions of what labor should cost," while proposed remedies are aimed not so much at ameliorating suffering as "diminishing the cost of the symptom."[96] In a similar way, *MKC* activities like "Stress Detective" put the burden of diagnosis on the military kid but target the symptom, for which the solution is an individualized form of "stress management," rather than its etiology. *MKC* also trains children how to distinguish between good and bad coping mechanisms: for the symptom of nausea, the choices range from "eat lots of candy" to "read a great book," and for legs that "want to move" uncontrollably, options include "kick the wall" or "ride my bike more." Stress, in this instance, is taken for granted as an inevitable and defining fact of military childhood, while military children are assigned the mission of controlling its effects. To be clear, I am not suggesting that military children should not play these games, or take deep breaths, or make Feeling Flowers, or identify with Elmo if those things are helpful. My concern here is with the ways that resources like this envision the emotional lives of military children and then represent them back to that very audience and, moreover, how the benefits of children's engagement with this media might ultimately accrue to the state.

The state's vested interest is clearly evident in *MKC*'s "Deployment Daily." This simulated advice column features exemplary problems faced by military kids before, during, and after deployment. The advice that circulates on the discussion boards is fairly pat and general: do your best to make friends, hang in there, trust that it will get better. Presumably, most discussion-board content comes from actual military kids. This means that it has been vetted but lacks the prescriptive force upon which the site relies. "Deployment Daily" fills that gap. Sometimes many paragraphs long, the responses typically diagnose or explain the problem while offering reassurance and guidance with an eye toward maintaining the functionality of the military kid. For example, one of the posts is titled, "Military Kid says, 'I'm Having a Panic Attack!'" Many of the hypothetical requests for advice in "Deployment Daily" evince this level of severity, but invariably, the responses operate at a much lower intensity, calmly explaining and often minimizing the problem. The answer to the "panic attack" begins, "You probably don't want to hear this but missing a parent when they go away is normal" and then outlines the reasons having a deployed parent might be anxiety-producing. But normalizing this reaction comes at the cost of directly addressing it—the "helpful tips" include reducing caffeine intake, keeping a journal, and building self-confidence with positive affirmations.

The post concedes that "if you feel really distressed it would definitely be helpful to talk to your parent," but the conditional form of that statement leaves it up to the child to determine how distressed is "really" distressed, and also leaves them to reckon with the costs of approaching the at-home parent, whom *MKC* consistently represents as taxed and struggling too. There is no indication here, or anywhere else on *MKC*, that the child should seek out the deployed parent for support. And the site repeatedly hints that the at-home parent will be ill-equipped to handle the stresses of deployment; this is a very different vision than the one conjured by *SSMF*, in which the at-home parent seems to do more, rather than less, for the child during deployment. So the child is left to calculate the costs and benefits of his feelings and assume all of the risk for getting it wrong.

SSMF defaults to an understanding of childhood, even military childhood, as essentially carefree and peaceful and tasks parents and communities with restoring it to that condition if there is a divergence. By contrast,

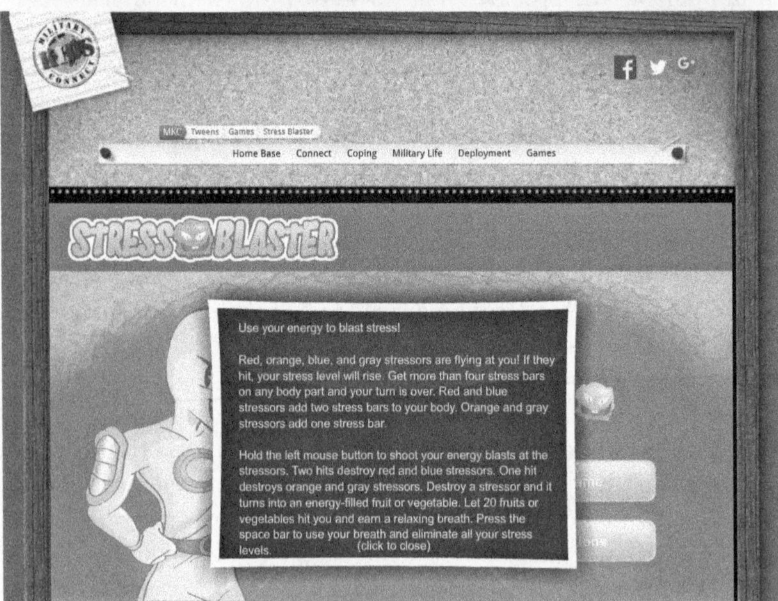

Ostensibly, *Military Kids Connect*'s "Stress Blaster" enables players to vaporize the things that leave them worried or anxious.

MKC takes for granted the opposite possibility and leaves it to military kids themselves to reconcile their circumstances. One of the games, called "At Ease," rhetorically asks, "Stressed? Emotions all over the place? So are these crazy shapes! Get them in line until you feel AT EASE." The game itself is patterned on the model of apps like *Candy Crush*, except "At Ease" players have to create groups of matching shapes (each representing a feeling, like a crying blue triangle or an angry red square) in groups of three or more and so make them disappear. If you make multiple feelings disappear at the same time, you get an "Explosion Bonus." I never tried, but I get the impression that one could play At Ease indefinitely, as there is no absolute ending or final level to clear.

And then there is "Stress Blaster." Players with an affinity for legacy games like *Pong* or *Arkanoid* might recognize their influence here. The center of the screen features a character vaguely reminiscent of Spiderman encircled by a device that launches little white balls meant to shoot the stress that comes from all directions. Stress here takes the shape of little monsters: hairy, fanged, multi-legged, fast and unpredictable in their movements. For every stressor that hits the body, one of his limbs turns partially red. A successful shot zaps a stress monster and transforms it into a fruit or vegetable, and if the produce hits the body of the protagonist, some of his stress is reduced. And if twenty fruits or vegetables make it to the hero in the middle, he earns a "breath" that, when redeemed, can eliminate all the red stress from the body. But if the body turns completely red from too many stress encounters, a screen pops up that says "Level failed," and grinning stress monsters bounce and taunt above the "Try again" button. I spent hours trying to figure this game out and never once earned a breath or survived past the first level. Again, this might just be a design flaw. But it is also a concession, howsoever unintentional, that there is no real cure for the stresses of militarization, which cannot be deflected, and so must be endured.

MKC takes for granted the possibility—generally unthinkable in *SSMF*—of a childish political subjectivity. It seems to regard this subjectivity as the consequence of an engagement with militarization; hence, it appears not as something that military children age into, but something that they simply have. This facilitates a different kind of address than that directed at civilian children or *Sesame Street* kids. It also belies a wariness toward its young audience, related to the broader cultural ambivalence around the "military brat," rooted in a sense that they might destabilize their families and hence the military as a whole. For example, children on *Sesame Street* can do virtually no wrong, acting only in ways that make

In practice, the game appears impossible to win. This may be a design flaw, but it is also an inadvertent confession that there is no real cure for the stresses of militarization.

them loveable or easy to sympathize with. But *MKC* includes content that features military kids misbehaving: being disrespectful to their elders, skipping school, shoplifting. The site places the blame for these transgressions on the children themselves, acknowledging no extenuating circumstance, with its own disapproval supplementing that of the parent who is unable or unwilling to discipline them properly, assuming a national interest in their good behavior.

What Should Jeremy Do?: Military Childhood as Insoluble Problem

The YouTube channel "Military Surprises" aggregates hundreds of videos, attracting tens of millions of views. Every video features a deployed servicemember surprising a loved one—often a child—by appearing, unexpectedly, in person. Nearly all of the scenes unfold in public, often at sporting events or in schools, locales that enable audiences to watch others watching and reacting to the child's own reactions, an endless mirroring and intensification of these sentiments. Each one is poignant in its way, but all follow the same narrative and emotional pattern, and there is no mystery to the appeal of a clickable happy ending in a period of perpetual war. But there is also a deeper pleasure to be had here, the restabilizing of childhood and family despite, or even through, militarization. Here, children appear as sweet, uncomplicated, and appropriately sensitive. These videos reduce the subjectivity of the military child to a desire for nothing more than reconnection with the formerly absent parent, whose military status hangs suspended in the scene; they often appear in uniform, but these garments serve only to amplify the emotional impact of their return home. Absent the military context, a parent-child reunion would scarcely be newsworthy. Militarization makes them matter, but deployment serves simply to intensify the affective charge of the reunion, never signifying as an experience that would negatively impact the child.

All of the examples that I analyzed here and in the previous chapter presume that both civilian and military children will feel something, and strongly, about militarization. A child that feels ambivalent, or feels nothing, is simply unthinkable. Berlant observes that in any situation, "being overwhelmed by knowledge and life produces all kinds of neutralizing affect management—coasting, skimming, browsing, distraction, apathy, coolness, counter-absorption, assessments of scale, picking one's fights, and so on."[97] But pursuing these behaviors would leave the child out of sync with the emotional community of the family, the affective mesh of the home; children must either be proud of their parents and hence self-sacrificially pleased about their deployment or discomfited by it and hence reasonably saddened by their absence. All of the interventions on behalf of the children function simultaneously to describe the feelings they are experiencing and prescribe the reactions that they should have.

MKC does this work without recourse to the image of the home as a bastion of security. Instead, the *MKC* home is imperiled by the combination of an absent parent and the likelihood that the present (female) parent will be left in less than ideal shape for the actual hands-on work of childrearing. To remedy this deficit, *MKC* relies upon the courage and sacrifice of the child, the compensatory value of their contribution to the overall mission. In confronting the subsequent dilemma of the parent unable to parent, *MKC* builds on the image of the military kid as mature and self-reliant, more balanced and capable than even the at-home adult. Many of the posts in "Deployment Daily" describe some kind of minor family dysfunction, with titles like "Why Is Everyone Yelling?" and "This Is Crazy—I'd Rather Be at School." But there are also posts specific to parental failings, like "Why Does Mandy's Mom Sleep All Day?" A different post, in response to a query about whether it's worth reestablishing intimacy with a father who is likely to be deployed again, says, "It's up to you if you want to take the risk of getting close again with your parent," which implies that the familial default is not love or intimacy, or that the love of the family is not enough to keep one of its members from going off the rails.

Most dramatically, a video in the "What Would You Do?" series, which features short vignettes about social and ethical dilemmas that young people might face and invites users to respond with ideas about how they would address them, tells a story about Jeremy, encountering a very different kind of "military surprise." The narrator tells us that his dad's deployment "has been tough on his mom, and she's been drinking and taking pills" (cut to a scene of a middle-aged white woman, apparently passed out in bed). His ne'er-do-well friends want to "take advantage" of the situation by raiding mom's stash ("Sleeping pills *and* painkillers? Man, these'll really chill you out!"), and Jeremy has to figure out how to keep them away. Later, he has to protect both his parents by covering for his inebriated mother when his father Skypes home from Afghanistan. Because Jeremy's story appears in the "What Would You Do?" section, the titular question goes unanswered, despite the desperate nature of his situation. But here, too, an inadvertent confession: there is no real solution in any case. Moreover, reflecting the silence on virtually all the peer-to-peer elements of

MKC, when I watched the video, there was no guidance in the "Comments" section either. Thus, the vision of the military kid as unsupervised and adrift, but bearing up, persists.

To speak of the military child is also to speak of the military parent, the military family. Figuring children is a way of figuring the family, the household, domesticity, reproduction, kinship, and various forms of filial loving-care. When outraged parents advocate on behalf of their child's troubled sensibilities, as I detailed in Chapter 1, they demonstrate that they have raised an appropriately sensitive child. Alternately, when their child appears to be insufficiently troubled, as in the case of the students too eager to draw their ISIS recruitment posters, parents step into the breach where their child's innocence is supposed to have been. During deployment, when parents who might otherwise speak up for their children are unavailable, the state resumes its role as protector of the military child. (Of course, when the state assumes custody of a minor like Omar Khadr, militarized by his father and corrupted at home, it adopts a different posture entirely.) Although *SSMF* and *MKC* make occasional references to families where both parents are in the military, the usual figure of the military child has one military parent, the father in the majority of cases, and one civilian, the mother. *SSMF* and *MKC* address themselves to military children, but also refer constantly, in ways both implicit and explicit, to their civilian parent, who is a mother in the majority of cases. Gender is crucial here. Alongside its provision of resources for the military child, *SSMF* delivers a normative affective pedagogy to the nonmilitary parent. *MKC*, on the other hand, speaks much more directly to children, acting on the assumption that the nonmilitary parent is unlikely to be helpful. The paradox here is that *MKC* is much more candid about the pressures and strains that militarization puts on families, but it tells this truth at the expense of military wives, blaming them for their failure to cope rather than acknowledging that their situations are often impossible, and encouraging their children to do the same.

CHAPTER 3

Recognizing Military Wives

The ticks arrived shortly after her husband left, deployed overseas for the first time. She was a new mother with an infant daughter, and they were outnumbered. Not just one or two ticks, or even a handful, but an actual infestation. She called an exterminator and followed all the advice he gave.

Treat the dog.

Bomb the house.

Get rid of the furniture.

And eventually, just seal off the room containing the worst of it and decamp to other parts of the home.

The exterminator, for his part, tried every solution he could think of. He was sympathetic to her and unsympathetic to the bugs, but the problem was bigger than ticks. It was deployment luck. This wasn't her only encounter with it (see also the middle-of-the-night gas leak caused by a malfunctioning dryer), and she's not the only one.[1]

Online forums for military spouses are filled with stories of house problems, car problems, illnesses, and sundry other mishaps attributed, with varying degrees of equanimity, to deployment "luck," "curses," and "gnomes."[2] Rationally, logically, objectively, there is no such thing as deployment luck. A skeptic might say that the same things happen to civilians all the time, but only seem more dramatic because the spouse is left to deal with them alone. But the discourse of deployment luck does more than make sense of random adversity; it also intimates, albeit jokingly, that a nefarious form of recognition targets military families, hurling bizarre problems at them during periods of acute vulnerability. Multiplied by the 1.1 million U.S. military spouses, deployment luck generates a staggering volume of misfortune.[3]

American military spouses already struggle for meaningful recognition—of their status, circumstances, and contributions—while being overdetermined as icons of heroism or victimization. In this context, the perverse recognition implied by deployment luck resonates viciously with systematic and structural ways of ignoring military spouses. The figure of the military spouse reveals the seamlessness with which affective recognition and material neglect can coexist in contemporary American militarism. Although programs and institutions designed to support military spouses have proliferated in recent years, they cluster less densely around the military spouse than around children. This differential corresponds to perceptions of military spouses as more agentic and less vulnerable, and hence less sentimentally appealing, than children. Moreover, while laudatory recognition of military spouses' resilience acknowledges their strength and ingenuity, it also obscures their hardships and downplays the failures of both government and military to meet their needs.[4]

The 1973 shift to an all-volunteer force (AVF) compelled the military to consider enlistees' children; adult spouses pose even more complicated dilemmas. The government and military exude much more confidence in their abilities to identify, meet, and manage the needs of military children than those of military spouses. Initiatives for military children, like the ones I described in the previous chapter, presume that children's needs are more straightforward and their political subjectivities much simpler, a certainty reinforced by children's inability to complain about their circumstances or their treatment. Military spouses, on the other hand, have the capacity to do both, but risk marginalization or worse—by the military and by the public—if they do.

Overall, the U.S. government has been relatively slow to acknowledge and respond to the needs of military spouses. Although programs like the Department of Defense's Military Spouse Preference Program, which gives priority consideration to qualified military spouses applying for DoD jobs, have been operative since the 1980s, provisions for military spouses have been scattershot at best.[5] While the Bush administration made some overtures toward military spouses in the early phases of the war, Laura Bush became a more outspoken advocate for military families after leaving the White House.[6] It was not until 2011, a decade after the War on Terror had begun, that Michelle Obama and Jill Biden launched their *Joining Forces* initiative to help military families with wellness, education, and employment.[7] And in early August 2017, women in the Trump administration hosted military spouses at the White House for a "listening session," but, as of the time of this writing, no official plans have been announced for better addressing their needs, besides a March 2018 executive order encouraging agencies to hire them.[8]

Military spouses occupy a range of contradictory positions vis-à-vis the institution that confers their status. Kenneth MacLeish observes that the descriptor "Army wife" implies a range of meanings: marriage to the institution itself, interchangeability with other women married to soldiers, or a wife who is also in the army.[9] As I noted in the previous chapter, there is a certain ideological awkwardness in the term "military child"; there is a similar tension operative in "military wife." On the one hand, the term sets these women apart from other, presumably civilian, wives. Even as it marks their proximity to the military, however, it also marks their distance from it, their auxiliary status. Like that of the military child and the military mother, the identity of the military wife bleeds across public and private spheres. Writing about military mothers, Wendy Christensen observes that this fractured identity can muzzle them, as the widely accepted "public/private divide that places [mothers] in the private and war in the public undermines mothers' legitimacy in speaking about the war itself."[10] The contours of this dilemma are different for military wives, whose conjugal relationships to their husbands do not engender the unquestioned purity of the love of a mother for her son; this means that acceptance of their behavior is more conditional yet. Often, their articulation of a political subjectivity costs them public sympathy, a dynamic similar to that surrounding the September 11 widows, who stopped being popularly likeable when they started questioning the government.

This trade-off between subjectivity and sympathy is operative across the materials I analyze in this chapter. I explore how they illuminate the paradoxical results of the combined recognition and neglect of military spouses: a circumscribed and contingent visibility as a function of their proximity to suffering, whether their own or their husband's, along with a chronic suspicion about their reliability and sincerity. The affective recognition they receive often substitutes for material support and is accompanied by additional prescriptive norms about how military wives ought to behave.

Like the military's stringent expectations for the behavior of military spouses, these public affective investments have exacting specifications. In short, military spouses are hypervisible, but narrowly. Their exposure increases in proximity to sacrifice, their own or their partner's, but they appear only through the screens of patriotic romance or the emerging trope of the victim of a traumatized soldier's domestic violence, a phenomenon I consider in more detail in the next chapter.

To contextualize the figuring of the military spouse, I begin with a sketch of two key histories: that of women's militarization during the War on Terror and that of the U.S. military's approach to military wives. From there, I describe the predominant affective investments in and expectations for military spouses. These are made explicit in presidential proclamations of appreciation for military spouses and their sacrifice, which are my first objects of analysis. Operational Security (OPSEC) materials, my second, reveal the other side of official regard for military spouses, which identifies them as vital but weak links in national security. Conversely, the American Widow Project (AWP), a network organized and maintained by military widows, offers an alternative to these official discourses, recognizing widows' sacrifices but also embracing a vision of widowhood that is independent and pleasure-seeking. I consider the politics of this strategy in the penultimate section. The work of the AWP intimates that, within the affective strictures established by prevailing systems for recognizing military spouses, their full visibility is possible only in the radical absence of their husbands. I conclude with a consideration of military spouse posttraumatic stress disorder (PTSD), an emerging line of inquiry that simultaneously maps and submerges the subject-position of the military spouse.

Women in the War on Terror

Despite the increasing participation of women in militaries worldwide, Christine Sylvester notes that "commonplace understandings of war today can still be starkly sex-differentiated: men do war and women suffer, support, or protest war."[11] These understandings both reflect and perpetuate the idea that women, especially Western women, and militarization are antithetical, in much the same way that the formulation of the "military child" links two statuses that would otherwise seem to be at odds. This dichotomy preserves both the supposed maternal gentleness of women and the masculine purity of war; thus, prevailing ways of imagining the military wife confine her action to the domestic sphere or the expanded domestic space of the base community, softening the "military" side of her identity. Realistically, of course, military wives participate actively in war making, if often in ways that are quotidian and nonspectacular: managing the household despite the ravages of deployment luck, affixing magnetized yellow ribbons to their cars, putting patriotic T-shirts on their children, staying awake in the wee hours to wait for a phone call from overseas.

At the same time, other militarized women have garnered an unprecedented visibility during the Global War on Terror. Making a concerted effort to brand the war as a feminist undertaking, both George W. and Laura Bush repeatedly emphasized the urgent need for women's liberation in the Middle East. The Bush administration leveraged the enlistment of women as a symbol of their empowerment, and Deborah Cohler describes the early years of the War on Terror as "historically noteworthy for the mainstreaming of media images of female U.S. combatants."[12] During the War on Terror, the sight of women in uniform has become far less remarkable, if not exactly commonplace or uncontroversial. Indeed, the 2013–16 process of opening all combat roles to women revealed lingering reservations, especially on the part of the marines' leadership, about women's capabilities while affording the Obama administration an opportunity to flaunt its progressive credentials.

The new visibility of American female military personnel boomerangs, inevitably, back to established ideas about femininity. Kelly Oliver argues that images of militarized women "haunt" spectators by toying with "age-old fears of the 'mysterious' powers of women, maternity, and female sexuality."[13] Confronted with potentially discomfiting images of women in the military, media outlets often domesticate them by reference to their husbands.[14] On the other hand, noncombatant military wives do not trigger those fears, and even assuage them, and these women do not garner the same kind of individualized attention as female combatants. Because service in the U.S. military is voluntary, a woman's decision to enlist grates against commonsense notions about inherent female pacifism.[15] Consequently, there is no generic archetype of the female soldier; rather, individual female soldiers become visible as aberrations of their gender, the U.S. military, or both.

Hence the iconic status of Jessica Lynch and her seeming obverse, Lynndie England. In March 2003, Lynch's supply convoy made a wrong turn that led them into an ambush; during the ensuing firefight, Lynch was seriously injured and captured by Iraqi forces. Ten days later, a team of U.S. Special Forces personnel rescued her from Saddam Hussein Hospital. The details of Lynch's ordeal remain contested, but despite or perhaps because of this, she became a celebrity. Every iteration of her story (except, importantly, Lynch's own) situated her as what Oliver describes as the embodiment of the "best of American womanhood."[16] Whether she was the patriotic small-town girl who valiantly enlisted to defend her country or the hapless damsel who needed to be rescued by heroic American men, Lynch became a singularly positive symbol of soldiering femininity, a mythologizing that rested heavily on her whiteness. Comparatively and predictably, the women of color who were present at the ambush (Lori Piestewa, a Native American who was killed, and Shoshana Johnson, an African American who was also captured) received less public attention and affective investment.[17] For Lynch, iconicity came at the cost of narrative control over the story of her captivity. She has repeatedly downplayed her status as a hero and contested reports that she was sexually assaulted by Iraqi military personnel, but these details fell quickly by the wayside, overshadowed by the image of her prone, smiling weakly, on a stretcher borne by her rescuers.

While Lynch's femininity was secured by her attachment to military men, England deviated wildly from this norm. Arguably the most recognizable face of the torture at Abu Ghraib (try to recall the faces of the American men captured in the photos), England participated in the sexualized abuse of male detainees and was romantically involved with another reservist, Charles Graner, which resulted in an out-of-wedlock pregnancy. With her ostensible appetite for violence, androgynous haircut, and nonheteronormative sexual practice, England became a stark counterpoint to Lynch's apparent delicacy, modesty, and virtue. The temptation to compare the women proved irresistible for commentators, but the details of their biographies are less revealing than the widespread urge to parse them. Tales of England's dirty job at a meat-packing plant or Lynch's desire to teach kindergarten do not tell us much about their political subjectivities; rather, they are ways of measuring their conformity to, or deviation from, ideals about militarized American women.

On the home front, Cindy Sheehan's liminal status as the civilian mother of a soldier became a vector for intense criticism. After her son, Casey, was killed in action in Iraq, Sheehan installed herself outside George W. Bush's Texas ranch in a durational antiwar protest. Sheehan used the moral authority of grief-stricken maternity to condemn the war, an affectively and politically polarizing strategy. Supporters embraced Sheehan's actions as pure and powerful, while critics denounced them as anti-American hysteria, condemnations revealing a certain tenuousness in the sentimental privilege otherwise afforded to the Gold Star mother. Whether one condones her methods or not, Sheehan's story illumines the policing of women's grieving over men lost in war; their grief is mandatory and expected, but permissible only below a certain threshold, above which it appears treasonous.

To date, no individual American military wives have gained equivalent celebrity, or notoriety, to that of Lynch, England, or Sheehan, despite the popular cultural fascination with military wives in general. This fascination is perhaps most apparent in the popularity of the Lifetime drama *Army Wives*, which at the time of its run was the most successful show in the network's history. Mary Douglas Vavrus describes it as "gendered military propaganda using the conventions of soap opera and serial drama . . . in an attempt to fix meanings around army family life" that works to "naturalize and normalize historically specific ideologies about army gender politics and the wars."[18] Accordingly, even as it followed the tribulations of its characters, the show affirmed the necessity and goodness of militarism, narratively disciplining the women who disagreed. For my purposes, the details of the show are far less relevant than the simple fact of its existence, an index of the intense sensationalizing and romanticizing of military life that extends to the women who inhabit it.[19]

Like Security Moms, but Sexier: Imagining Military Wives

Military wives have some access to the popular admiration and gratitude directed at the troops (about which much more in Chapter 4) and to what Catherine Lutz describes as the "supercitizenship" afforded to enlisted military personnel.[20] But unless she herself is enlisted, her supercitizenship is derived from that of her husband, and if she is enlisted and a mother, exhortations about her commitment to country are tinged with doubt about her dedication to family. Military wives are widely represented as accidental or even reluctant conscripts to the war effort; figuring collapses their complex political subjectivities into affiliations that are derivative or by proxy. In her study of military mothers' online practices, Christensen describes how they claim "noncivilian" identities in order to garner rhetorical authority. She writes, "The paradoxes of public/private boundaries are especially salient for military mothers, for whom war is both personal/private (they have children who are in the military) and public/political (their own experiences of war are shaped by public political processes)."[21] Military wives occupy a similarly fractured position. Politicians and military leaders invoke them and sometimes even trot them out as exemplars of what is best about the nation, including its solid heteronormativity, but relegate them to the private when they become liabilities or demand different forms of recognition and support. To negotiate this simultaneous hyper- and invisibility without running afoul of the military itself, military wives must delicately calibrate belief and action. The sheer number of self-help books, magazines, and online forums that military wives create attests to both the complexity of the role and their resourcefulness in managing it.

While the political subjectivity of the military wife is often regarded suspiciously, militarized civilian women in the War on Terror, the so-called security moms, enjoyed widespread approbation from supporters of American militarization. According to the conventional wisdom, after September 11, the moms who had previously been concerned with getting their kids to soccer pivoted to security and foreign affairs and became hawkish on those fronts. Profoundly agentic, in Bree Kessler's terms, security moms would eagerly "initiate the policing of themselves, their family, and those around them through surveillance techniques."[22] Inderpal Grewal theorized security moms as symptomatic of the neoliberal fusion of public and private, arguing that they act at the edge of state power and extend it into new territories.[23]

Although security moms elicited criticism from antiwar feminists, pronounced curiosity in the media, and tactical attention from presidential candidates, polling data suggests that they might not actually exist. Laurel Elder and Steven Greene debunked the myth of the "security mom" (and her counterpart, the "NASCAR dad") but argued that the idea of the security mom retained traction because of the "'man bites dog'" appeal of a story that "defied long-held stereotypes about women, mothers, and military action."[24] Because she enacts a quaintly militaristic and feminized geopolitical sensibility firmly localized to the home, the fiction of the security mom might have defied those stereotypes but did not really challenge them. The idea of security moms supporting defense spending and aggressive militarism provided a vision of maternal plenitude that licensed the state to pursue its ambitions but demanded nothing in return. Indeed, it seems that the state so desired this kind of female citizen that it had to concoct her where she did not truly exist.

Of course, compared to military wives, "security moms" do not generate quite the same *frisson*, as they lack the charge drawn from an attachment to military men. While mythic security moms were motivated by an instinct to protect their children, the figure of the military wife embodies a militarized heteronormativity: a sexier alternative to security motherhood. Yet if the civilian security mom is imagined as fully supportive of an aggressive U.S. military, the military wife, paradoxically, can seem more suspect, her support for militarization more conditional. And so she must be managed. Lutz describes the military as a "total institution" in both ambit and practice, the primary determinant force in the lives of its members.[25] This power diffuses onto their families, channeled by the servicemembers themselves, enacted through policies that apply to spouses and children, and replicated through networks of socialization.[26] MacLeish writes that for military spouses, the army dictates "privacy, comportment, and conduct," along with "access to the rest of the world" and their "sense of what's possible."[27] By bringing spouses and children into the fold, the military engages in the business of heterosexuality.[28] But the state cannot militarize heterosexuality alone; it requires and incentivizes the participation of military spouses themselves.

Both the military and the family operate as "greedy" institutions, placing copious and often nonnegotiable demands on their members and potentially absorbing all available resources of time and energy.[29] Inevitably, this leads to conflict between them. For much of its history, the U.S. military managed this dilemma by pronouncing families incompatible with military service. But in the late twentieth century, as the U.S. military became an all-volunteer force with the elimination of the draft in 1973, the practical exigencies of recruitment and retention forced it to adapt. All of the policies on marriage and family that the military has enacted since are meant to address these demands. They do so without addressing the fundamental incommensurability at the core of military marriage—namely, that active military service, by imperiling the life of the servicemember, jeopardizes the very security and futurity upon which the heterosexual family form is based. It falls, therefore, to military spouses to resolve these dilemmas as they manifest in countless ways.

The pre-AVF U.S. military maintained its incompatibility with the family in ways both official and not. Until the middle of the twentieth century, as Lutz observes, a soldier wanting to marry would need the permission and approval of his first sergeant.[30] Soldiers themselves circulated the conventional barracks wisdom that "if the Army wanted you to have a wife, it would have issued you one." At the level of policy, the U.S. armed forces continually retooled the relationships between marriage, fatherhood, and the draft; a likely noncoincidental spike in weddings followed the revision of the 1940 draft law to include an exemption for married men.[31] Although the particularities of deferments and exemptions for paternity and marriage varied over time, from at least the era of World War I, the U.S. government accepted that a man's marital status should matter in the determination of whether or not he would be conscripted.[32] Currently, the Selective Service System retains that logic, promising deferment for "men whose induction would result in hardship to persons who depend upon them for support" in the event of a hypothetical draft.[33] These seemingly compassionate provisions also reveal a persistent skepticism about whether military personnel benefit from marriage and family.

Servicemembers no longer need the authorization of commanding officers to marry, but tight and regulatory social networks ensure that new military spouses learn and adhere to a range of expectations that assimilate them into military life. Among the ranks of military men, MacLeish documents a lingering misogyny toward military wives, noting that many of the soldiers he encountered were skeptical of women's motivations for marrying military men. The attitude was sometimes overtly hostile, with soldiers joking about "Standard-Issue Military Wives." Sensing the widespread adoration of military men, others suspected that women's interest in them was either opportunistic or purely, even fetishistically, sexual.[34]

In fact, the popular idea that military personnel make especially desirable mates has a relatively short history. Lutz notes, for example, that in communities surrounding military bases, families actively sought to keep their young daughters away from soldiers, despite army efforts to allay their fears. During World War II, for example, the army released a "barrage of images of soldiers in ads, news stories, and military promotional material that showed every soldier, enlisted or not, as clean, tidy, ironed, sexually innocent and nonaggressive, and good-humored" that ultimately did little to dislodge suspicions that they were quite the opposite.[35] Whether or not these public relations campaigns improved soldiers' marriage prospects, they surely revealed the state's growing investment in the heterosexual propriety of its armed forces.

In a meticulous archival study of the federal regulation of homosexuality, Margot Canaday documents the many ways that the U.S. military sought to reacclimatize its men to domestic life, a negotiation evident in the complex sexual politics of the G.I. Bill.[36] With its emphasis on education, marriage, and homeownership, Canaday contends that the Servicemen's Readjustment Act was "directed at settling men down" after their wartime adventuring. Although the bill entitled unmarried male and female veterans to benefits, it did so partially and grudgingly. Canaday demonstrates that it allocated the most support to male heads of households with wives and dependent children, establishing them as the "most deserving citizens."

The G.I. Bill marks one of the most significant thaws in the military's orientation toward domesticity, and its affinity for heteronormative arrangements now extends to active-duty servicemembers as well. Following the end of the draft, the U.S. military changed its institutional relationship to romantic relationships, accepting married enlistees and parents and absorbing their spouses and children. Currently, more than half of military personnel have "family responsibilities for spouses, children, or other dependents."[37] In fact, new enlistees are more likely than their civilian age-mates to have spouses and children when they join up. Presumably this trend can be explained by the promised stability in income and benefits afforded by a military career. These lures would likely be extra appealing to those with familial responsibilities, particularly for young people of color and those from less affluent backgrounds with fewer other avenues for attaining comparable security.[38]

When the AVF era began, the military had the luxury of the post–World War II baby boom and its resultant large pool of young men.[39] Over time, however, the demographic math changed, compelling the military to be more aggressive and strategic in its recruiting. To entice the most qualified men to enlist, the Department of Defense had to position itself as a viable competitor with other employers.[40] Accordingly, it began making accommodations for their needs, developing career paths that promised more predictable and regular work schedules, softening the demeanor of drill sergeants, perhaps to make them more like civilian managers, and relaxing regulations about appearance.[41] And because enlisted women tended to leave the military once they got married and had children, in the late twentieth century, the DoD began branding itself as an alternate route for women to the stability that would otherwise be provided by marriage.[42] In sum, according to Lutz, once marriage and family were "no longer officially discouraged . . . money and effort poured into recruiting and retaining not just the solider but the whole family."[43] Instead of continuing to operate as a fraternity of untethered bachelors, the military hewed to a vision of heteronormative family life, providing more remuneration and benefits to personnel who were married and thus incentivizing marriage over more transient affiliations or cohabitation.[44]

Just as the military was adjusting to its new entanglements, American culture more broadly was trying to make sense of the sudden compatibility of military service and family life. In the previous chapter, I queried the trope of the military brat. Military brats are, of course, begotten by at least one military parent, raised in military families, and the clinical diagnosis of "Military Family Syndrome" exists alongside the lay imaginary of the military brat. Rooted in the post-Vietnam concern over veterans' mental and emotional struggles, mental health professionals turned to the rubric of Military Family Syndrome to explain their interpersonal ramifications. As a diagnostic, Military Family Syndrome envisioned the dynamic of an authoritarian father, abetted by a timid mother, who bullied children into compliance with his every unreasonable demand.[45] Today, this theory is largely discredited, with the emerging expert consensus being that military children are more psychologically and behaviorally similar to their civilian counterparts than they are different.[46] But the stigma of the dysfunctional military family persists, even among mental health professionals.[47] It also circulates in popular culture, and it is worth questioning why this fantasy lingers and how it might resonate with ideas about masculinity, femininity, and militarization.

Despite practical incompatibilities between military service and family life, the U.S. armed forces actively rely on military spouses.[48] Liz Montegary notes that "while draftees in the twentieth century imagined service as a release from their everyday lives and a detour within their heteronormative trajectories, recruits today are encouraged to think about a military career as reconcilable with and perhaps even enriched by a wife and children."[49] The military promotes this line of thinking because it is convenient. Given the tremendous investments of time and resources entailed in recruiting and training personnel, the military garners its best returns if they choose to reenlist, and the military knows that families, especially spouses, play a crucial role in that decision. Consequently, the Department of Defense works hard and studiously to understand the family dynamics that bear on enlistment decisions, employing sociologists and other experts to study rates of reenlistment, divorce, domestic violence, and overall measures of the happiness of military wives.[50] This interest in the affective stems from the military's concern for the "morale" of its troops. Morale, in short, is a willingness to persist in the war effort despite overwhelming hardship, and this unquantifiable sense became, as Ben Anderson notes, the "object for specific techniques of power" during World War II.[51] Increasingly, military leaders have come to recognize the variables of military spouse and family morale and endeavor, of necessity, to keep them intact.

Beyond combating the cosmic vicissitudes of deployment luck, military wives confront a range of tangible obstacles as well. As a whole, young military personnel and their families are massively indebted, often relying on credit cards and high-interest short-term loans to supplement the pay afforded to low-ranking enlistees.[52] Because they must relocate frequently and manage households and children singlehandedly, as well as supporting their own military personnel and volunteering to serve the larger community, many military spouses struggle to advance their own careers.

This compounds financial difficulties and requires that wives be enterprising about finding work.[53] Especially during deployment, military wives do all this beneath clouds of worry and insecurity. These are issues that policies like preferential hiring and *Joining Forces* are meant to address, but their success so far has been partial.

Military dependence on spouses is concrete and profound and manifests in part through an established culture of their volunteerism. But the language of volunteerism understates the pressure on military wives to participate, which emanates both from base social networks and, often, from their husbands' commanders. During the twentieth century, women often volunteered through military auxiliary units as well as through the Red Cross, the USO, and the sale of war bonds.[54] In addition to its expectation that women perform duties like these, the military today often outsources the work of caring for military families back onto the families themselves, expecting that they will support others in times of need and aid in the recuperation of injured military personnel. Christensen points out that the DoD actively "mobilizes mothers" to aid recruitment and reenlistment, to provide support to their children during deployment, and in the event of an injury, to provide rehabilitative health care after.[55] Expectations for military spouses are even more numerous. In a *New York Times* op-ed published five years into the War on Terror, Tanya Biank, a journalist and army wife (her writing inspired *Army Wives*) encapsulated the situation as follows:

> Almost 30 percent of all Army wives volunteer in a formal capacity. . . . Our work is expected, underappreciated and often goes unnoticed. We volunteers save the Department of Defense millions of dollars that would otherwise have to be spent on consultants, accountants, social workers, publicists, counselors, fundraisers, program managers, administrative assistants, advisers, class instructors and event coordinators.[56]

In Biank's assessment, the DoD both depends upon military wives and disavows this reliance, banking on their contribution but refusing to recognize its true value.

Moreover, the demands of military service and deployment tax their relationships, frequently rendering them conflictual as a matter of course. Even absent the strain of deployment, being married to someone in the military is difficult; Biank characterizes the army as the "other woman" in her marriage.[57] And this other woman often outranks the wife, who must accept her secondary role because virtually everything about her family depends on her doing so. MacLeish describes a pervasive sense of marriage in crisis in the Fort Hood community, as his informants cited divorce rates (though they are not confirmed by actual statistics) of 60, 70, and 80 percent.[58] MacLeish observes that these couples feel the weight of the army pressing constantly on their marriages and consequently experience their love as either more fervent or more fragile or both. This fragility settles disproportionately on military wives. Precarious as their economic situations might be during marriage, divorce, with its attendant risk of losing access to military resources, would probably make them worse—likewise if a man is dishonorably discharged from the military because of a domestic violence conviction and loses his benefits. Indeed, when military wives experience domestic violence, they often confront the extent of their invisibility: as Cynthia Enloe notes, abusers' commanders often ignore these issues lest they distract from waging war, while the American news media is reluctant to cover the stories lest it seems unpatriotic.[59]

If their husbands return suffering from the "signature wounds" of the Global War on Terror, PTSD and traumatic brain injury, military wives may find themselves responsible for the day-to-day management of their symptoms and coordination of their care, even when they have access to rehabilitative services. This dynamic is further complicated by an ethos that valorizes recovering from combat stress without recourse to professional therapeutic intervention. David Kieran, in a study of recent veterans' memoirs, observes that they often concur with an official approach that "places the onus for recovery" on veterans themselves rather than clinicians or the military.[60] This distribution of responsibility implicates their families as well, especially because many of these authors do not attribute their trauma to combat experience, instead identifying the "primary locus of trauma as lying outside of the combat zone in the veterans' predeployment or domestic life."[61] But this coping strategy potentially jeopardizes the mental health of military spouses, who fare better, both individually and relationally, when they understand their husbands' trauma as connected to a specific cause, like combat exposure.[62] The memoirs that Kieran analyzes consistently identify the home as a space of healing, where veterans promote their own recovery through a "recommitment to a normative domesticity."[63] By design, this domesticity requires other actors, a wife and children, thus obligated to fashion the home as a sanctuary, working invisibly to attain this end.

The various institutions that figure military wives also unilaterally reclassify their suffering as patriotic and necessary and insist that it is stoically and willingly borne. The recognition they afford is symbolic compensation, while the admiration and gratitude they bestow function normatively and prescriptively. And so it is that military couples in general are less likely to divorce while one spouse is enlisted and more likely to divorce afterward, presumably because the military provides a supportive environment for managing the stresses that might otherwise become the undoing of the relationship.[64] But the pervasive valorization of military personnel and the derivative mythologizing of their wives might also conspire to make separation seem unthinkable.

Of course, military spousehood is not only, or inherently, miserable; the remedy for the idealization of military wives is not to overcorrect by portraying them as victims, an inverse of the figuring that constructs them as heroes. Many practical benefits inhere in the role of military wife, including senses of purpose and connection and financial security.[65] And then there is the matter of love. In her consideration of the ways that love becomes a vector of terrorism, carving a path for it to travel along, Asma Abbas asks, "How have we loved in order to suffer this way? How might our life in and with terror, and our status as subjects of this terror, be connected with the inherited modalities of love and suffering insinuated in this subjectification?"[66] This question takes on an added urgency in the case of military wives, for whom love and suffering are intimately predicated on one another and so beget variegated political subjectivities. Military marriage cannot simply be explained as a form of "cruel optimism," Lauren Berlant's term for attachment to one's pursuit of a "good life" that is actually antithetical to one's well-being. Rather, the more pernicious optimism is the one imposed on military wives and military personnel from outside, which downplays the complexities of their decisions and the intricacies of their dilemmas.[67]

For civilian audiences, the figure of the military spouse makes militarization comprehensible, familiar, and romantic.[68] Recent proclamations of military spouse appreciation days, which I analyze later in this chapter, have explicitly referenced military casualties and the spouses who care for wounded warriors or mourn the loss of husbands killed in action. But paradoxically, the figure of the grieving military widow might also deflect attention from the fact of wartime casualties and the state's obligations toward them. In her study of photographs of mourning women worldwide, Marta Zarzycka describes their function as follows: "Easily accessible, undemanding in its familiarity, and well-suited to mass-mediated collective memory, the trope [of the mourning woman] proves particularly powerful in the case of the coverage of atrocity: it replaces the un-picturable character of trauma and loss with a recognizable scene."[69] In this way, when military wives become recognizable as individuals, they simultaneously facilitate public nonrecognition of the larger causes of their griefs and the role of the state in engineering them.

Love, Militarized: Affective Constructions of Military Wives

Despite the deliberate attempt at inclusiveness embedded in the official language of the military *spouse*, the term functions, both practically and conceptually, as a near-perfect synonym for military *wife*. All branches of the U.S. military remain predominantly male, and the vast majority of military spouses are women. MacLeish observes that military *husbands* are uncommon and essentially unacknowledged, even more discursively illegible than female soldiers.[70] Presumably, the same can be said of the same-sex spouses of military personnel, who acquired a modicum of official legitimacy with the repeal of Don't Ask, Don't Tell. The official rhetoric of the "military spouse" simultaneously invokes and erases women while implicitly reinforcing the heteronormative social structures and cultural practices that remain central to military life. Indeed, the admiration, gratitude, pity, and anger operative in the figuring of the military spouse are expressly gendered and hinge on the femaleness and wifeliness of their objects. MacLeish describes the "Army family" as more than a demographic; it is, he argues, a "lived affect."[71] In the following analysis, I explore how various entities, among them an American public eager to know how love feels when militarized, have sought to identify and regulate that affect.[72]

Relative to the military children I considered in the last chapter, military wives are figured as both less pure (always the dilemma of the married woman) and more noble (because they have, ostensibly, chosen the sacrificial path of the military family). The conjugal ties that bind the military wife to the military exclude affection from popular forms of figuring. When it sends her husband off to war, the nation-state adopts a custodial posture toward the military wife, offering minimal sustenance until her husband comes back. Iris Marion Young argues that the United States has, in its War on Terror, relied on affectional bonds to establish new forms of citizenship. She observes that this new arrangement is patterned on the family, with the citizens in the position of obedient women and children dependent on the chivalrous protection of the masculine state, a softer expression of power, paternalistic rather than dominant.[73] The state directs a much more targeted form of this power—often expressed as care and concern—toward military wives. And so it is that George W. Bush averred, just before the end of his speech marking the 2008 observance of Military

Spouse Day, that legislation he sent to Congress to expand services for military spouses "should send a clear message that we care for you, we respect you, and we love you."[74] But for an outsider to feel affection for her would be to transgress against the fact of her marriage and, by extension, the military itself. It would also raise the specter of the stereotypically unfaithful military wife.

The military systematically commandeers love for its own purposes. Lutz and MacLeish document both the army's reliance on and operationalizing of love to sustain morale while interfering with it as a matter of course. In sum, Lutz writes, the "Army creates separations and new loves on a mass scale."[75] By participating in these loves and separations, the military wife becomes an object of the other affects I consider here. First, admiration for her stoicism and fidelity. This form of regard is profound but intensely moralizing, quickly rescinded if she defaults on either measure. She can be tearful, occasionally overwhelmed—this is, after all, the mark of ideal emotional womanhood.[76] But she must always bear up. Anything less might read as weakness, disloyalty, or worse. There is no leeway at all on her fidelity in this affective investment. Gratitude has lately emerged as the necessary corollary to admiration, an evolution I trace in my reading of the presidential proclamations, which derive from a broader cultural thankfulness for her husband's service and sacrifice. This gratitude stands in complex relation to pity, extended on the assumption and condition that she is lonely, lovelorn, pining for her deployed serviceman. Importantly, the figure of the military wife does not draw pity because she has to manage a household or children singlehandedly or forsake her own career; the idealized figure of the military wife cannot register those losses. In this affective regime, she can occasionally express her frustration, but cannot voice any stronger discontent than that. Overtly angry military wives risk censure, ostracism, the careers of their husbands, and by extension, the security of their families if they criticize the mission or the institution publicly. Their anger does not, cannot, appear in this imaginary. Others may feel angry on their behalf, particularly when they are widowed, but can direct this hostility only toward the enemy or the general tragedy of war, not at the state or the military itself. The political subjectivities of military wives are largely inaccessible, therefore, because of the institutional mechanisms that constrain their expression and dictate the terms of their appearance.

Thus, as an object of sentiment, the military wife is capacious but contingent. All of these affective bestowals—by the government, by the military, by the media, by various charitable organizations—depend on her emotional performance. They expect her to be effusive about her romantic love for her husband, accepting of its costs, and always a- or even antipolitical. Military leadership depends, Enloe argues, on the military wife embracing that identity to the exclusion of others. The goal, according to Enloe, is that she will "imagine that her greatest contribution to the country's security is mediated through her role as a wife" and realize in turn that "women's value as citizens and as patriots is embedded in their roles as mothers of soldiers and wives of soldiers, *not* as spokespeople on foreign policy."[77] Moreover, these affective investments hinge on a range of fantasies about military marriages: that they are happy; that all nondeployed spouses experience deployments as heartrending (rather than, for example, liberating); and that these relationships are sustained by love or loyalty rather than convenience or a lack of alternatives. Simultaneously, their participation in a militarized, nationalized form of heteronormativity constructs them as objects of a sentimentality that sanitizes out both desire and unpleasantness, but never regards them as full political subjects or trustworthy partners in the military enterprise.

"Responding to the Call of Duty": From Recognizing to Appreciating Military Spouses

Military researchers have consistently found that military children and families fare better when they have a sense of "broad societal appreciation for the value of military service."[78] But I argue that multiple interests, beyond those of the military spouse, are served by this kind of officially mandated appreciation, which functions as an affective means for the consolidation of state power over military families. Deborah Gould identifies the centrality of affect for expressions of power. She writes,

> Power certainly operates through ideology and discourse, but it also operates through affect, perhaps more fundamentally so, since ideologies and discourses emerge and take hold in part through the circulation of affect. Those seeking power and control sometimes can bypass the realm of ideas and attempt to influence, manipulate, or harness affective states to the desired objectives of a leader, the state, capital.[79]

The state stands to benefit from a population that appreciates military spouses. With public professions of admiration and gratitude for military spouses, the state generates a convincing alibi for other forms of systematic neglect and disenfranchisement of them while cultivating more support for militarism by hitching it to the sympathetic figure of the loyal military wife.

The origins of official recognition for military families are sandwiched between an act permitting the National Park Service to scout for new lands and a resolution authorizing the construction of a memorial to Haym Solomon, a Revolutionary War–era financier. Senate Joint Resolution 115, passed on June 23, 1936, designates the last Sunday in May as "Gold Star Mother's Day." It offers the following justifications:

> Whereas the service rendered the United States by the American mother is the greatest source of the country's strength and inspiration; and Whereas we honor ourselves and the mothers of America when we revere and give emphasis to the home as the fountainhead of the state; and Whereas the American mother is doing so much for the home and for the moral and spiritual uplift of the people of the United States and hence so much for good government and humanity; and Whereas the American Gold Star Mothers suffered the supreme sacrifice of motherhood in the loss of their sons and daughters in the World War. . . .[80]

In and of itself, S.J.R. 115 is relatively toothless, merely a request that the president issue a proclamation instructing the American public to observe this holiday. Though it provides instructions on what forms that observance might take, S.J.R. 115 leaves it up to the president to make the yearly request. It was not until 1984, roughly a decade after the transition to an all-volunteer force, that President Reagan took the liberty of proclaiming a day for military spouses, and in this section, I focus on the official discourse that mandates and shapes their recognition.

The senators behind S.J.R. 115 were not the first politicians to suggest that military families deserved special recognition. For example, in 1865, Abraham Lincoln ended his second inaugural address with an exhortation about widows and orphans that the Veteran's Administration would subsequently incorporate into its mission statement:

> With malice toward none, with charity for all, with firmness in the right as God gives us to see the right, let us strive on to finish the work we are in, to bind up the nation's wounds, to care for him who shall have borne the battle and for his widow and his orphan, to do all which may achieve and cherish a just and lasting peace among ourselves and with all nations.[81]

Notably, Lincoln's address explicitly tasks the nation-state with active care for widows and orphans, while S.J.R. 115 simply recommends that it should spare a thought for them. Over time, the differences between actual care and thoughtful recognition have blurred, with recognition becoming the preferred and predominant service provided to military families.

Gold Stars as a symbol for maternal loss of a child during war date to the era of World War I. Typically reserved for birth mothers, Mary Clark notes that the U.S. government granted the earliest Gold Stars in the 1920s to mothers willing to have their sons buried in Europe rather than having their remains repatriated.[82] In this way, Gold Stars signified not only the mother's sacrifice but also her willingness to relinquish final control of her child's body to the state. Today, the Gold Star remains a singular icon of the death of a child. But celebration of Gold Star Mother's Day has been intermittent. President George H. W. Bush proclaimed the first Gold Star Mother's Day in 1989. It has been observed steadily ever since, and in 2009, President Obama renamed the day more inclusively, in honor of "Gold Star Mothers and Families," and proclaimed it as such in all subsequent years in office.[83]

Presidential proclamations matter, but in a qualified way. They emblematize an administrative commitment, a sovereign infusion of value in a particular object. But they are not overdetermining, and they originate from such a height that they might trickle down onto actual citizens only in very diluted form. Elisabeth R. Anker theorizes the relationship between official discourse and political life as follows: "To presuppose that subjectivizing processes are the same as or are exhausted by discursive intent would be to assume that political discourses equal political subjects, that discourses determine psychic life."[84] Because these proclamations are nonbinding, they encourage participation, but do not require their audiences to undertake any specific action. Instead, they invite their audiences to imagine themselves as the types of people who would be inclined to act in accordance with their terms. Thus, presidential proclamations may not bear in direct or predictable ways on the lives of particular citizens or on the political subjectivities of military spouses themselves. Rather, they encapsulate, descriptively and prescriptively at once, a broader sentiment; I read across this corpus to track shifts in key affective and ideological markers.

Ronald Reagan used his presidential authority to proclaim the first "Military Spouse Day" for April 17, 1984. In both tone and content, Reagan's proclamation sets a pattern that his successors would follow:

> Since the early days of the Continental Army, the wives of our servicemen have made unselfish contributions to the spirit and

> well-being of their fighting men and the general welfare of their communities.
>
> Throughout the years, as the numbers of our married men and women in uniform have grown and as their military missions have become more complex and dispersed, their spouses have made countless personal sacrifices to support the Armed Forces. In many instances, they subordinated their personal and professional aspirations to the greater benefit of the service family. Responding to the call of duty, they frequently endured long periods of separation or left familiar surroundings and friends to reestablish their homes in distant places. And there they became American ambassadors abroad.
>
> As volunteers, military spouses have provided exemplary service and leadership in educational, community, recreational, religious, social and cultural endeavors. And as parents and homemakers, they preserve the cornerstone of our Nation's strength—the American family.[85]

There is an ambiguity in the "call of duty" to which these spouses responded, as they are not the direct recipients of orders to mobilize. Theirs is a second-order obligation, one that is enforced by their love for or commitment to their spouses and families, yet the phrasing of the proclamation endows it with a quasi-legal status. This emphasis on selflessness and sacrifice, the enumeration of the work that military spouses do both domestically and abroad, and the slippage between "spouse" and "wife" provide the template for subsequent proclamations. Yet within these parameters, over time, the degree and orientation of their affective focus and their assessment of the nation's obligation have evolved.

Since 1984, the observation of Military Spouse Days or their equivalent has come into and out of presidential fashion. Reagan never proclaimed it again. George H. W. Bush renewed the tradition with an adaptation, issuing proclamations yearly from 1989 to 1992, this time in the name of military families, and Bill Clinton followed suit in 1993 and 1994. After that, the tradition lapsed until George W. Bush renewed it in 2007 for military spouses, and the day has been proclaimed in early May ever since. In 2009, President Obama began proclaiming November "Military Family Month," a designation that seems to honor both military families themselves and the idea of the military family in general.

Most of the proclamations urge Americans to actively show support for military spouses and provide varyingly specific instructions on how to do so. Obama's are unique because they also enumerate what his administration did on behalf of military families, a bureaucratization of appreciation

Table 3-1. Presidential Proclamations on Behalf of Military Spouses and Families

Year	*Day Proclaimed*
1984	Military Spouse Day
1989	National Military Families Recognition Day
1990	National Military Families Recognition Day
1991	National Military Families Recognition Day
1992	National Military Families Recognition Day
1993	National Military Families Recognition Day
1994	National Military Families Recognition Day
2007	Military Spouse Day
2008	Military Spouse Day
2009	Military Spouse Day
2010	Military Spouse Appreciation Day
2011	Military Spouse Appreciation Day
2012	Military Spouse Appreciation Day
2013	Military Spouse Appreciation Day
2014	Military Spouse Appreciation Day
2015	Military Spouse Appreciation Day
2016	Military Spouse Appreciation Day
2017	Military Spouse Day

that also reveals how political capital can be obtained by professing it.[86] Typically, the text of the proclamations mentions actual military excursions only obliquely; George H. W. Bush, in the years around the Gulf War, made the most explicit references. Otherwise, most of the proclamations decontextualize the work that military spouses do, a generalized endorsement of militarism that obfuscates the strains that deployments place on families while underscoring the official expectation that they equip themselves for the demands of any geopolitical exigency. The Pentagon speaks regularly of the importance of military family "readiness," which corresponds to the vision of a "family whose members were willing and equipped to handle—uncomplainingly—the stresses of sudden, repeated, and long dangerous deployments."[87] Moreover, the repeated invocation of extended but vague histories of supportive military spouses depoliticizes the specific demands that each new war places on them. And quite predictably, when the proclamations enumerate the hardships faced by military spouses, they omit mention of marital difficulties or domestic violence, imagining the military family as one strengthened and drawn closer by shared suffering.[88]

Often, the language of the proclamations implies a collective national connection with military spouses. The word "our" appears frequently. In earlier proclamations, the plural possessive marked public relationships to things like the military, the Nation, ideals, freedom, gratitude, and military families, with each day proclaimed in the "year of our Lord." The authors of these proclamations regularly use the possessive to describe the nation's relationship to military personnel, as in the 2016 invocation of "the spouses of our men and women in uniform." They also refer quite freely to "our" military families, but they are less possessive of military spouses—only 2007 and 2018 references to "our Nation's military spouses" and "our military spouses" in 2012, 2017, and 2018. I suggest that the lack of reference in the proclamations to "our" military spouses also introduces a distance between the state and the military spouses, enabling a denial of collective responsibility for their well-being while preserving the intimacy and sanctity of marriage, making no claims to ownership on a serviceman's wife.

Yet even as the proclamations smudge the state's obligation to military spouses, they also increasingly make connections from military spouses to larger entities. There are predictable references to "our Nation's character" (2009), "our flag" (2014), and "our casualties" (2010). Then, in 2011, references appear to "our communities," enriched by the efforts of military spouses, and "our businesses" (2013), which were instructed by Obama to hire military spouses—an admirable idea that nonetheless transfers responsibility for military families from the state to the private sector. Alongside these gestures, presidential language about military spouses has also become more normative and pedagogical. This begins in 2009 with references to "our obligations" (2009) and "commitment" (2012) to them, and then becomes more direct with "our debt of gratitude" (2014) and "sacred promise" (2015) to repay all that they have done on "our behalf" (2015). The language of obligations and debts is contractual, binding, which implies the possibility (even likelihood) of default and establishes a subtly agonistic framework.

Steeped in the rhetoric of admiration and gratitude and tinged with pity, these proclamations route those affects through the sacrifices that military spouses make. The details of the civilian responsibilities that they assign change over time. What begins simply as a "day" for military spouses evolves—minus the unadorned interval from 2007 to 2009—from a period of recognition to an affective mandate for appreciation. In practical terms, this is consonant with a proliferation of programs designed to support military spouses and families during the Global War on Terror.[89] And it reflects the professed commitment by the First and Second Ladies of the Obama administration to provide services to military families; Michelle Obama in particular emphasized the importance of actively demonstrating national gratitude, especially to military children.[90] But it also points to a relationship to military spouses reconceived as both affective and compulsory.

In the presidential lexicon, recognition connotes something less active, more neutral than appreciation. The semantic shift toward appreciation marks a change in the affective positioning of the military spouse, though—crucially—this change does not necessarily translate into tangible benefits for them. We can recognize someone or something without much effort or even being conscious of the processes that result in the sense of recognition. Recognition can be an end in itself; it does not necessarily imply any other action. In an argument that refers specifically to people of color in the United States but neatly encapsulates the limitations of grievances rooted in appeals for recognition, Herman Gray contends that these operate on "a cultural field where private differences remain, but collective claims on the state do not."[91] Recognition, in other words, does not necessarily beget any kind of structural change and is often positioned as an end in itself. This analysis resonates with the long-standing practical neglect of military spouses by the U.S. government and underscores the contingent nature of recognition, as evidenced in the inconsistent observances of these days for military spouses.

As a mandated gesture of inclusion, the act of recognition reifies exclusion. References to "our" military are ostensibly gestures of solidarity; however, they also serve as an objectifying expression of power over and separateness from military personnel. The creation of a specific day for the recognition of military spouses devalues their contributions relative to those of their active-duty partners (as presumably every day should include appreciation of military personnel). It also undermines their claims to militarized identities of their own; after all, many army wives understand themselves as actively serving alongside their husbands.[92] Even when the proclamations recognize this service, they also set it apart. Although the federal government has, since the 1980s, eschewed the language of "military dependent" (a category that included wives and minor children) in favor of "military family member," the act of recognizing a military spouse bears traces of the same subordinating logic.[93] If recognition is too little, however, appreciation might be too much. The tilt toward appreciation diverts more attention toward its recipient, intensifying her visibility and the attendant scrutiny.

Among presidential proclamations in general, "appreciation" appears rather infrequently, and in this voluminous archive, very few things have been officially appreciated since 1984. There have been a handful of people: Eugene Ormandy (1985), China-Burma-India Veterans (1988), the Rose Fitzgerald Kennedy Family (1990), and teachers (1986, 1988, 1990, 2015, and 2016). The Actor's Fund of America got presidentially appreciated twice (1988 and 1989). And American wine got a week's worth of appreciation in 1993. Military spouses got their first appreciation day in 2010; since 2009, the only other object of appreciation, besides teachers in 2015, has been African American music. Clearly, there are two distinct meanings of "appreciation" operating in this pairing: enjoyment of the music (an aesthetic experience) and a feeling of thankfulness for military spouses (an affective and ethical one). But the repetition of the word is revelatory. It indicates, first, that the appreciation of military spouses can gratify appreciators by affording them an opportunity to enact an idealized form of citizenship. Second, it intimates the possibility that this process can consume and objectify its target.

Appreciation implies corresponding valuation (the ascription of worth to the thing being appreciated) and evaluation (the assessment of whether or not a particular thing meets the criteria for being worthy of appreciation). The appreciation-worthiness of the military spouse is often defined by her capacity to perform specified affective work, to feel certain ways about her deployed loved one and to demonstrate that feeling in ways that are publicly legible. All of the proclamations emphasize the love that military spouses feel, both for their servicemembers and for their country, often citing these feelings to introduce the proclamation. For example, the military spouses envisioned by these proclamations:

> "sen[d] love, love, prayers, encouraging words, and care packages to their loved ones stationed around the globe" (2007)
>
> "endure separations that are filled with worry and anxiety" (2009)
>
> "constantly wonde[r] what kind of dangers lie ahead for their loved ones" (2015)
>
> "must brace themselves for the uncertainty that comes with goodbye" (2017).

All of these things may well be true. But these descriptions also prescribe a set of affective tasks while claiming a national interest in their completion. Although military spouses, at least in the abstract, inhabit a discursively privileged position, these official decrees of their worthiness make it contingent on them performing that role in a narrowly specified way while inviting their civilian observers to hold them to it, gratefully but firmly.

At 11:00 a.m. Eastern on May 12, 2017, the Friday before Mother's Day, *MilitaryOneClick*—a popular source of news, information, and advice for military families—published a story about what Donald Trump had not done: issue the now-customary proclamation designating that day for the appreciation of military spouses.[94] At 4:31 that afternoon, the author posted an update, indicating that the Press Secretary's Office had since issued the proclamation, which reverted to the original name of "Military Spouse Day." The White House did not provide any explanation for the change in nomenclature, though the content of the proclamation's text was essentially indistinguishable from those that preceded it. In this instance, I don't know whether we can divine much from the omission of "appreciation" in the name of the day, given that the proclamation itself may well have been a hasty afterthought; Trump used the same title in his 2018 proclamation, this time issued the day before. But nonetheless, the ease with which "appreciation" was subtracted provides an important insight into the ultimate fungibility of this affect.

"Keep Your Eyes Open and Your Mouth Shut": Operational Security for Military Spouses

In early 2015, hackers who claimed to be affiliated with ISIS launched a series of small but targeted cyberattacks. News outlets began reporting that the group was monitoring the social media profiles of military wives. Noting that a group calling itself the "Islamic State Hacking Division" had posted photos and addresses of 100 U.S. military personnel online, the reports warned that ISIS was encouraging supporters in the United States to target military personnel and their families, who would be identifiable by their social media profiles and by things like supportive decals affixed to their vehicles.

Publicly, the military did not say much about the credibility of the threat. The news media sensationalized it—what more perfect example of ISIS's villainy could there be?[95] Neither approach made it easier to evaluate the veracity of the stories. But military families cannot take chances. Many of the news stories featured quotes from military spouses who asked to remain anonymous for fear that they would be targeted, lending credibility to the threat and derealizing it at once. These women described frantically redacting their online profiles, changing their settings and preferences to those that would keep them least visible, most secure.[96] They recounted how they stripped their homes and vehicles of any symbols that might betray their military connections. Even as the news stories kept the blame squarely on ISIS and generally lauded the spouses' quick responses, their angle intimates that the exuberance of military wives to share their status incites malefaction in a dangerous world. This is a digitally mediated variation on an old theme.

"Some birds talk too much." So says a poster circulated by the U.S. Office for Emergency Management during World War II. To illustrate its reminder that "silence means security," the poster features the profile of a large black bird with its orange beak tied shut.[97] While the cartoonishly sinister-looking animal embodies the danger that loose lips pose in general, the slangy subtext of "birds" intimates that women in particular are not always trustworthy. Playing on the stereotype of the female chatterbox, the poster amplifies her threat beyond irritation to national security, a motif that recurs throughout wartime publications. In all such content, the notion that military spouses are essential to the security of the nation coexists with a concern that they might also imperil it.

Along with the Department of Defense, many branches and commands of the armed forces have developed OPSEC (Operational Security) training materials for spouses (read: wives) and families, identifying them as potential leaks of information that might jeopardize deployed American troops. Here, I trace the origins of this perceived threat, beginning with World War I, in an effort to map the anxieties surrounding military spouses. The converse of the acclamation I described in the previous section, this skepticism of military wives—dubiousness about their self-restraint, common sense, and understanding of geopolitics—lays bare the ambivalence operative beneath official proclamations of gratitude for them. It reveals that the military's regard for them is essentially calculated tolerance for the added liability that they bring, mostly but not completely offset by the value of the services they provide. In this model, a woman secures herself by attaching to a military husband, but at a potential cost to the security of the nation-state, a troublesome dividend of its heteronormativity. These materials position military wives as insignificant to the war effort while simultaneously emphasizing the magnitude of the threat they might pose to it.

When associated with the safety and tranquility of the domestic, women can serve as a comforting symbol during wartime. Many scholars have observed that, during the Global War on Terror, homemaking women were hailed as key nodes of securitization, frequently through advertising. Patricia Ticineto Clough and Craig Willse, for example, have described how security itself became a gendered commodity in post–September 11 advertising.[98] Grewal, discussing the image of the security

During World War II, the Office of War Information produced countless posters alerting citizens to the dangers of "loose talk." This bird with its beak tied shut, I suggest, makes an elliptical reference to the chatty sociality of women. National Archives and Records Administration, identifier 513597.

mom, discerns "interconnections between the war on terror, hyperconsumption, suburbanization, and new spaces of 'community.'"[99] And Marita Sturken, in her study of kitsch and comfort cultures in the aftermath of terrorist attacks, notes that advertisements for high-end durable goods like SUVs targeted women, enticing them to protect their families and, metonymically, the nation-state.[100] Alongside the military wives envisioned in the presidential proclamations, whose duties center on the home, immediate community, and possibly the workplace, these women cozily fortify the home front with their purchases.

Always in OPSEC materials, the problem arises when they seek to connect with the larger world; the issue is women's communication. Generally, the public speech of military mothers and wives is tightly regulated, whether by the military, the media, or the women themselves. For example, the military mothers that Christensen studies repeatedly remind one another that they should express public support for military campaigns, even if they have private doubts.[101] And Enloe observes that the media and the public are quick to brand military wives as disloyal if they criticize an ongoing war.[102] The OPSEC materials do not address women's political speech, focusing instead on a smaller target, which might otherwise seem inconsequential: women's interpersonal communication.

Public information campaigns from World War II portrayed women either through the tropes of the femme fatale spy or the overeager gossip. By contrast, current OPSEC materials typically operate on the presumption that women's breaches are accidental and attribute mistakes to a lack of awareness, a naïve failure to discriminate between types of information and auditors. In this way, they traffic in stereotypical representations of dimwitted women, the Facebook-happy descendants of the lady chatterboxes of the mid-twentieth century. But the World War II posters that envisioned women deliberately sabotaging the war effort attributed a form of political agency, however nefarious, to them. By contrast, contemporary OPSEC materials seem to assume that women will compromise security simply because they do not know any better. They also imagine that these small miscalculations can have outsized consequences, including failed missions, lost equipment, and American casualties. These are updated versions of the axioms about what loose lips (which were often represented as female) can do. With their short chains of causation, they affirm Oliver's contention that women have long been imagined as "secret weapons of war" for better or worse, who are possessed of uncontrollable and "'mysterious'" powers.[103] Earlier warnings about women's talk focused mainly on direct communication, whether they intentionally shared information with someone whom they did not recognize as dangerous or an enemy eavesdropped on an otherwise safe exchange. Here, the emphasis is on mediated communication, particularly but not exclusively via new media, being intercepted by a dangerous third party.

An iconic poster from World War I begins, "Hello! This is liberty speaking. . . ." "Liberty" here is a woman dressed as the Statue of Liberty, holding a phone with her lips slightly parted, looking imploringly outward in a bid to sell war bonds, reminding viewers that "billions of dollars are needed and needed *now*!"[104] By the era of World War II, however, the image of a woman on the phone would have signified much differently. Propaganda posters circulated in both the United States and the United Kingdom repeatedly identified loose-talking women (and, to a lesser extent, loose-talking men who spilled too much information around them in a misguided attempt to impress them during courtship) as a major threat to the war effort.[105] "Careless talk," specifically that of men, occupied a prominent place in public communication about the war effort, seeking to police risky connections between civilians and the military. In the United Kingdom, the pervasive instruction was to "keep mum." Some played on a double entendre of mother and silence, urging them to "be like dad, keep mum."[106] A variant on this theme featured British military officers drinking, smoking, and apparently sharing important operational information as a lithe blonde woman in an evening gown reclines in their midst. The men do not acknowledge her, and she appears to be ignoring them as well, but the poster warns, "Keep mum, she's not so dumb." This underhanded compliment reveals a sense that women, otherwise incidental, can have a devastating impact on the prosecution of a war: that women might matter.

Whether they are itching to boast about their exploits, keen to impress the ladies, or inclined to underestimate them, military men, it seems, needed to be reminded of this. Many posters feature couples *in flagrante*, apparently a risky juncture for exuberant young recruits. In one, a young woman, heavily made up, leans in toward a young sailor in uniform who looks at her sidelong: "Sailor beware! Loose talk can cost lives." Elsewhere, a strawberry blonde woman in a low-cut sweater wraps her arms around a sailor and inclines her head as if listening. Here, the awkward posture of the sailor suggests a reticence, but the caption ("Tell nobody, not even her . . .") implies that he would have cracked eventually. For men who feel they must say something, a cartoon drawing of a G.I. with a buxom young woman perched on his lap, kissing his cheek, advises, "Tell her you love her/That's all she needs to know," as a tiny warplane flies overhead trailing a banner that reads, "Keep your eyes open and your mouth shut."

This poster from the Office of War Information provides a script for military men who might be tempted to divulge too much. It also presages the government's subsequent turn toward appreciation for military spouses, in the absence of more concrete forms of support. National Archives and Records Administration, identifier 515551.

Although the "tell her you love her" injunction is clearly aimed at individual servicemen, it reads like a predecessor to the presidential proclamations. "That's all she needs to know" presages contemporary profusions of gratitude largely disconnected from substantive action on behalf of military spouses. In this exchange, "she" is reduced to an accessory, easily satisfied by a sweet word or two, rather than someone who might actually contribute to the war effort.

In the event that men failed to heed these warnings, posters tasked their female consorts with the job of protecting the mission in their stead, reminding them sharply of what might happen if they act as carelessly as the men did. Some seek to discipline their sociality. "Telling a friend may mean telling the enemy," a poster produced in the United Kingdom, depicts a chain of conversations that begins with a sailor chatting with his girlfriend, who in turn tells another woman, who tells another, who tells a Teutonic-looking man in a suit just three frames later. A "careless talk costs lives" poster produced by the Works Progress Administration places a white X over an enormous pair of shiny, red lips. In another instance, the U.S. Government Printing Office produced a fake wanted poster with a photo of a lovely woman in the center; her beguiling expression contrasts sharply with the text proclaiming, "WANTED! For murder," because "her *careless talk* costs lives."

Although these scenes are rendered somewhat playfully, there is no ambiguity about the lethality of the stakes. One poster features a mournful-looking spaniel on the back of an armchair, the distinctive collar of a sailor's uniform draped over and a Gold Star flag hanging on the wall behind him, "because somebody talked." Yet while the dog registers loss poignantly in its mournful face, the women drawn in these posters seem to act heedlessly of the consequences, whether because they are subversives who deliberately elicit sensitive information or sweethearts who are cute but oblivious.

That type of woman is on display in another "loose talk" poster centering on a park bench where a soldier regales a young woman. She cocks her head toward him, her expression attentive but noncomprehending, as one hand flutters near her mouth and another by her lap. Next to them, reading the paper in a poorly cut suit, Hitler (with an overgrown ear) eavesdrops. Despite being dressed patriotically in a red skirt and white sweater with a blue scarf tied around her neck, the woman does not appear to register the significance of what is transpiring. With her lips and her knees slightly parted, her body language suggests a dangerous kind of openness.[107] As an alternative to this unregulated talk, the military provided a channel for women who desired to communicate safely with soldiers: V-Mail. One promotional poster promises, "Be with him at every mail call . . . ," as a smiling woman sits at a desk penning a letter promised to be "private, reliable, patriotic." Her blouse is buttoned to a respectable height, and the desk is located squarely inside the safe confines of the home. Behind her, faintly visible as if she is daydreaming them (rather than interacting directly with them), a group of soldiers in the field happily collect their letters.

While the U.S. military promoted this form of letter-writing as a way to maintain intimacy and bolster morale, today more instantaneous, digitally mediated forms of communication provoke the most intense OPSEC concerns. They also change the family experience of deployment. The

Although the gabby soldier is the main security threat in this scene, his female companion, with her uncomprehending face and parted legs, is portrayed as a less-than-ideal audience. In a historical period where women were consistently represented as threats to the war effort, the poster may also be implying that she and Hitler are in cahoots. Designed by John Holmgren for the British and American Ambulance Corps, © Victoria and Albert Museum, London.

ubiquity of digital communication, as well as its instantaneity, helps keep families close during deployment, but may also increase their anxiety, especially stateside. MacLeish characterizes information as the "main vector" of vulnerability for military families, whether they avoid the news or watch it voraciously.[108] He writes, for example, that wives become habituated to semiregular phone calls home and then panic when their husbands deviate from these patterns. Simultaneously, Patricia Lester and Lt. Col. Eric Flake note that it can make things difficult for the deployed serviceman by "bring[ing] family problems to the battlefront."[109] While the military portrays women's maintenance of the home front as insignificant compared to the men's work, its worry about this reveals the importance and complexity of all that women manage. By identifying its stressors as potentially severe enough to distract a soldier in battle, the military concedes that domesticity might be complicated, even politically significant.

Contemporary OPSEC materials are primarily concerned with how military wives communicate with people other than their husbands. Reprising the themes of loose lips and sunken ships, this instructional content for families reiterates that information they might deem insignificant could be valuable to the enemy and identifies spouses as custodians of potentially crucial facts. "An OPSEC Family Guide" produced specifically for Special Forces families—for whom communication can be especially vexed given the highly classified operations that Special Forces often carry out—notes succinctly, "You have access to critical information."[110] Yet many of the documents also acknowledge the leakiness of the military itself. The same guide says, "Information about the daily operations of a unit is fairly easy for our adversaries to gather (we often give it up voluntarily)." The United States Africa Command describes sensitive information as "laughably easy to find if you know where to look."[111] Claims like these leave military spouses responsible for redressing the military's shortcomings or the ingenuity of adversaries in uncovering such information, tasking them with a mission that is potentially impossible but also absolutely essential.

The category of seemingly insignificant information recurs throughout the documents, which often invoke the metaphor of a "puzzle" that enemies patiently piece together from fragments and scraps. In this model, no bit of information is too small to be useful; OPSEC requires military spouses to micromanage their own communication, and the trainings remind them that any miscalculation could mean defeat or death for their loved ones. Consequently, many of the OPSEC materials suggest, whether implicitly or explicitly, that the less information the spouse has, the safer everyone will be. Military OneSource reminds spouses that "it can be frustrating not to receive any specific information about your service member's location or return date, but it's for the safety of everyone involved."[112]

Dutiful OPSEC-savvy spouses accept their lack of information as necessary and keep close to home. Describing such forums as "open sources for spies," Special Operations Command advises spouses not to "communicate critical information over the phone, Internet, email, or in public places." No doubt, these are important reminders. However, they radically circumscribe the interpersonal world of the Special Operations Spouse so that she can discuss her experience with very few people, only in person, and exclusively in secure locations. In fact, Military OneSource hints that even staying home might not be enough. Their guidelines alert spouses that "it is important to continue your usual routines and maintenance of your home to disguise the absence of your service member." Here again, commonsense advice translates into extra responsibility for the military spouse; for example, if the wrong parties notice it, even an unmowed lawn might make her, her family, and her deployed husband vulnerable.

Elsewhere, Military OneSource depicts the home as dangerously penetrable, noting that over the course of a typical day or week, a householder might open the door for the "pizza delivery boy, appliance repairman, parcel delivery service, door to door salesman, or the kid from next door whose ball landed in your backyard." During deployment, however, she should always ask herself beforehand whether she is expecting anyone and be mindful that any of these characters might be preying on her being alone. Indeed, much of this advice is heavily gendered. "As good OPSEC practice," Military OneSource suggests that spouses "always give the impression that you are not home alone and never reveal that your service member is deployed. Don't be afraid of seeming rude."

Cumulatively, instructions like these reinforce military sociality, reminding wives that the safest, most trustworthy interpersonal networks are those prefabricated by their husbands' service. For military spouses who do not wish to participate in these networks (which many women describe as insular or oppressive) or do not have access to them, the only truly secure alternative is isolation. For example, the Special Operations Command warns that representatives of the enemy might "befriend [a military spouse] for the single purpose of gathering critical information." Although the OPSEC materials do not evince much trust in military spouses, they also expect them to serve as active participants in protecting and advancing the mission, using their imaginations in order to "think like the bad guy." Because, according to Special Operations Command, bad guys are "opportunistic and focus on easy targets," it is up to military spouses to fortify themselves. An army briefing on social media safety tells users to "always assume the enemy" is reading everything they post online.[113] In this way, they are conscripted into the mission, even as they are kept at a remove from it.

As if anticipating a military spouse's objection that she simply cannot manage one more thing, Military OneSource warns that "criminals can use your distraction against you." The article continues, "No matter how hectic things are, remember that you're never too busy to be conscious of what you say and where you are." OPSEC materials advise spouses to guard dates of deployment and return with special care. Yet these are also the events around which military spouses are likely to need extra support from family and friends. The materials often assume that military spouses will be inclined to talk because they are proud of their servicemembers and want to share this with others. Against this instinct, Military OneSource recommends that they do not engage in practices like displaying yellow ribbons outside their house and thus communicating an absence to the larger world. In this way, the experience of deployment is made intensely private and isolating.

With varying degrees of subtlety, OPSEC materials for families shift a significant portion of the burden of troop safety off the state and onto the spouse. Unlike OPSEC posters from World War II, which often apportioned at least part of the blame onto the indiscriminate talk of soldiers and sailors, these materials place it squarely on the spouses. A conversational PowerPoint from the Department of Defense poses a rhetorical question to the spouse: "Did you know you're half the team?"[114]—this after an assurance that "your loved one has the training, leadership, and equipment needed to perform the mission and come back home to you." Alongside this inventory of specialized weapons and vehicles and military expertise, a full half of the responsibility for bringing the loved one home alive falls to the spouse, and more precisely her ability to keep quiet and remain invisible.

"It's Okay to Smile": The Work of the American Widow Project

As the history of the Gold Star shows, military mothers and wives first garnered visibility through their losses, recognition in exchange for their griefs. Yet that recognition is contingent on them expressing that grief in normatively appropriate ways: displays of fidelity, love, and sadness that are reasonable and adequate but not excessive and never critical. Military widows are held to a different kind of silence. They encounter a representational landscape on which they are hypervisible in rich public imagination but ultimately liminal, as the death of a husband complicates the terms of the military wife's already provisional inclusion. Despite the widespread valorization of military wives, Lutz notes that "in a society that values independence and individualism, a dependent has an ambiguous status, perhaps even less than full cultural citizenship." In this way, military spouses are both "supported and disempowered" by the systems that order their lives.[115] Given the intensity with which the military regulates the identity of "military wife," when a woman effectively loses both of those statuses at once, she is consigned to illegibility. She is linked to the military only by her past decision to affiliate with it, whatever financial benefits it affords, and the rhetorical force of her own claims of belonging. In these circumstances, the military widow is excluded from its forms of recognition by precisely the circumstances that locate her as the living embodiment of its ultimate sacrifice.

Perhaps the greatest visibility afforded to widows during the War on Terror was in the aftermath of September 11 as the women widowed by the attacks appeared as indices of all that had been lost. Cohler describes the "9/11 widows" as key icons of the event.[116] The sentimental comprehensibility of the grieving wife, she suggests, was readily absorbed into the larger militarizing project of "nationalist feminism," whereby the Bush administration promoted its wars in Afghanistan and Iraq as crusades on behalf of women in the United States and the Middle East.[117] But in the allocation of visibility and sympathy, not all widows are created equal. Ruby Tapia discerns a difference between media representations of 9/11 widows who were mothers (especially those who were pregnant when their husbands died) and those who were not. She writes, "Heterosexual *mothers* provid[e] a constantly regenerating site on which the American public can project their feelings of loss, fear, anger, and recovery." On the other hand, the childless widows resisted seamless incorporation into such affective undertakings, especially when they criticized the government's failure to prevent the attacks.[118] Indeed, presidential proclamations single out Gold Star mothers and active-duty military wives, which suggests a struggle to place military widows, who are included primarily as an afterthought, particularly those who do not have children.

The 2015 proclamation of Gold Star Mothers and Family's Day notes, "The depth of their sorrow is immeasurable, and we are forever indebted to them." In addition to taking the measure of this eternal debt, such presidential proclamations also envisage the devastated home front. The 2008 proclamation invokes "empty dinner tables," an image that recurs in 2010 as "an empty seat at the table" and in 2013 through "empty seats at family dinners and folded flags above the mantle." The 2014 proclamation references these "families' front windows, [hung with] blue-turned-gold stars." The proclamations also consistently reference the unknowable depth and enormity of Gold Star grief, portraying it as incomprehensible and unimaginable, an unthinkability echoed in the relatively limited official imaginings of what a home looks like after such a death. Yet beyond pat references to this abiding sadness and occasional affirmations of the grit they display by getting on with their lives, it is unclear from these proclamations precisely what a Gold Star family member ought to do. At least in the title of the proclamations, the Gold Star wife (or widow) goes unnamed and largely unrecognizable.

Enter the American Widow Project. Founded by Taryn Davis, who was widowed in 2007 when she was twenty-two years old, the AWP is a nonprofit organization that provides peer-to-peer support for "the new generation of military widows." The AWP is open to all military widows regardless of age, branch of service, or their spouse's cause of death; they include women widowed by suicide, and the website specifically mentions suicides precipitated by TBI. The organization focuses on providing education, support, and—most uniquely—fun to military widows. The AWP runs WidowU, which offers courses aimed at helping widows make practical choices about what to do next, like going back to school or starting a business. Volunteer members (not counselors) staff a 24/7 hotline for other widows needing support. But the marquee aspect of the AWP is its coordination of vacations and retreats for widows and their children.

Overall, the AWP stands in vexed relationship to the military. Beyond Davis founding it with money from her husband's life insurance, her organization does not have any material connections to the military. Moreover, both Davis and the other women who give personal testimonies on the AWP's website regularly comment on the impersonality and uselessness of the official guidance that the military provides. Consciously deviating from the conventional "seminar and speaker" model, which Davis identifies as sterile and unrelatable, the AWP aims to meet the practical needs of a particular demographic: very young widows, many of whom had very new marriages when their husbands died. Recognizing that many of these women had to radically readjust life plans that they had barely begun to make, the AWP focuses on providing them "the opportunity and tangible tools available to help rebuild" their lives. In this way, the organization deviates sharply from images of widowhood as sackcloth-and-ashes in perpetuity, but also skirts the edges of normative expectations about grief, femininity, and sacrifice, alternately refusing and reshaping attendant forms of recognition.

The AWP evades political declarations, either for or against war, and generally seems disconnected from the military as a whole. Its rituals are tied more toward individual than nationalized sacrifices. As an impassioned and photogenic spokeswoman for the plight of widows, Davis often explains her work from the premise that widows are culturally invisible and young widows even more so. Prevailing depictions of widows, as she frequently notes, generally conjure up images of elderly women who lose their husbands after long marriages and now have nothing much left to do but wait for death themselves. As evidence of this, Davis often tells the story of searching online for the word "widow" shortly after she became one, only to have Google suggest that perhaps she meant "window" instead. Thus, part of what the AWP endeavors to do is to make widows and the fact of widowhood visible and recognizable. If, in the cases of presidential proclamations and OPSEC materials, recognition of the military wife is a derivative of recognition for the servicemember, the relative autonomy of military widows claimed by the AWP points to the chilling possibility that, within the militarized economy of recognition, full visibility of the military wife is only possible in the radical absence of her husband.

The transition to an all-volunteer force heightened the distinction between civilians and military personnel and freighted both the decision to enlist and the choice not to enlist with new meaning.[119] But after their husbands die, the women who chose to marry men who chose to enlist now inhabit an unchosen position of widow, and so hover between volition and misfortune. Generally, the visibility of military spouses corresponds to their proximity to sacrifice; when husbands are actively deployed, the sacrifice is easier for outsiders to imagine as active and ongoing. When the husband has been killed, sacrifice may start to look past-tense and passive: a continuous experience of lack. Moreover, the identity of "military widow" is contingent on the identity of her dead husband and thus routed through a signifier whose meaning is fungible and always prone to slippage. Clark notes that, in legal terms, dead bodies are classified as *res nullius*: "'a thing owned by no one.'"[120] Yet both the state and the family make claims on them, to them, and about their meaning that may or may not align; in the event of a divergence, the identity of "military widow" begins to fracture.

For the military, war dead are primarily a bureaucratic (and sometimes public relations) issue.[121] Over the course of the twentieth century, the U.S.

government increasingly claimed authority over dead military personnel, particularly in terms of repatriation, funeral procedures, and interment, and stripped this control away from the mothers and wives who might otherwise have claimed it.[122] In tandem with these processes, the federal government also determines what forms kinship might take in and after death, restricting burial alongside military personnel to minor children and dependent spouses who do not remarry.[123] This stipulation that the military widow actively retain her unmarried status to receive this final benefit enforces her fidelity, even posthumously, through the promise of a morbid reward.

Thus, the figure of the vibrant young widow is discordant, ineffably disturbing; it does not square easily with prevailing ideas of mourning. Accordingly, AWP materials include a common lament that their losses are often overshadowed. The AWP mission statement observes,

> While the service member's sacrifice is often acknowledged by society, many simply forget or fail to recognize the sacrifice of the spouse who is now left a grieving widow. Oftentimes the invisible wounds of military widows are disregarded due to age or a simple lack of knowledge and understanding.

Rhetorically, this is a bold strategy, a reassertion of their presence in the national community, a reminder that they make and remake their "ultimate sacrifices" daily, and a demand for recognition on those grounds.

The AWP infuses its demand with patriotic iconography. Their website includes a photo of a solitary young woman wrapped in an American flag, staring off into the distance and smiling faintly in apparent reminiscence. The group's logo is a heart that is half solid red and half blue triangle with white stars evocative of the folded flag presented at military funerals. But it also claims a right for military widows to start over, which is not at all guaranteed or encouraged by official discourses and policies. The AWP is not expressly critical of the war or the state but derives part of its mission from the inadequacy of the military's provision for widows beyond the return of the deceased's personal effects and the administration of death benefits, even as it relies on patriotic signifiers for its legitimacy. In some ways, this is a variation on the tightrope antiwar position that criticizes militarization while still claiming to "support the troops."

Violently detached from the systems that treat military personnel and, by extension, their spouses as modular, disposable, and replaceable, the AWP provides participants with a chance to repersonalize their lives through adventure and (subsidized) consumption. The AWP fills a year-round calendar of weekend getaways; these constitute the majority of events listed on their website. Some are tailored to widow-child families, and options include an "AWP Gives Back" getaway for a weekend of volunteer work. With the exception of the give-back weekends, where the destination is determined by need (like the December 2015 trip to New Orleans to help rebuild homes damaged by Hurricane Katrina), the trips focus on pleasures: days at the beach, white-water rafting, hot-air ballooning, gourmet meals. Outside donations offset the costs, so that most of these events are offered to participants for less than $100, excluding airfare. Unsurprisingly, given their affordability, the appeal of the locations, and the need that the AWP fills, most of these events sell out, according to the website, in "30 seconds."

By contrast, media coverage of the AWP's work conforms to more traditional expectations of what military widowhood entails. News stories often include shots of Arlington National Cemetery or footage of military funerals, yoking the visibility of widows back to their husbands. For example, an *ABC News* story describes them as "exceptional young women" who are "forever linked to the military" yet moving on "despite their grief" to find other widows in "common sorrow."[124] This is not an inaccurate description, but it is a partial one that starkly contrasts with the content that the AWP generates for itself, which asserts that widows deserve to be vivacious and forward-looking.

In her acceptance speech for a 2012 CNN Heroes award, which was preceded by a video montage of vignettes about how various widows' husbands died, Davis asserts that although "widowhood is a lifetime process," it is not a totally mournful one; "it's okay," she says, "to smile." While this ostensibly gives permission to other widows, it also telegraphs something to those who might be inclined to police them and their expressions of happiness. In her foundational analysis of the story of Antigone, Judith Butler mines the radical political potential of the defiant act of grieving a life that might otherwise be classified as ungrievable.[125] The eminently grievable life of a servicemember, on the other hand, instantiates strict normative expectations for how people, especially their wives, should mourn them. Both presidential proclamations and popular discourses of military widowhood valorize coping and provide a rather austere vision of survival as sustaining the minimum (home, work, children) of what is necessary for a meaningful life. In contrast, the AWP makes a radically different claim that these women deserve a life beyond grieving, asserting a new and layered political subjectivity in which it is also okay to smile.

The AWP eschews prescribed existential meagerness by insisting that widows, particularly young widows, deserve to do more than eke out a life. Butler writes, "Antigone is caught in a web of relations that produce no coherent position within kinship. She is not, strictly speaking, outside kinship or, indeed, unintelligible. Her situation can be understood, but only with a certain amount of horror."[126] The widows, especially the childless ones, have had their links to the state and its preferred forms of kinship violently severed. Simultaneously, the status afforded by their previous marriage to military men tethers them to a particular discursive frame that predicates recognition for their loss on their appropriate embodiment of it. Refusing to consign widows to a life of austerity, the AWP rewrites the terms of this exchange, but only partially, relying on patriotic iconography to legitimize its assertions that military widows "deserve" another chance at life.

Significantly, the AWP does not rely on the language of entitlement by suggesting that the widows have "earned" the right to fun because they have suffered. Abbas notes that liberal societies typically mandate recognition of claims for rights, justice, or reparations. Alternatively, the AWP simply presents the grief of the widow without transforming it into currency and claims its pleasures without apology.[127] It entertains no doubts about whether widows deserve a future and asserts unequivocally that they do. Abbas makes a case for "acts of (and in) abundance" for oppressed populations.[128] She makes this assertion on the grounds that they are "acts of love and acts of suffering that defy the imposed logics of the kinds of intimacy, knowledge, and premises that the structure wants from us; they are exceptions to the prescribed modes and unfoldings of love and suffering that do not fit within a liberal utilitarian calculus."[129] The AWP, with its skydiving trips and spa retreats, enacts precisely such a practice.

AWP widowhood is intensely social and almost entirely homosocial among women.[130] The AWP avows that its programs exist to "unify," "educate," "empower," and "assist" its constituents and declares that "no military widow should feel alone in her grief." AWP networking is, in many ways, an inversion (or perhaps a reversal) of the socialization that new military wives undergo, often under the tutelage of expert peers. Enloe notes that these women operate on unofficial behalf of the military, offering newer wives "advice, support, and warnings" on "how to support her husband, how to help him win promotions, how to contribute to the smooth running of the military base community," "pack, unpack, and repack" and "when to socialize and when to keep silent."[131] By contrast, the AWP guides women who are trying to restructure their lives outside of these rituals, often when they have barely had a chance to get accustomed to them in the first place.

At the same time, the AWP provides a substitute for the heteronormative sociality that marriage would have provided, and so there is something compensatory about these affiliations—a rerouting of the widows' desire. Berlant defines desire as "a state of attachment to something or someone, and the cloud of possibility that is generated by the gap between an object's specificity and the needs and promises projected onto it."[132] Given the unclosable gap between the widow and her husband, the desire is redirected outward into elaborate networks of care maintained by the AWP. It relies on widows for its 24/7 hotline not because they are credentialed counselors, but on the assumption that they can instinctively relate to the difficulties of their callers, their intimate familiarity with the mechanisms of all the ungrantable wishes and unfulfillable wants that widowhood entails. This lateral form of recognition has the potential to evade the normative traps of emotion and expectation inherent in other forms of assistance.

From my outsider perspective, something about the AWP feels hyperreal: photos of women grinning broadly, exuberant testimonies from members about healing and community, lists of events that range "from surfing to ziplining." Much of the AWP rhetoric emphasizes the importance of military widows recapturing the "ordinary lives" that they might have had if their husbands had survived. Here, however, the ordinary is hypertrophied, almost fluorescent in the brightness of its vision. MacLeish argues that the circumstances of military wives require new interpretive frameworks that trade notions of agency and resistance for attunement to the "ability to defer, endure, make oneself forget that one is waiting, or maintain oneself in the empty space between irregular episodes over which one has no control."[133] He suggests that military wives form a kind of political subjectivity through this endurance, which is cemented and affirmed upon the eventual return of the husband. When the husband will never return, the result for the widow is a kind of abandon, a life beyond these signifying systems predicated on a future.

By emphasizing the positively transformative power of crisis and advocating a recourse to luxury consumption, the AWP veers toward a neoliberal response to grief. Despite my awareness of this, I find myself reluctant to critique the AWP and wager that the practical needs it meets outweigh my ideological concerns; I proceed in my analysis with that tension unresolved. After all, AWP neoliberalism might simply be reflecting the pervasiveness of neoliberalism in American life. Without steady sources of funding, and aside from the start-up money provided by Davis's life insurance settlement, the AWP depends on private philanthropy and grant funding. The promotional video showcases Davis working at a home office, a mournful entrepreneur. And even as the AWP deliberately eschews the format of the therapeutic seminar, its WidowU offerings center on self-improvement and equipping widows with skills necessary for success in the marketplace.[134] "Pursuing Your Passion" helps widows identify the right "path" for their future, including opening their own business. "Overcoming Obstacles" uses physical and mental challenges to help widows find "the full depth of [their] resilience and power," emphasizing individual fortitude in the face of systemic challenges. These courses promote lifestyles that blur the distinctions between labor and pleasure, work and not-work, even as they provide potentially useful and necessary skills and guidance to their participants.

Interestingly, this is a vision of "the good life" that does not include marriage (or, apparently, remarriage). The course description for "Overcoming Obstacles" promises to teach participants ways to navigate "past the road blocks that may be holding you back from living the life you deserve. This course is aimed at those willing to push physical and mental limits in order to find a path to living a more fulfilling life." But what, following Berlant's pointed query, does it mean to "have a life"? She asks:

> Is it to have health? To love, to have been loved? To have felt sovereign? To achieve a state or a sense of worked-toward enjoyment? Is "having a life" now the process to which one gets resigned, after dreaming of the good life, or not even dreaming? Is "life" as the scene of reliable pleasures located largely in those experiences of coasting, with all that's implied in that phrase, the shifting, diffuse, sensual space between pleasure and numbness?[135]

The vagueness operative in the AWP's rhetoric of "paths" and "fulfillment" points to the uncharted space beyond the heteronormative fantasy. It signals the widespread unthinkability of a woman's life persisting beyond her husband's death and beyond the comforts and reassurances of the nation-state, whose protection could not be secured even by ultimate compliance with its demands.

The right to happiness that the AWP claims for military widows apparently coexists with mourning rather than transcending it.[136] In this way, the AWP operates in complex relation to Butler's contention about the radical potential of grieving as an alternative to the state's determination of what lives and deaths are worthy of recognition.[137] Their postmortem, postmarital *joie de vivre* intimates that such mourning might not be sustainable over the long term. In the case of military widows, both the imperative to mourn and the process of mourning itself are inextricably entangled with ideas about normative femininity and the living out of heteronormativity on the grid of militarization.

Davis, who lives outside Austin, Texas, plays on that city's adopted slogan and says in the AWP's "Join Us" video that she wants to "keep widows weird," because, she says, "we *are* weird and we're proud of it." She does not elaborate on the content of that weirdness, though it seems to be a shorthand for their affective desynchronization with expectations about how widows ought to behave. Shortly before her advocacy for weirdness, Davis notes that when she talks to a new widow, she understands that "the last thing she wants to hear is 'I'm sorry' because, God, how many times can you hear you're sorry?" Instead, she suggests that widows want and need concrete information (especially from other widows) about how to move forward. But this possibility can scarcely be countenanced within prevailing affective regimes. And this kind of moving forward—notwithstanding the individual recognition that Davis has gotten for her social entrepreneurship—comes at the cost of refusing the admiration, gratitude, pity, and vicarious anger underpinning the "I'm sorry" that she finds so exhausting.

Given the AWP's unconventional version of widowhood, I was initially surprised to learn of the many plaudits it has received, including favorable mentions in major media sources ranging from NPR to the *Today* show. Jasbir Puar's argument about the absorption of select homosexual subjects into the national community after September 11 is instructive here. She writes that the exigencies of the War on Terror compelled the United States "to temporarily suspend its heteronormative imagined community to consolidate national sentiment and consensus."[138] In such an environment, the praise of mainstream media outlets for the work of the AWP neutralizes the threat embedded in its otherwise unthinkable implication that there can be a life, sometimes a really joyful one, beyond the daily practice of heteronormativity.

Acknowledging widows' losses while embracing a version of singleness that is resilient, independent, even radiant, the AWP operates in complicated relationship to heteronormativity. It does not imply that widows earned their spa time and self-improvement because they have sacrificed their husbands. Instead, it asserts that widows deserve to rebuild their lives

These scenes from an informational video by the American Widow Project depict its founder, Taryn Davis, as a mournful entrepreneur who crafts a new future for herself after the sudden death of her husband in combat.

without justifying that claim by recourse to grief. Thus, it departs from the logic of indebtedness that circulates in forms of recognition like presidential proclamations. On the other hand, the organization also intimates that those abbreviated marriages were a necessary step on the way to the fulfilling lives toward which they are now headed. This history authorizes their participation in these activities. It frees them for pursuits that might otherwise resemble the chilling stereotype of the merry widow, whose cheer cancels out whatever devotion she might have felt before and the sympathy she might otherwise have been afforded. They had to have the experience and demonstrate their mastery of that perfect heteronormative love to qualify for something transcendent.

Searching for the Traumatized Military Wife

The dangerously loquacious military wife who compromises a mission and the freewheeling military widow who grieves while whitewater rafting hint at the possibility of more complicated subject-positions than traditional forms of recognition might allow. The ideal for the military wife is of a woman who positions herself as only the mute recipient of presidential appreciation or the weaker helpmeet of a deployed husband.[139] But the "signature injuries" of contemporary American wars—PTSD and TBI, which are the focus of the next chapter—introduce perturbations in this model. They do this by repositioning military wives as permanent caregivers for injured husbands, who often display symptoms that fracture the romantic ideal of marriage to a military man. Even in the absence of PTSD or TBI, a husband's deployment can have deleterious consequences for the at-home spouse's physical and mental health.[140] And most researchers surmise that military spouses underreport their levels of stress and the severity of their psychological and bodily reactions to it to "minimize the perceived impact of military service."[141] There are strong institutional and informal pressures not to "complain" about the burdens of deployment, reflected and perpetuated, for example, in presidential proclamations that depict military spouses as stoically fulfilling their duties.

Domestic violence affords some military wives a perversely refracted form of visibility, particularly when that violence is precipitated by combat injuries like PTSD and TBI. As in the case of military widows, such victims of domestic violence can become visible because their husbands are absent, but in this case the absence is mental, emotional, or cognitive rather than physical. If he survived deployment but is, in the common parlance of PTSD and TBI, "not himself," it is as if he is not truly there, at least in this discourse. If he aggresses toward his wife, he abdicates his role in the militarized romance of the military couple. In this case, there is no coherent, legible, masculine subject-position to overshadow that of the military wife; this absence enables her to appear more fully, but only in victimized relation to him.

Increased attention to PTSD and TBI in military personnel has recently begotten a new line of inquiry into their consequences for military spouses, as emotional changes and interpersonal difficulties are key symptoms of these conditions.[142] The fact that researchers are considering women at all is a positive development, but the research began from the premise that the psyche of the military spouse is simple, malleable, and receptive. As clinicians began noticing that the spouses of veterans with PTSD and TBI were also presenting signs of mental health disturbance, early explanations for this phenomenon proposed a fairly direct transmission that led wives to react to their husband's new, and often unpredictable or difficult, behaviors.

More detailed research has revealed that the women are not suffering simply because their husbands are. In fact, wives' outcomes are worse when their husbands do not acknowledge the severity of their own mental health issues, and they generally fare better when they can discern clear etiologies for their husband's symptoms, like combat experiences.[143] For this reason, some experts advocate educating wives about common PTSD symptoms like emotional numbing and interpersonal withdrawal to help them better understand what they are witnessing.[144] Here too, interest in military wives is instrumental, as most clinicians emphasize the importance of strong relationship ties for the recovery of the person with PTSD: the male servicemember.[145] Furthermore, both the visibility of military wives and their access to care are contingent on the severity of their situations. One study found that even subclinical symptoms of PTSD in veterans diminished their wives' well-being. The researchers suggested further that these less severe conditions might actually be more deleterious for wives because symptoms at this level left them without recourse to clinical intervention.[146]

An investigation of the incidence of PTSD in military wives found that 21.6–41.6 percent of them had symptoms indicative of the disorder. Notably, however, when the researchers asked the women about the causes of their distress, more than half said that their symptoms arose from their own life experiences, not their husband's combat experience.[147] In this way, the study provides empirical evidence that women have inner lives that are not coterminous with their status as military wives, a possibility that is otherwise systematically discounted.[148] To enact that role in accordance with its imagined form, military wives must be sensitive enough to miss their husbands, correctly oriented in their desire for a military man, but selectively insensitive enough to deflect, ignore, or manage the costs that desire entails.

CHAPTER 4

Economies of Post-Traumatic Stress Disorder and Traumatic Brain Injury

In the summer of 2016, if a shopper for patriotically themed athletic clothing were so inclined, he could have surfed over to the Under Armour website and outfitted himself head to toe in gear branded for the Wounded Warrior Project (WWP).[1] A tactical beanie ($24.99). Tactical "Ranger" sunglasses ($74.99). A long-sleeved tech shirt ($34.99). Or, as an alternative, a short-sleeved shirt announcing that its wearer is "proud to be American," above a line of text that provides a brief inventory of what that entails: "Honor, Stars, Stripes" ($24.99). "Raid" shorts (marked down to $22.99 from $29.99). "Assert" running shoes (a steal at $69.99). If he got cold, he could add a tactical hoodie identifying himself as "Property of the WWP" ($59.99). And to carry it all around, a tactical backpack ($24.99), for a total of $337.92, before tax. During this period, the WWP commended Under Armour's "commitment to supporting the brave men and women who put their lives on the line to protect the American public," a mission to which the company promised a minimum of $5,000,000 through the end of 2016.[2] In turn, buyers got the assurance that they were paying for "an official Wounded Warrior Project® licensed product."

Corporate social responsibility campaigns like this one are hardly novel, and Under Armour is far from the only business that seeks to capitalize, albeit charitably, on the patriotic sentiments of American consumers. Indeed, the appeal of these arrangements for all parties has proven quite durable, as ongoing controversies at the WWP appear not to have damaged its brand identity, which remains popular despite widespread publicity about its mismanagement and diversion of funds away from the wounded warriors it purports to serve.[3]

For my part, I am most interested in the conflicting logics of quantification operative across Under Armour's WWP line. There is, on the one hand, the luxury commodification of the wounded warrior by association with the pricey and professional-grade Under Armour brand. But on the other, the WWP collection is quite a bit cheaper overall than most other Under Armour products; among the dozens of options in its large array of men's running shoes, only three cost less than the "Assert." The WWP logo worked a contradictory commodity magic here, both increasing and decreasing the worth of the products to which it is applied. By late summer 2017, when I searched Under Armour's website again for WWP-branded products, only one item came up, and it seemed to be an error: a men's tactical quarter-zip pullover ($39.99) that didn't reference the WWP at all.[4]

The fungible and apparently transitory value of the injured servicemember lies at the core of my inquiry in this chapter, which explores the fluctuations of gratitude, pity, and anger around this figure.[5] Zoë Wool writes that, as the image of the "fighting soldier body" has traveled across institutions and discourses, it "has been disciplined and ennobled, decried and obscured, displayed and defaced."[6] In the WWP logo, it has also been commodified and reduced to a cartoonish essence, a Rorschach-y and expressly masculine silhouette of a wounded soldier draped limply over the shoulder of another who spirits him away, presumably to safety. This graphic condenses the experience of battlefield injury to an outline made intelligible by the accoutrements of a helmet and a gun and inserts the promise of rescue (and, by extension, rehabilitation and survival) directly into the scene.[7] Likewise, the imaginative figurations I analyze in this chapter incline toward the reparative as they work to reconcile the fact of military injury with an unflagging commitment to militarism.

Despite capacious rhetoric about widespread gratitude for and commitment to the troops, this figuring relies on an actuarial sensibility that takes exacting and sometimes dubious measure of the injuries it is meant to offset. The relative invisibility of the injuries in question—post-traumatic stress disorder (PTSD) and traumatic brain injury (TBI)—complicates this process.[8] The WWP's wounded warrior is drawn so abstractly that his injury is unknowable, signified only in his absolute dependence on the heroism of his buddy. In this chapter, I demonstrate that the broader figuring of PTSD and TBI works in similar ways, rendering these conditions visible to unafflicted civilian publics, but only sketchily. After all, the injuries themselves are generally undetectable to casual observers, manifesting as they do in clusters of symptoms that can be unpredictable or puzzling. PTSD and TBI do not lend themselves readily to representation compared to more obvious afflictions like amputations.[9] This means that recognition of veterans with PTSD or TBI occurs through the interpretation and evaluation of their behaviors, most of which concern affect and their ability to connect with others.

Epidemically prevalent, PTSD and TBI have become the "signature injuries" of U.S. forces in the Global War on Terror.[10] PTSD, in particular, serves as an evocative shorthand for the emotional consequences of militarization. Wool describes PTSD as "ubiquit[ous] . . . in public accounts of soldiers' combat-connected transformations."[11] The familiar cultural motif of the deranged veteran relies on a simplified backstory of overexposure to the horrors of war and translates it directly into socially and emotionally aberrant behavior. On the other hand, the all-volunteer nature of the U.S. military brings with it a concomitant expectation that enlistees are self-selected and better equipped for combat, an assumption that enables denial—at cultural, clinical, and policy levels—about the incidence and severity of these conditions.[12] Indeed, even the experts can only estimate. Department of Defense statistics for TBI cite a total of 361,092 servicemember diagnoses from 2000 to mid-2017.[13] The Department of Veterans Affairs reports that 11–20 percent of veterans from Operation Enduring Freedom and Operation Iraqi Freedom "have PTSD in a given year."[14] The goal of figuring the veteran with PTSD or TBI is to make these phenomena comprehensible, manageable, and resolvable in the context of continued militarization.

In this book, veterans with PTSD and TBI operate as a hinge, a switchpoint between unknowability and unthinkability. While civilian and military children and military spouses are relatively easy targets for affective investment, these figures often threaten to refuse or exhaust the gratitude, pity, and anger that they elicit. Signature injuries cause damage to subjectivities that often cannot be quantified or specified, introducing an element of unknowability, which becomes more pronounced in the case of the detainees and nearly absolute in dogs, the foci of my next two chapters.[15]

These investments align most closely with the state's agenda, which grafts from overwhelmingly positive feelings about soldiers to maintain support for militarization.

All of my objects of analysis in this chapter mediate combat trauma in some way, reshaping the people who have experienced it into receptacles for gratitude, pity, and anger. I begin with a brief history of PTSD and TBI as diagnostic categories and targets of an administrative calculation that spans roughly the last century. Then, I explore how these conditions have become sites of affective investment for gratitude, pity, and anger and note the shift from the unknowable subjectivities I discussed in the preceding chapters to the unthinkable ones characterizing the figures I will consider thereafter. Because the mandate to "say thank you to the troops" is so ubiquitous and so frequently unquestioned, I provide a fuller history of the militarization of gratitude in the section after that. I then turn to a consideration of the various ways that gratitude for veterans has been materialized through charitable organizations and the conditions according to which they dispense their care. To focus more specifically on the figure of the veteran with PTSD or TBI, I turn to the exacting standards by which the Department of Defense awards Purple Hearts for TBI but refuses them for PTSD. In stark contrast to the DoD's decidedly unsympathetic approach to PTSD, the next object of my analysis—David Finkel's bestselling work of nonfiction, *Thank You for Your Service*—provides a poignant account in which these conditions play out, often violently, in domestic spaces. *Thank You for Your Service* generally eschews scientific descriptions of PTSD and TBI, making them intelligible instead through intensely emotional scenes. By contrast, in the penultimate section, I consider research efforts to make TBI clinically legible by looking for specific signs of the injury on posthumously donated brain tissue. The cases I analyze in this chapter evince a wish to render these subjects knowable, thereby sliding the veteran with TBI back into a schema that makes the injury intelligible and compatible with ongoing militarization. The example I consider in the concluding section offers a different vantage on TBI, and I end the chapter with a brief reflection on veterans' own efforts to make their brains visible to others.

What Combat Trauma Costs and What Combat Trauma Is Worth: Pricing PTSD and TBI

The American history of care for disabled veterans began in 1636, when Massachusetts established a lifetime pension for them. In 1776, the Continental Congress formalized these arrangements on behalf of the fledgling republic, and in the interim before the Civil War, the government offered recompense in money and, when federal reserves were low, bequests of farmland.[16] In 1862, the government adopted a general pension scheme for veterans that included gradated bonuses for injuries.[17] The current U.S. military system for adjudicating disability preserves this logic, but citizenries and administrations have varied in their ideas about what, exactly, the state owes its disabled veterans. For example, after the Civil War, Victorian animosity toward people with physical and mental disabilities eroded popular support for such measures.[18] This skepticism, which seems so incomprehensible today, persisted well into the twentieth century.

Injuries that impact the veteran's mental or emotional health are among the most confounding and so have drawn more scrutiny and speculation than conditions that appear visibly on the body. By now, it is axiomatic that some form of combat stress or trauma has always attended the experience of combat, though symptoms and nomenclatures have varied over time. Favored terminology for combatant malaise included "Swiss disease" in nineteenth-century Europe, "irritable heart," "soldier's heart," or "nostalgia" in the Civil War, "shell shock" or "combat neurosis" during World War I, and "battle fatigue" or "operational fatigue" during and after World War II, which gave way to "combat stress" during Vietnam and "PTSD" thereafter.[19] Although these phenomena share a common genealogy and correspond to similar categories of experience, they are not identical conditions.[20]

Just as clinical frameworks for interpreting these conditions change over time, so too do levels of sympathy for and understanding of these diagnoses. For example, Civil War "soldier's heart" often manifested as palpitations and profound homesickness, and military authorities typically viewed these psychiatric casualties as malingerers.[21] By contrast, shell shock drew much more sustained and serious attention during World War II.[22] And in 1980, the DSM-III included PTSD for the first time. This inclusion was largely the result of activism by Vietnam veterans as they lobbied for official clinical recognition; accordingly, diagnostic criteria for the condition were drawn extensively from veterans' accounts of their experiences and symptoms.[23] This paradigm held that PTSD was expressly a psychological condition resulting from emotional and cognitive overload in the direct encounter with militarized violence. In this model, accompanying physical symptoms like insomnia or elevated heart rate were understood as physical manifestations of a mental condition.

The idea of PTSD was legible and persuasive in part because it resonated with familiar ideas about the "shell shock" endemic to earlier wars.[24] In February 1915, Charles Myers, a British physician, published an article in the *Lancet* that included the first recorded use of the term.[25] Although "shell shock" implied a specific and singular etiology for the disorder, Myers himself was vague about the mechanisms by which it took hold.[26] Shell-shocked soldiers might display a collection of symptoms that Tracey Loughran characterizes as "bewilderingly diverse," including spontaneous deafness, speech impairments, nightmares, and paralysis.[27] Very early scientific explanations of shell shock framed it as a physical condition, an affliction of the brain or spine caused by proximity to an explosion, but over time, the consensus shifted to favor emotional explanations.[28] This paradigm had a practical appeal. After all, if shell shock was purely emotional, then therapeutic rehabilitation could resolve it, while an internal physical injury causing the symptoms would have been all but impossible to locate and repair. With this agreement about the emotional origins of shell shock widely ensconced, rehabilitation of the shell-shocked soldier became a national priority. As Annessa Stagner writes, "By the end of the war Americans' optimism regarding their own abilities to restore minds and improve the behavior of individuals and nations fueled a sense of their country's international distinction and even superiority."[29] Popular media accounts, she observes, "described it as an injury experienced by healthy, mentally strong soldiers," which in turn reasserted the underlying fitness of the national body politic.[30]

During my first efforts to think through this chapter, in the fall and winter of 2014, I anticipated focusing primarily on PTSD, with TBI as a supplemental concern. But as an explanatory frame for the emotional and intellectual harms of war, TBI has since become ascendant.[31] It is important to note that PTSD was not simply a misnomer for or misdiagnosis of TBI; innumerable veterans experience nightmares, flashbacks, and hyperarousal in the absence of blast exposure. But TBI has begun to eclipse PTSD in descriptions of the health crises caused by the wars and as an explanatory framework for the behavioral changes that mark these signature injuries.[32]

If the concept of shell shock has helped to make war seem more broadly interpretable, TBI is somewhat less familiar, more difficult to imagine. After three decades of intellectual predominance, the emotional explanation for PTSD, particularly in cases where it is comorbid with TBI, has begun to recede. The new consensus resonates with the earliest physiological explanations for shell shock, evidenced, for example, in the title of a June 2016 piece in the *New York Times Magazine*, "What If PTSD Is More Physical Than Mental?" Alison Howell characterizes this movement as a "reverse trend" that relocates trauma in the body, a paradigm shift facilitated by the "increased authority of biomedicine and especially neurology."[33] Compared to the alluring simplicity of psychological explanations for combat trauma, this neurological turn opens up a range of new questions, problems, and uncertainties. Options for imaging the brain are somewhat limited, particularly when the person being evaluated for TBI is still alive. And there is no scientific consensus about how, exactly, blast exposure damages the brain; current hypotheses include changes in pressure and motion of the brain inside the skull.[34]

And the full scope of the TBI problem may still be unknown, or unknowable. The U.S. military has become more systematic in documenting blast exposures and screening for concussions, but reporting remains imperfect, particularly from earlier phases of the wars in Iraq and Afghanistan, and the record-keeping problem is complicated by the many symptomatic resemblances between PTSD and TBI.[35] Additionally, virtually all of the statistical research takes for granted that military personnel are likely to underreport injuries and understate their symptoms. The article in the *Times* magazine offers a seemingly commonsense explanation for this: "So many who have enlisted are too proud to report a wound that remains invisible."[36] Such descriptions may accurately reflect military culture and socialization but can also become prescriptive, reiterating the link between honor and stoicism.

But then again, blast exposure itself, instantaneous and fundamentally jarring, can render sense-making systems useless; how can one report something indescribable? Most personal accounts of being blown up emphasize the overwhelming everywhere-ness of the experience. In his ethnographic work on combat, J. Martin Daughtry describes it as "full-spectrum sensory onslaught"[37]; blast exposure compresses this onslaught into milliseconds. Moreover, the physics of blasts are still only incompletely understood. Every explosion unfolds in four stages, but the "primary" stage remains largely mysterious in both mechanics and effects on the human body, though experts believe this stage includes the phenomena that cause TBI.[38] Overall, rudimentary technologies like improvised explosive devices cause many blast injuries and create a powerful leveling effect against the more technologically advanced U.S. military. Current body armor and helmet technologies provide minimal, if any, protection against brain injury from blast explosions. Avoiding blast exposure in the field is nearly impossible, and treatments for TBI remain experimental.

The Veterans Administration estimates that 22 percent of all combat injuries from Afghanistan and Iraq could be brain injuries.[39] Of these, the most common form of TBI by far is "mild."[40] The TBI nomenclature of mild, moderate, and severe refers only to the circumstances of the injury and does not indicate severity of the symptoms. Civilians suffer MTBI, too, usually as the result of car accidents or falls. But military MTBI often progresses differently, entailing a longer and less linear recovery.[41] The VA's overview includes three types of symptoms: "somatic (headache, tinnitus, insomnia, etc.), cognitive (memory, attention and concentration difficulties), and emotional/behavioral (irritability, depression, anxiety, behavioral dyscontrol)."[42] Alongside these clinical descriptors, Kenneth MacLeish offers a more impressionistic account of TBI symptoms, describing them as "oblique, complex, debilitating, yet often unmarked by physical pathology."[43] The RAND Corporation's 2008 report, *Invisible Wounds of War*, was one of the first and most substantive efforts to capture the scope of military TBI. The authors explain that "returning servicemembers suffering from mental disorders report problems restraining negative emotions, especially anger and aggression," and cites this lack of restraint as a probable explanation for the "distressingly high" rates of domestic violence in Vietnam veterans with PTSD.[44] This narrative of causation implies that the impulse toward domestic violence is natural but that veterans who do not have these conditions can restrain it, an explanation that reduces military domestic violence to an individual condition rather than a systemic problem. More generally, as medical confidence in the likelihood of an organic cause for TBI symptoms increases, so does the variety of traits included under this rubric. Howell notes that many clinicians now list "memory difficulties, irritability, mood swings, suspiciousness, and even guilt" among the roster of "'neurobehavioral symptoms.'"[45] Subjectivity, in such a framework, becomes merely an expression of a physical condition.

MTBI imprints on subjectivity in complex ways. People with MTBI exhibit variations in duration and severity of symptoms, but a defining characteristic of the condition is its disruption of core functions commonly understood to comprise subjectivity: memory, emotions, and thought processes. Simultaneously, MTBI impinges on the functions from which people derive much of their identities, like communicating, relating interpersonally, and maintaining jobs. Spouses and family members of veterans with MTBI often describe a sense of absence despite their physical presence, because they have become affectively unrecognizable.[46] As opposed to moderate or severe TBI, both of which entail longer losses of consciousness and more profound amnesias, in MTBI, some of these neural bases remain intact while others are profoundly altered.

Current clinical protocols for TBI generally consist of pharmacological treatment of symptoms, along with various forms of therapy aimed at improving memory and physical function.[47] Unlike therapeutic interventions for PTSD alone, which focus more on processing, integrating, and defusing traumatic memories, TBI interventions target the symptoms rather than the injury. Unsurprisingly, such "invisible" wounds incur very tangible medical costs. A 2009 accounting estimated that treatment for mild TBI would typically cost $27,000–$33,000 per patient, with more complicated cases at $270,000–$408,000.[48] Moreover, because the symptoms of TBI are often chronic, overall medical costs are difficult to tally. All totaled, disability costs for the wars in Afghanistan and Iraq could exceed one trillion dollars.[49] This, along with functional concerns about force readiness, explains the military medical establishment's recent interest in researching and preventing TBI.

Even these aggregate cost figures, staggering as they are, do not account for the other, less quantifiable sequelae of such injuries. In their discussion of its impact on military families, the authors of *Invisible Wounds* identify two routes by which TBI can compromise their functioning: "direct" in the emotional and relational problems TBI can cause, and "indirect" in the ways that TBI can limit employment possibilities and thus result in additional financial stress.[50] Moreover, because military TBI casualties are often young and otherwise healthy men, the work of managing their recovery can impose a "substantial, and usually unexpected, caregiving burden" for their wives. The authors confirm that "secondary traumatization" is common among the wives of veterans with TBI.[51] And, because the work of TBI caregiving can be quite consuming, it can interfere with wives' pursuit of their own careers.

In addition to calculating how much these injuries cost, the government and military must also determine how much they are worth. Bureaucratically, this pricing takes the shape of military and VA adjudication of disability claims. Recognition of trauma has always been shadowed by questions of what might constitute adequate compensation for it.[52] Yet on a cultural level, military trauma seems priceless. This valuation, which fetishizes the singular significance of combat trauma above all, inspires popular efforts to recognize and reward the troops for their service. These gestures channel effusions of gratitude, pity, and anger into both symbolic actions and material objects.

Gratitude, Pity, Anger: Feeling for PTSD and TBI

The gratitude, pity, and anger that orbit around veterans with TBI and PTSD distill a range of historical, political, technological, and cultural dynamics into these easy affective investments. Of course, the veneer of innocence laminated onto children and spouses cannot adhere fully to these veterans. Yet there is also a pleasure to imagining the interiority of a servicemember: the thought that someone might be willing to fight, kill, and die on *my* behalf . . . astonishing. These affective investments begin with the premise that veterans comprise a discrete and specially deserving group and that injured veterans constitute a more special subset of that population. Both of these ideas are relatively new, and the awareness of and interest in empathizing with veterans' suffering have varied over time.[53]

Since the era of the Gulf War, the professed responsiveness to veterans' suffering has been translated into the ubiquitous mandate to "support the troops." Although it was common, even acceptable, in the aftermath of earlier conflicts for politicians and civilians to regard returning veterans ambivalently, if not skeptically, tolerance for criticism of U.S. military personnel has declined precipitously since the era of Vietnam. By the time the United States was mobilizing for the Gulf War, media and politicians generally portrayed any opposition to militarization as "deviant."[54] Roger Stahl argues that, increasingly, American political and public discourse about war has moved away from earlier paradigms of good versus evil to justify military missions. Instead, he contends, public deliberation has collapsed into variations on a pro-war position based on supporting the troops, a framework that "works as a regulatory mechanism for disciplining the civic sphere itself."[55] The result, he suggests, is a "tautology" whereby it appears that the "military exists to save itself," which is a curiously demilitarized version of militarism.[56]

Support for the troops, as it is currently configured, begets a range of imaginaries about them.[57] In his philosophical meditation on drone warfare, Grégoire Chamayou critiques Western nation-states' commitment to "hyperprotection of military personnel," which aspires to eliminate any risk that they might be injured or killed.[58] In this logic, the military bodies that are so precious to the nation-state must be protected at all costs. In practice, however, this protection happens much more at the level of imagination than in battlefield reality, a discrepancy confirmed by high rates of wartime injury. MacLeish surmises that a combination of "military logic," "civilian imagination," faith in new technologies, and military discipline create the abiding fantasy of soldiers "who, thanks to their insensitivity to pain and their immunity from danger," can be sent safely to war.[59] Consequently, as David Gerber observes, news reports of American casualties are often delivered with a "tone of surprise."[60] The discourse of "sacrifice" by the troops smooths over these dilemmas, attributing to them an absolute agency that inspires their will to "put themselves in harm's way" on behalf of their fellow citizens.[61] This, in turn, locates the injured veteran in an affective market that trades largely in shows of sympathy and gestures of recognition.

The affection and admiration that I discussed in the previous two chapters drop away here behind the weight of the pity, gratitude, and anger foisted onto the figure of the veteran with PTSD or TBI. In other circumstances, affection might be directed toward the eroticized figure of the (uninjured, hypermasculine) soldier. But when the soldier is injured, this fetishization gets swamped by the mixture of pity, fear, and anxiety prevalent in modern Western responses to disability. Such responses, Gerber notes, are especially common in confrontations with "neuropsychiatric disability."[62] Often mysterious in origin, unpredictable in manifestation, and directly contrary to liberal ideals of self-control and self-awareness, such impairments thus militate against the kind of admiration that might otherwise be uncritically bestowed on military personnel.[63] Broadly, however, as David Serlin argues, disabled veterans encounter more cultural acceptance than those whose disabilities result from congenital conditions; they are celebrated as heroic, even modern, as opposed to being condemned as monstrous.[64] But because PTSD and TBI often entail relational difficulties, sometimes severe, "affection" for veterans with those conditions, as I demonstrate in my later analyses, is often framed as a sacrifice in itself and largely unrewarding, if not dangerous.[65]

Gratitude, by contrast, establishes a much less threatening relationship, although its role in the civilian-soldier-state nexus has varied from conflict to conflict. Minimally, the extension of gratitude requires widespread recognition of veterans as members of both a "status group" and an "entitlement group"—that is, members of a demographic identified as deserving particular treatment from the government.[66] Contemporary articulations of gratitude for the troops have their origins in the Civil War, though the line from that era to the present is neither straight nor unbroken. Jennifer Mittelstadt argues that social welfare programs for veterans who fought in conscript armies like those of the Civil War and World Wars I and II were largely understood as "rewards for faithful service or compensation for loss."[67] Gerber describes the introduction of an emotional component to this transaction; precisely because the Civil War was fought with a massive army assembled via an unpopular system of conscriptions, many civilians and politicians began professing a "feeling of gratitude to the saviors of the Union." Prior to this, pensions for veterans were regarded as forms of benevolent "welfare" and then a "right" that veterans had earned. With the subsequent influx of gratitude, they began to look like a debt to be paid for their service and also a nonnegotiable obligation of the state.[68]

However, good will toward veterans eventually dissipated into resentment for the special treatment they were receiving, which seemed—in the eyes of their detractors—to encourage laziness, criminality, and unemployment. These charges recurred after World War I.[69] Then, the burden of gratitude shifted onto injured veterans who were now the beneficiaries, willing or otherwise, of systematic and bureaucratized rehabilitation efforts. According to Ana Carden-Coyne, these "veterans were supposed to appreciate the new systems of medical and government attention by recovering rapidly. By physically and mentally adapting to retraining, disabled men were thought to be responding appropriately. Recovery was thus a way of performing gratitude with the body."[70] Following World War II, the rehabilitation movement was shaped by what Wool describes as the pervasive "moral desire and economic need to return men to work."[71] Over time, however, the onus for gratitude relocated onto the civilian population. Contemporary expressions of gratitude adapt this history for new conflicts, relying on the conventional wisdom that an ungrateful American public exacerbated the trauma of combat for returning veterans of the Vietnam War. These processes have transformed gratitude, as an ethical or affective position, into a popular obligation.[72]

Pity stands in complex relationship to gratitude here, and these affects are rooted in a nexus of injury, innocence, and heroism. Pity, as I noticed in previous chapters, requires an attributed innocence. But innocence is a thorny matter in this case, because the essence of military heroism as it is commonly understood is a willingness to injure or kill. In the case of the injured veteran, however, his own suffering serves as an offset, expunging whatever harm he might have inflicted on others during the war. And both pity and gratitude are contingent on the veteran demonstrating an appropriate attitude toward both his injury (stoic, noble, uncomplaining) and his healing (tough, committed, agentic).[73] Reflecting on a discourse that emerges in the rehabilitation movement of World War I, MacLeish observes that for individual soldiers, "the experience of being injured is decisively shaped by the idea of 'healing' as a willed undertaking for which soldiers are responsible."[74] Thus, the ideal figure of the soldier is endowed with a conditional agency that he expends only in the pursuit of noble military objectives and, subsequently, in the service of his own recovery.[75]

While the figure of the injured veteran provides a clear target for pity, which then encompasses his spouse and children, the anger that surrounds him is much more diffuse. This multidirectionality arises from the complex status of the soldier who is, in MacLeish's terms, "at once the agent, instrument, and object of state violence."[76] Ushered along by pity and gratitude, anger could be channeled into antiwar or anti-state platforms. Alternatively, anger might be directed at the enemy who caused the injury. At the level of public discourse, gratitude and pity generally neutralize anger that might be directed at military personnel themselves. Military personnel with PTSD or TBI who commit acts of violence, domestic or otherwise, do pose an acute challenge in this affective system. Even in those instances, however, pity blunts outrage, and anger gets diverted toward the governmental and military systems that failed to detect and treat their conditions.

In addition to anger felt on behalf of the veteran with PTSD or TBI, there is also the matter of the veteran's own anger, which is difficult to reconcile with their status as objects of sentiment. Wool describes the "soldier body as an icon of normative masculinity."[77] This iconicity, I argue, makes these figures particularly appealing for affective investment and those investments especially adherent. According to John M. Kinder, over the sweep of American military history, two "competing visions" of the reintegration of disabled veterans oscillated into and out of prominence, so that the men appeared "as an index of the United States's ability to enter the global arena and return home functionally, if not aesthetically, unscathed" or "portents of a terrible new age of unparalleled violence."[78] Yet if injuries like missing limbs or disfigurement prompt a national reckoning with how militarism looks, the affective damage done by PTSD and TBI force a consideration of how militarism *feels*.

Alongside the expectation that the injured veteran be a worthy object of gratitude and innocent enough to be pitiable but not so weak as to be pitiful, representations of these men as erratically and explosively angry circulate widely. Since the World War I–era adoption of the rehabilitation paradigm for injured veterans, the ideal subject of these regimens was to be both compliant and cheerful.[79] PTSD and TBI, however, with their attendant emotional symptoms that can include both the flattening of some emotional responses and the exaggeration of others, confound this image. Representations of angry veterans today echo the stereotype of the bitterly enraged Vietnam vet, a staple of depictions of that war in popular culture.[80]

Given the military's insistence on discipline and toughness, it is possible that anger was and is among the only emotions deemed acceptable to express.[81] Yet most of those representations coded veteran anger as political, provoked by the betrayals of state and fellow citizen alike.

The anger that attends some cases of PTSD and TBI, however, is rarely represented as having such an origin and often manifests in media representations as senseless and/or confined to the domestic.[82] For example, in 2016, after the domestic violence arrest of her son Track, who had served in Iraq, Sarah Palin elliptically implied that PTSD was to blame for his actions. This rhetoric minimized domestic violence and exonerated him, but also echoed common portrayals of veterans as, in Nate Bethea's terms, "ticking time bombs, as damaged beings primed to harm."[83] These representations deprive veterans of their agency, which may be an expedient strategy when they are charged with a crime, but also reduce their subjectivities to collections of misfiring reflexes. Through all of this, the veteran's anger remains an unwieldy affective liability in the effort to imagine these figures as perfect embodiments of national virtue. Simultaneously, the public's understanding of itself as grateful to and supportive of the troops forecloses any thought that military personnel might harbor any animosity toward them or any hesitation at all about going to war on their behalf or in their stead.

Saying "Thank You" to the Troops: On the Militarization of Gratitude

"Get Some," the first episode of the 2008 HBO miniseries *Generation Kill*, features a mail call for the marines, who find themselves, yet again, to be the beneficiaries of unsolicited but earnest correspondence from American schoolchildren.[84] One little boy has written that he hopes the marines will be able to come home without fighting because peace, in his youthful estimation, is always "better than war." One character sneeringly composes a reply aloud, as his mates encourage him along:

> Dear Frederick, thank you for your nice letter, but I am actually a U.S. Marine who was born to kill, whereas clearly you have mistaken me for some sort of wine-sipping Communist dick-suck. And although peace probably appeals to tree-loving bisexuals like you and your parents, I happen to be a death-dealing, blood-crazed warrior who wakes up every day just hoping for the chance to dismember my enemies and defile their civilizations. Peace sucks a hairy asshole, Freddy. War is the motherfucking answer.

Another chimes in, his face radiant in mockery of sincere appreciation, "But thanks for writing anyway." With a sigh, the first marine encapsulates their predicament:

> Aww, man, every motherfucker in this camp is just waiting for packages of dip, Ripped Fuel, porn mags, batteries, hash chunks, dirty-ass jerk-off letter from Suzy Rottencrotch. . . . But no, all we get is this happy-day fucking horseshit from Miss Cunt Lips' fourth grade class. Can you fucking believe this shit?

This, clearly, is a perversion of gratitude: the childish expression of appreciation that offends rather than pleases its recipient, the marine's utterly disingenuous thanks for the letter, and the shameless litany of the commodities that they really want.

Yet there are also things worth hearing in this character's complaint. It is, essentially, a critique of the patriotic ritual of "saying 'thank you' to the troops," and it demonstrates how the collectively imagined entity of the "troops" erases the needs of individual soldiers.[85] Their benefactors' radical misreading of their wishes reveals how gratitude can maintain the distance between civilians and the military by the very actions that seem to shorten it. This complaint also contains a lament, however crass, about the double insufficiency of this kind of gratitude, which is both too big to be personally resonant and too small to meet the simplest wartime bodily needs of its recipients. This mismatch can be amusing at best (as when the *Atlantic* cataloged the "bizarre" and "unsatisfying things" that soldiers often receive in care packages, like a stack of AARP magazines), or profoundly inconsiderate at worst.[86] Because the instruction to "say thank you to the troops" is so ubiquitous and seemingly irrefutable in the landscape of contemporary American militarism, I want to linger a bit on the interface of gratitude and militarized violence.

Every day, concerned strangers compile and mail care packages, make donations in money or in kind, and organize homecoming events, and countless citizens, from schoolchildren to the president, participate in rituals by which they offer their personal thanks. These and other similar practices have become commonplace, almost reflexive, but are fraught with complexities and contradictions. All of these thanks circulate during a conflict where war has scarcely entailed any meaningful civilian sacrifices. Indeed, as Andrew J. Bacevich points out, the tax refund checks mailed out by the Bush administration early in the War on Terror effectively paid us for our "acquiescence" to war.[87] Gratitude offers an appealing remedy for the tension provoked by awareness of the discrepancy between military and civilian experiences of war, a form of emotional compensation.[88]

Any true expression of gratitude requires a preceding feeling of gratefulness.[89] To feel gratitude, someone must have the sense that someone else has done something to benefit them directly. Mark Jonas distinguishes between "debt" and "recognition" forms of gratitude. "Debt" models of gratitude, he argues, establish hierarchies where the gracious doer occupies a position superior to and powerful over the grateful beneficiaries. These hierarchies also imply that the original deed is something that must be repaid, that the beneficiary has a duty to "discharge" this obligation. Jonas suggests that this is an unstable, perfunctory form of gratitude. Alternatively, the "recognition" form of gratitude is much less mercantile, marked by genuine appreciation and culminating in a simple expression of thanks.[90] In the current context of militarization, these forms of gratitude commingle. There is, on the one hand, a sense that gratitude to the troops must be acted out, preferably in some way that is concrete or commoditized, so that gratitude becomes a thing to do or buy rather than a feeling to be experienced. On the other, because the current wars require so little of American civilians, the feeling or expression of gratitude often qualifies as sufficient proof of one's patriotism.[91]

A key dimension of these gestures of thanks is asymmetry between the enormity of the sacrifice being repaid and the relative smallness of the thing given in exchange. This is not to imply that the people who receive these tokens do not appreciate them; indeed, military personnel often report that seemingly little things have outsized impacts on morale or quality of life.[92] Though there is something exceedingly, even dismissively, optimistic about the idea that the hardships of militarization can be offset by a note or a box of cookies shipped overseas, the problem is not simply one of scale. After all, true parity ("I am going to enlist on your behalf because you enlisted on mine" or "I am going to kill someone for you because you killed someone for me") is unrealistic and perhaps undesirable. And in any case, the act of thanking someone for his service is also a way of intimating that he has done this work so we did not have to. In all of this, militarized gratitude starts to seem obligatory, and my concern here is with the propulsive force these norms provide to various kinds of militarism and violence. Indeed, the state itself seems to recognize its obligation to enact at least a semblance of gratitude for its military personnel; the standard script for military funerals is recited "on behalf of a grateful nation," and the folded American flag is given over to the family "as a symbol of our appreciation."

Such symbolic gestures of gratitude by the state comprise an alternative to more substantive, material forms of recognition that might be foreclosed by budgetary constraints and financial crises. Indeed, the state has not always understood its fiduciary obligation to veterans as a given. For example, President Hoover deployed an army regiment to disperse protesting veterans in their "Bonus March" on Washington in 1932.[93] Subsequently, President Franklin Roosevelt cut veterans' bonus and disability pay, refusing their claim that they had unique entitlements.[94] Currently, congressionally mandated budgetary sequestrations can trigger cost-saving measures like base closures and consolidation and the reduction or elimination of benefits, positions, and salaries. In a rare moment of agreement before the government shutdown in October 2013, both houses of Congress passed the Pay Our Military Act unanimously and so guaranteed paychecks to military personnel and their civilian support at the DoD, signaling a belief that there were some debts that could not be forgiven or deferred. But this largesse extended only to active-duty personnel; claims were delayed for those who had already served and were seeking tuition or disability benefits, as well as for survivors of military personnel killed in action.[95] And, as Brienne Gallagher notes, antipathy persists toward veterans who seek to actually avail themselves of the benefits to which they are entitled, particularly for mental health care related to PTSD. She argues that military personnel are widely regarded as the most deserving recipients of the welfare state *until* they make demands, at which point they often come under suspicion of malingering and fraud.[96]

When the state fails its military personnel, the private performance of grateful citizenship becomes a way of assuming the abandoned or outsourced obligation of the state. Much of the practical work of expressing national gratitude has fallen to individual citizens or corporations, who have devised their own rituals for it. When other forms of remuneration are scarce, gratitude seems like a cheap and limitless resource to be offered as payment for the work of military personnel. Elisabeth R. Anker argues that post–September 11 American citizenship has been defined by intense identification with the state. She writes, "State identification idealizes what the subject 'should have been'—sovereign, self-making—and what it now desires to be 'like.'"[97] Here, citizens inspired to imitate the state perform militarized gratitude according to its model or in the breach created by its failure to uphold these obligations.[98]

All of this gratitude pivots on the reasoning that military personnel have deliberately and willingly sacrificed on behalf of their fellow citizens, a belief central to contemporary American militarism.[99] This reading of military sacrifice is pervasive, moving both downward from the government and upward, or laterally, from citizens. President Obama, for example, relied heavily on the rhetoric of military sacrifice to mobilize support for the continued war in Afghanistan.[100] The feeling of being personally sacrificed-for motivates individual expressions of gratitude as well. Both Wool and MacLeish's ethnographic accounts of military life include depictions of awkward professions of thanks to soldiers from strangers. MacLeish critiques these expressions on the grounds that they express appreciation for a gift that is "not freely given," writing that "the soldier's giving is at once the ultimate gesture of will and an act that is truly 'selfless' by virtue of its enactor's ambiguous selfhood."[101] Wool objects on the grounds that the expressions of gratitude construct the receiving soldier as a "flattened figure" whose own interpretation of his wartime actions gets nullified in the process. She recalls that for many of the soldiers she knew, those "unqueried tasks of war or acts of violence . . . had seemed more like the workplace injuries of a life-or-death job than patriotic sacrifices."[102] Such thank-yous operate performatively and prescriptively, imposing a narrative onto the servicemember's actions; the thanker sets the terms of the exchange and relocates the servicemember into a subordinate position. This constrains in advance the servicemember's options for reply, lest he be perceived as rude or ungrateful; after all, even the marines in *Generation Kill* did not actually send the letter that they drafted aloud. By demurring or accepting the thanks humbly, the servicemember suggests a willingness to have done it and to do it again.

Consequently, the thank-you may actually function to reinstate the servicemember's obligation rather than marking that it has already been fulfilled. Jacques Derrida insists that a true gift requires the giver to forget the giving, lest the gift be tainted by the expectation of a thank-you in return.[103] In this instance, the publicly bestowed thank-you shifts a burden back onto the servicemember, who is obligated by conventions of politeness to respond with some variation on "you're welcome." That "you're welcome" relieves citizens of any further responsibility while reifying the civilian/military separation and thus keeps open the possibility that the servicemember will fight again on their behalf. So the thank-you establishes a militarized line of credit. Mauricio Lazzarato's definition of credit captures its perils; credit, he writes, is "a promise to pay a debt, a promise to repay in a more or less distant and unpredictable future, since it is subject to the radical uncertainty of time."[104] In this case, however, the servicemember as creditor bears all the risk while the citizen has purchased, very cheaply, unlimited access to the reward.

"A Wish May Also Be Denied for Any or No Reason": On the Materialization of Gratitude

In June 2013, President Obama signed the Stolen Valor Act, which makes it a federal crime to fraudulently represent oneself as the recipient of military medals or decorations "with intent to obtain money, property, or other tangible benefit."[105] The act, which sailed through its House and Senate votes and into law, replaced a 2005 iteration of the Stolen Valor Act, which the Supreme Court declared unconstitutional—as an infringement of the right to free speech—in 2006. Stolen Valor is both a legal and cultural phenomenon: for example, the popular website StolenValor.com acts as a clearinghouse for stories of "stolen valor," and a YouTube search for stolen valor videos returned nearly half a million results, most of which focus on exposing would-be thieves of valor. The 2005 version of the law made the lie itself a crime; the 2013 version instead criminalizes the intention of the lie, so that this kind of fabrication is tantamount to fraud. The valor in question is "stolen" from those who have rightfully earned it, effectively cheapening it, but the implied victims of the fraud itself are those who fall for it and offer something in exchange. For my present purposes, Stolen Valor matters because it reveals, indirectly, the thoroughgoing commodification and monetization of military status. In other words, attempts to steal valor only matter, or work, in a context where valor has a value.

To take a seemingly trivial example of this valuation: in the long aftermath of September 11, airlines used the pretext of increased security and recovery from the terrorist attacks to reconfigure their business models. The resultant monetization of every aspect of air travel created a multitude of petty hierarchies, including the process of boarding the plane. To the ranks of priority boarders like families with small children, people with disabilities, the elderly, and well-to-do or frequent travelers, airlines added uniformed military personnel. Unlike more spontaneous shows of thanks like rounds of applause for servicemembers commanded by enthusiastic flight attendants, this practice has become essentially standard across the industry. Yet these practices feel "good" precisely to the degree that they are directed at servicemembers who remain largely passive in the exchange. Consider, for example, how it would feel if servicemembers got early boarding privileges by elbowing their way to the front of the queue and demanding them.[106] Culturally, we seem to prefer to "give" things to veterans over listening when they ask for attention. For example, when World War I veterans became too enterprising and tried to capitalize on their uniforms for pan-handling or preferential treatment in hiring, the federal government moved to prohibit such "'commercialization' of the uniform."[107] The contemporary glut of charities working to provide things for "warriors" or "the troops" or even "any soldier" operate largely by channeling the unfocused good will of benefactors and rather less frequently by responding to the wishes of their recipients.

To mobilize popular support during World War II, Franklin Roosevelt promised that victory would ensure a perpetual "freedom from want" that would more than offset the sacrifices it necessitated. In the interim, a truly communal mobilization would be required, built on public sacrifices widespread and quotidian enough to be felt and hence create solidarity, but not so overwhelming as to erode popular morale.[108] That campaign modulated the war's demands on civilians, but the contemporary civilian of economy of conflict in the United States is much different. Currently, war in the United States is marked instead by surpluses. There is virtually no direct impingement on consumer liberties alongside the proliferation of outlets where good will might be routed toward the troops. Whether this good will takes the form of thank-you notes or cheap consumer goods, there is a lot of it.[109]

Some volunteer organizations exist to coordinate displays of thanks. For example, a group called Hugs for Our Soldiers welcomes back single soldiers with care packages of toiletries and friendly signs in their barracks, often provided by their unit's Family Readiness Group, which draws its membership from among married soldiers' wives. In my home state, Operation Welcome Home Maryland coordinates groups of people to be at Baltimore-Washington International Airport when planes with military personnel land. They promise that "when the troops arrive through the doors of the International Arrival terminal, we shower them with praise, cheering them on, hugging them, offering them a 'goodie bag,' and thanking them for their service." The group is emphatic about the importance of a warm homecoming, and the website includes suggestions on details like what to wear in order to facilitate that. Yet it also advises volunteers to be prepared to rein in their exuberance, reminding them that the military personnel are likely hungry, thirsty, and tired and so might not be as enthused for these displays of gratitude as one might hope. Even in the videos posted on the group's website, there are traces of awkwardness in the juxtaposition of the effusiveness of the welcoming gauntlet and the weary disorientation of the troops. But their explanation of this behavior, while thoughtful and considerate, oversimplifies the complex dynamics operative in these welcomes, avoiding the possibility that the personnel being welcomed feel ill at ease with or disconnected from the display.

While networks like Operation Welcome Home emphasize personal, if fleeting, contact with soldiers, other groups operate at a distance. Among these, care packages are the preferred form of communication. But before I offer a critique of this practice, I want to say that I get it. I have never written a thank-you note to an unknown member of the military, sent an enlisted stranger a holiday card, or assembled an anonymous care package. For a time, I did these things for someone overseas that I knew and loved. So, I think I understand what could incline a person to load a big-box retail shopping cart full of goodies for someone that she will never know. For my part, I remember deliberating in the snack aisle, how ordinary commodities like laundry detergent were infused with significance because someone fighting in a war had asked me to buy them, how the task of buying spiraled out when I asked if anyone else over there needed anything and then found myself pressing a quizzical-looking gas station cashier for a very specific brand of chewing tobacco. I remember the jigsaw work of fitting it all in the box, lettering the FPO address in all caps as I had been instructed, and the dizzy freedom of being unconcerned about the cost of postage because my mission was so noble, and what was money, after all? I remember visualizing the package's progress across the ocean. I remember waiting for my thanks. I remember getting it. I remember starting my shopping for another package on my way back from dropping one in the mail.

While the packing and mailing of favorite items from home is a key deployment ritual for military families, civilian families who want that experience can work with organizations like Anysoldier.com. Givers generally create a package suitable for "any soldier" (or "any female soldier") and address it accordingly. Other soldiers volunteer to act as intermediaries to distribute them, giving first to soldiers who do not regularly receive mail. Anysoldier's list of selected gifts includes things like cleaning kits for M-4s, which I would have expected the military itself to provide. But perhaps there is a pleasure, for givers who identify strongly with the military, in the notion of actually equipping them for battle. Anysoldier's instructions also alert overzealous but budget-minded shoppers that low-quality items from dollar stores or Walmart don't last, implying that such goods are more for the sender's benefit than the receiver's. Yet the group also suggests that the sender's letter is the most important element of the care package: "Better than spending a bunch of money that you don't have" on commodities. In essence, the site caters to those who have a profusion of gratitude and nowhere to put it. It jokingly warns of "support withdrawal" in the interval between the compilation of the last care package and the assembling

Anysoldier caters to people who have a profusion of gratitude for military personnel and no place to put it. Some of these people are so-called support junkies who cannot seem to get enough of the ritual of shopping, boxing, and shipping care packages to servicemembers. These little vignettes are part of Anysoldier's "You Know You're a Support Junkie When . . ." feature.

of another and provides special content for so-called "support junkies" who cannot stop shopping, boxing, and shipping. Although actual soldiers ostensibly provided input for Anysoldier's list of suggested purchases, the operation does not have a mechanism that allows soldiers to make specific requests: this enables Anysoldier to accommodate gratitude on a massive scale while insulating that gratitude from any reply, restraint, or consequence.

These charitable souls, who appear mostly to be women, are not alone. A group called A Million Thanks (AMT) has overseen the delivery of more than 9,100,000 thank-you notes to the troops. To be eligible for carriage by AMT, letters must be "generically addressed": no messages for specific military personnel.[110] Letters must also convey only "positive messages," and the group ominously elaborates that "any negative messages will be discarded." Likewise, Operation Welcome Home Maryland notes on its website that "we always need hand-made thank-you cards!" but reminds creators, "PLEASE: Keep it positive and supportive—nothing negative." AMT encourages its senders to be creative, embellishing their letters with color and drawings, and instructs them that because soldiers need "encouragement," they should be "kind and uplifting" in their sentiments. Both Anysoldier and AMT target generic soldiers, and so their guidance is based on a rather minimalist view of what they might want and need: playing cards, beef jerky, good socks, a cheerily illustrated note, candy or Beanie babies to pass out to local children.

When I first began researching AMT in 2013, their ambit included only letters; since then, it has expanded to include the "Fund A Scholar" campaign, a scholarship program for children whose military parents were killed as well as its "Grant a Wish" initiative. AMT has accumulated a range of corporate partners, but many of these, like Southwest Airlines, seem only to serve as nodes for collecting letters rather than providing resources directly to the organization or its intended beneficiaries. Its "Grant a Wish" feature is different, however, in that corporate sponsors sometimes provide goods directly, like a $12,000 bed from SleepNumber for a veteran with chronic pain or a fishing boat from Bass Pro Shops for a veteran whose PTSD makes his need for solitude especially acute. There is a massive disjuncture between the number of letters that AMT sends and the amount of material support it provides; at the time of this writing, AMT was working to meet the scholarship needs from just over thirty students and had granted or was soliciting donations to grant wishes for roughly the same number of veterans.[111]

While the other mechanisms for gratitude seem to imply that soldiers' primary needs are for tiny luxuries like Nutella or small infusions of positive sentiment, AMT's "Grant a Wish" establishes a different economy of desires, both in terms of what the injured veterans say that they want and the mechanisms by which AMT adjudicates those requests. Alongside the expressive guidance that AMT provides to well-wishers who write letters, it tightly regulates the wishes of servicemembers themselves. Veterans provide narratives to justify their requests, stories that reveal elaborate clusters of needs, often directly linked to their combat experience. For example, one man wanted, and received, a pool table because he loves to play but cannot go to bars (recovering alcoholic) or be comfortable in recreational pool halls (PTSD). As of this writing, the vast majority of the wishes on the website come from veterans with PSTD or TBI.[112]

Individuality in these narratives can only appear when articulated through the language of trauma, which in turn must be tied to a promise that a specific commodity will help the wisher to recover from it. If average "any soldiers" seem to need only snacks and socks to keep them going, being injured apparently launches the veteran into a whole different economy, where more expensive commodities are required. Generosity on this scale, measured in hundreds or thousands of dollars, requires a performance of suffering that the AMT decision-makers, who remain unnamed on the site, find persuasive and legible. AMT notes that if their wish is granted, recipients will be "required" to share their injury stories on the website, along with photos of themselves.

In its FAQ for this program, AMT outlines the terms of its generosity. These include the stipulations that veterans must have served specifically in the Global War on Terror and cannot ask directly for cash to pay off debts.[113] Although military families are themselves among the most indebted in the nation, AMT says:

> A Million Thanks' Grant A Wish campaign is designed to improve the overall quality of life and is not intended to grant cash to individuals for assistance with debt. Therefore, the wish that is requested should be one that will assist with improving individual and/or family life, but not debt relief.

They continue, in response to a hypothetical question about what a veteran *can* ask for, "Special devices, career improvement, special vacations, and living area improvement are some examples of the wish areas that may be granted."[114] Ultimately, AMT exercises absolute, sovereign discretion over whether or not these wishes are worthy and grantable. In a rigid summation, AMT identifies the limits of this good will: "The eligibility, and criteria for granting a wish is the sole and final decision of the A Million Thanks organization. A wish may also be denied for any or no reason." Institutional caprice here sets the limits of generosity.

In all of these instances, the expression of gratitude is constrained and mediated. With the exception of Operation Welcome Home, where the contact is intentional but also brief and contrived, all of these organizations keep the grateful citizen and servicemember apart. Both Anysoldier and AMT broach the issue of the reciprocal obligations that the servicemember incurs by being thanked, but in fundamentally divergent ways. One of AMT's "Do's" for letter-writers cheerily instructs them to include a

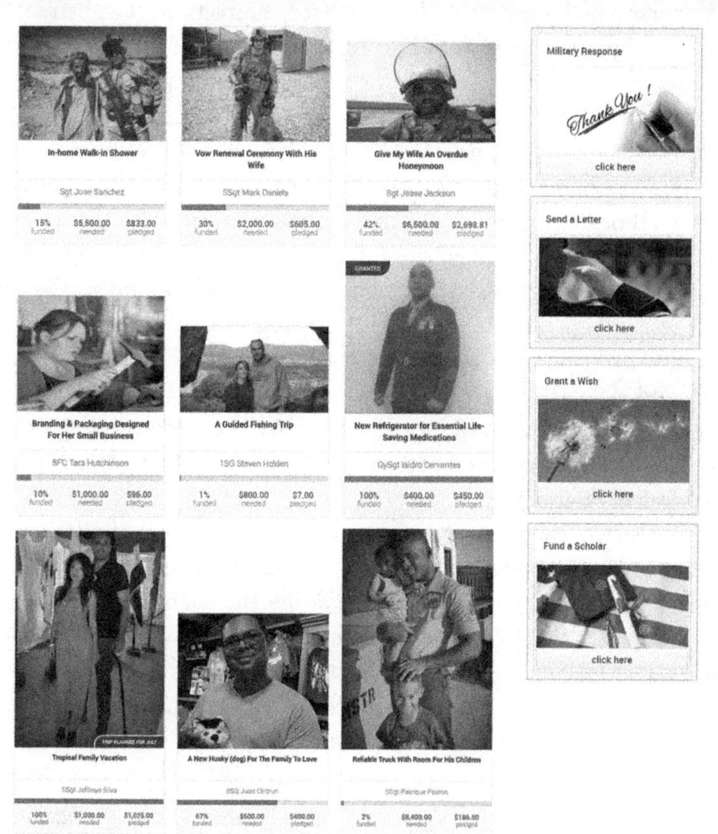

A Million Thanks began as a clearinghouse for thank-you cards to military personnel but has since expanded its operations to include more tangible forms of support. Its "Grant a Wish" initiative invites military personnel to narrate their struggles and needs for specific items. AMT uses its discretionary power to determine whether or not these needs will be met.

mailing or email address if they wish because "most military will write back to you!" On the other hand, although Anysoldier.com posts expressions of thanks from military personnel, they also chide speculative givers for even wondering if they will receive a response. Consummate with the grittier aesthetic of the website overall, they provide this answer: "The Soldiers are under no obligation to reply. They can't be," citing the danger, hardship, and privation of military life. This admonition seems to relieve the servicemembers of a burden, but it also keeps them at a distance and shields the givers from any kind of consequence, so that the care package retains its magical status and the servicemember remains an ideal figure.

My own care-packaging was animated by my desire to make a very specific person feel a very specific way, but generosity in the abstract is complicated, because these acts are so intimate. The food will be eaten and metabolized into energy, the clothes will be worn, the toiletries will be used to wash or lather or shave or moisturize a very specific, very real, but very unknowable body. It is worth thinking about the wishes, imaginative processes, and financial considerations that aggregate into the giver's decision to buy all of these things, to wonder about the variables that result in the purchase of those cookies, these razors, that shampoo as opposed to the countless other varieties of the same type of product, and all the tiny little feelings expended or intensified in the process of assembling them. And I can only imagine what it is like to open up a box full of these things, carefully and expectantly arrayed, how uncanny it would be to discover that a stranger had guessed and sent exactly what you were craving, or, alternatively, how lonely and disheartening to find things for which you have no use, desire, or appetite.

In practice, gratitude is messy, even if it is not so hyperbolically cruel as it appears in *Generation Kill*. But these carefully choreographed expressions of gratitude enable the people doing the thanking to leave their visions of the military intact while sanitizing the violence embedded in military service itself. Derrida has categorized the true gift as an aporia.[115] To be a true gift, it cannot be understood as such, lest it get pulled into an economy of exchange and obligation.[116] He says:

> The gift is precisely . . . something which cannot be reappropriated. A gift is something which never appears as such and is never equal to gratitude, to commerce, to compensation, to reward. When a gift is given . . . no gratitude can be proportionate to it. A gift is something that you cannot be thankful for. As soon as I say "thank you" for a gift, I start cancelling the gift, I start destroying the gift.[117]

The expression of gratitude for military service idealizes and devalues it while presuming total volition, absolute agency, and complete consent on the part of the people who give the presumed gift of enlistment. This presumption obscures the tangled series of choices and compulsions that underlie the decision to join up.

These expressions of gratitude are predicated on an understanding of citizenship as debt that is only called under certain circumstances. Military personnel become legible and visible primarily when they are risking their lives, when they are, or potentially will be, sacrificing. And it may be, problematically, that this kind of sacrifice is the only way that certain bodies can gain recognition as valuable members of the polity, recognition that might be differentially bestowed.[118] At the same time, understanding military service as sacrifice places all civilians in perpetual indebtedness, organizing the body politic according to obligations it can never relieve. Saying "thank you" to the troops seems to offer a form of expiation. But this damages in every direction.[119] Violence, with its avenging logic, goes beyond the common economic corruption of the gift, replacing reciprocity with retribution.

The state, which is organized by force, is inimical to true expressions of gratitude,[120] while the compulsion to say "thank you" to the troops establishes a form of national belonging that hinges on death and injury, collectivized.[121] In an elegant refusal of the claim that all Americans are responsible for the torture at Abu Ghraib, Timothy V. Kaufman-Osborn asserts that such an argument diffuses guilt to the point of meaninglessness. In so doing, it exonerates the state that is actually to blame while perpetuating the fiction that the state acts on behalf of its citizens, thus licensing further state-sanctioned atrocities.[122] There is a similar dynamic operative in the militarization of gratitude, which is predicated on the idea that the soldiers are sacrificing for everyone. By obfuscating the state's role as the entity that decides when, where, and how military personnel will be deployed, this action may contribute to militarism more generally, indirectly guaranteeing that military personnel will continue to be called upon to sacrifice in subsequent wars. This becomes even more complicated in the circumstance of subjectivity-altering conditions like PTSD and TBI. If such a person no longer acts like or is recognizable as "himself," then the person who served is, in many ways, different from the person who is being thanked for his service. The expression of gratitude, however, stabilizes the veteran's identity long enough to constitute him as an object of gratitude and so downplays the severity of the injury that ostensibly invited the thanks in the first place. Given this, what would happen if we resisted the reflex to say "thank you" in these circumstances, accepting the resultant discomfort of leaving the books unbalanced? How might we accept the possibility that some debts cannot be repaid and are perhaps better left unsettled?

Clearly Justify / Clearly Do Not Justify: Debating Purple Hearts for PTSD and TBI

In 2009, as public and legislative awareness of the pervasiveness of military PTSD was increasing, the Pentagon decided that PTSD did not merit a Purple Heart. The decision was and remains controversial, but the quasi-official Military Order of the Purple Heart (MOPH) supported it. Describing itself as "very sympathetic" to the plight of veterans thus afflicted, the MOPH reasoned nonetheless that it would be hard to find anyone who did not have "'some form of PTSD'" after combat. It further distinguished PTSD from other injuries on the grounds that it could be both faked and treated.[123] Relatedly, a member of the Veterans of Foreign Wars leadership clarified the rationale for such decorations, arguing that "Medals aren't awarded for illness or disease, but for 'achievement and valor.'"[124] The DoD's decision remains in force and reveals the narrow terms on which the government will recognize the suffering of its own military personnel.

Both government and veterans' organizations have worked actively to shape the meaning of war-related disability. During World War I, for example, the British Army Council differentiated between Shell Shock-Wounded and Shell Shock-Sick to indicate whether the cause was enemy action or not, with preferential and more sympathetic treatment afforded to the former.[125] In the United States, veterans' organizations have been active in the public sphere and influential in policymaking since the era of World War I, though they have not always pursued the same agendas or strategies. Both the American Legion and Disabled American Veterans worked to "frame war-related disability as a civic ideal" after the Great War, but relied on different textual and visual strategies to make their claims.[126] All such campaigns, including criticisms and endorsements of the DoD's recent Purple Heart decision, utilize the image of the injured veteran to press a claim on government recognition, appealing to governmental and popular imaginations and their conceptualizations of wartime suffering.

Conferring the Purple Heart entails a complex and sometimes contradictory form of recognition. On the one hand, military personnel are recognized as individuals for acts that go beyond the baseline normative expectations for battlefield conduct. On the other, on bestowal of the medal, in Wool's experience, "*a* soldier is transformed into *the* soldier."[127] The Purple Heart recipient becomes exemplary of the ideals of bravery and sacrifice that adhere to that figure. And the injury then becomes a visible sign of those otherwise ineffable traits, a purpose that the ostensibly invisible injury of PTSD cannot serve.

Purple Hearts are reserved for serious injuries, which are easy enough to identify when the damage is physical. They are also given only for militarily meaningful ones, sustained in battle rather than, for example, at home or on base. For these reasons, they serve as symbolic compensation, a prosthetic replacement for what has been lost. For an injured veteran to receive a Purple Heart, the Pentagon and the relevant branch of service must agree to confer it, and both weigh the extent to which the injury was caused by enemy action. If an enemy plants an explosive that detonates and nearby personnel subsequently are diagnosed with TBI, the role of the enemy in the injury is clear. On the other hand, so goes the argument, it is impossible to prove that enemies intended to cause military personnel to develop PTSD. In April 2011, the DoD released its Purple Heart standards for TBI, after some deliberation but with little hesitation. The marines had already begun instructing their commanders on how to assess Purple Heart eligibility for such injuries, while the army's standards were still in the works. Among the allowances made by the DoD is a provision allowing for a retroactive determination of the seriousness of the injury. According to these guidelines, even if the servicemember was not treated for concussive symptoms at the time of the injury, a medical officer later can certify that she or he would have needed treatment if it had been available.[128]

A background report prepared for Congress as the debate about PTSD was unfolding describes the Purple Heart as the "oldest and most recognized" of military decorations.[129] Its history begins with a Badge of Military Merit conferred by George Washington in the early 1780s and then lapses until 1932, when it was resurrected under the auspices of General Douglas MacArthur. By the era of World War II, Purple Heart eligibility criteria were tailored to those injured or killed in combat. In the ensuing decades, presidents have modified and expanded the criteria to include consideration for people held as prisoners of war and hostages; peacekeepers; victims of terrorist attacks; and casualties of violence by other military personnel, as in the shootings at Fort Hood and in Little Rock.[130] The report points out that "invisible wounds" complicate the evaluation of Purple Heart requests but also indicates that the increasing medical verifiability of TBI has compelled new recognition for that class of injuries.

The Purple Heart is different than other awards because it is based on entitlement, rather than recommendation. Consequently, the injury itself becomes the criterion for decoration. The Human Resources Command of the U.S. Army provides an exhaustive list of the criteria for the Purple Heart, apparently prompted by questions about MTBI.[131] The army reiterates that the injury must be incurred during engagement with the enemy. It also distinguishes types of conditions that "clearly justify" or "clearly do not justify" recognition in the form of a Purple Heart. The former category, after relatively succinct references to injuries from enemy bullets, mines, or chemical weapons, offers a much longer explanation for TBI:

> Mild traumatic brain injury or concussi[on] severe enough to cause either loss of consciousness or restriction from full duty due to persistent signs, symptoms, or clinical finding, or impaired brain functions for a period greater than 48 hours from the time of the concussive incident.

Among the conditions that "clearly do not justify," however, it lists:

> Mild traumatic brain injury or concussions that do not either result in loss of consciousness or restriction from full duty for a period greater than 48 hours due to persistent signs, symptoms, or physical findings of impaired brain function.

Those nonjustifying injuries are classified alongside maladies like frostbite, food poisoning, battle fatigue, first-degree burns, post-traumatic stress disorders, and self-inflicted injuries not sustained during the "heat of battle." Subsequent guidelines clarify symptoms that would justify granting a Purple Heart for MTBI, as well as the kinds of testing that could be used to identify such an injury. The army's distinction hinges on the severity of the injury, measured by duration or, more specifically, the duration of the time that the soldier's brain or bodily function was impaired to the point that fulfilling his or her duties was impossible.[132]

The debate about the Purple Heart and PTSD unfolds according to a complex politics of recognition. One prominent psychiatrist argues that the distinction between PTSD and other forms of bodily injury is baseless, both because PTSD can be debilitating and because the brain is an organ.[133] NAMI, the National Alliance for Mental Illness, issued its own report arguing that PTSD should qualify for a Purple Heart, making a rights-based claim in the language of "parity for patriots." Importantly, this report does not mention TBI at all; NAMI is arguing, instead, for an understanding of PTSD itself as a disability.[134] While the MOPH has held to its position about PTSD, it also issued a public statement urging the U.S. military to consider PTSD diagnoses in reviewing discharge update requests. In doing so, it reminds readers that World War I veterans with shell shock were often maligned for cowardice and draws extensively on the consequences of PTSD for Vietnam veterans (and their families, in a reference to divorce rates). But the statement makes no mention of veterans from the current war. In this way, PTSD becomes not an entitlement to recognition for valor, but rather a mitigating circumstance for which a punishment ought to be softened.[135] Importantly, veterans with PTSD are eligible to receive disability pensions from the Veteran's Administration, in which the state extends financial recognition for suffering while withholding the symbolic recognition of the Purple Heart, hinting that these forms of gratitude operate according to different economics.[136]

New findings about TBI have the potential to radically alter the terms of the debate over the Purple Heart. In a development I will discuss at more length in the chapter's penultimate section, researchers are beginning to identify specific changes in the brain that, when present, can verify that a TBI has occurred. This research has the potential to be transformative for TBI patients but may also engender further marginalizing or delegitimizing of military personnel who have PTSD symptoms in the absence of a verifiable TBI. The prospect, which is gradually being realized, of making TBI itself visible on the brain holds out the enchanting possibility that clinicians might be able to *see* the telltale signs of damage and in turn provide definitive support for the claim of injury. The relative ease with which the DoD and branches of the armed forces agreed that TBI warranted a Purple Heart while PTSD did not suggest that the state endows the servicemember's body with a value it does not attribute to other dimensions of their personhood.[137] And by deeming only longer lapses of consciousness worthy of the Purple Heart, it affords, chillingly, a special recognition to those who lose themselves completely, if temporarily, in its service.

Gooey Eyes, Orthodontist Smiles: Feeling for PTSD in Thank You for Your Service

The debate about whether or not veterans with PTSD should be eligible for Purple Hearts proceeds according to an exacting gradation of suffering and recognition. These standards endeavor to cordon off emotional responses and refuse to bestow governmental recognition on purely emotional symptoms. By contrast, Finkel's widely acclaimed *Thank You for*

Your Service approaches recognition in expressly emotional ways. This embedded journalistic account of soldiers returned home from Iraq makes a sly commentary on the profusion of thanks for veterans. The book is a standalone sequel to *The Good Soldiers*, Finkel's account of their experience in Baghdad. None of the men he follows returned unscathed, and in their stories, the leitmotifs of PTSD and TBI (though in some cases *avant la lettre*) unfold across backdrops both domestic and bureaucratic.

Among other laurels, *Thank You for Your Service* was named a best book of 2013 by both the *New York Times* and the *Washington Post*.[138] Finkel's account provides a searching view of the families he profiles. He chronicles the fear and bitterness of wives coping with their husbands' new illnesses and treatments, the bafflement of the soldiers trying to navigate military and governmental benefits systems, and the exhaustion of the officials who try vainly to help them.

The title's sarcasm echoes the scorn of one of the book's main characters who finds himself alienated at a free hunting trip organized for veterans. Finkel channels his internal monologue as follows:

> Those people who drive around with "We Support the Troops" signs on their cars, as if a sign on a car makes any difference? The ones who have never been to war and will never go to war and say to soldiers, "Thank you for your service," with their gooey eyes and orthodontist smiles?[139]

Those rhetorical questions vex the book's popularity: does a nodding assent with these criticisms permit readers to believe that theirs is a savvier, more appropriate form of gratitude for the troops? If so, then the book knits them into a powerful irony, as reading it comes to stand in for a more substantive ethical reckoning with the meaning of military service and the costs to one's subjectivity that it might entail.

Throughout, *Thank You for Your Service* queries the limits of gratitude, revealing how soldiers struggle and fail to reconcile their wartime actions, how bleak their employment prospects are upon return, and how burdensome the process of recovery can be.[140] Focusing on the intimate details of their lives—Wool describes soldiers in general as inhabiting a "kind of life publicly renowned but practically unknown"[141]—Finkel provides a counterpoint to images of heroic soldiers and seamless honeymoon homecomings. Such grim realism is not unprecedented; films in the aftermath of World War II, for example, invoked the character of the maladjusted veteran as a threat to "postwar harmony." In those stories, as

Beth Linker and Whitney Laemmli observe, "women appeared to shoulder the burden—practical, emotional, sexual—of reintegration," an echo of women's advice literature from the period, which encouraged them to study up on veterans' issues.[142] The women in *Thank You for Your Service* do this work as well, but their caretaking is often tinged with resentment, malice, or fear.

At the outset, Finkel provides a mathematical view of the problem, multiplying the prevalence of PTSD and TBI by the number of military personnel who have deployed to Iraq and Afghanistan and estimating that the overall toll of the wars could amount to "some five hundred thousand mentally wounded American veterans." But the book itself operates on a much smaller scale, tracking the melodramas of PTSD and TBI as they unfold mostly on domestic stages.[143] Throughout, Finkel adopts the voice of an omniscient narrator. He shifts between the inner monologues of the men, their spouses, their providers, and another voice whose vantage is both knowing and detached, as when, after a key character accidentally drops his infant son, he writes that the father "is sorry. He is always sorry now."[144]

Their relationships disintegrate as the men become more and more dependent on the women in their lives. In this way, the book also dramatizes a tension that disabled veterans must continually negotiate of maintaining their independence while making rightful claims on the systems designed to aid in their recovery.[145] In the book, these systems are marked primarily by their failure. The bureaucracies that the men navigate are the artifacts of a century-long history of rationalization in medical approaches to war injury that begins in World War I. As they established protocols for care, early-twentieth-century clinicians proceeded with confidence that war injuries could be known and understood and in turn approached their duties with a mass-casualty framework, assuming that injuries would be similar across the ranks rather than manifesting idiosyncratically in individual soldiers.[146] Yet in *Thank You for Your Service*, injury appears as mysterious, intensely personal, and largely incurable. It falls primarily to the women to compensate for health care systems that are overburdened, health care providers who are flummoxed by the conditions that veterans present, and their partners' noncompliance with treatment regimens.

Indeed, in the years after World War II, marriage appeared as a common solution in popular culture to the problem of how to reintegrate and remasculinize disabled veterans.[147] Subsequent depictions, however, departed from these fairy-tale visions of tireless and patient caregiving women whose ministrations magically restore husbands physically and emotionally. Instead, contemporary works by and about veterans, as David Kieran notes, often reprise Vietnam-era discourses that utilize veterans' "failures as partners and parents as evidence of the wars' psychological impact."[148] Along these lines, *Thank You for Your Service* narrates their debilitating incapacity for marriage, fatherhood, and partnership as a sign of a larger betrayal by the state and the military, a defalcation on the promise of heteronormativity so central to the experiences of children's and spouse's militarization that I described in previous chapters. In the book, home becomes a space of violence and danger for wives, children, and the men themselves.[149] These men drop their babies, trash their apartments, beat their partners, get divorced, profess to be "grateful" that their daughters do not have to be subject to their tempers, and attempt suicide in their children's bedrooms.[150] As Gallagher notes, military personnel seeking recognition and treatment for conditions like PTSD must often overcome a "burden of proof" to verify that they need and deserve care. Often, in order to garner proof that is satisfactory and persuasive to the institutions that determine who has access to medical care, the servicemember must engage in destructive behaviors like domestic violence, substance abuse, and suicide attempts.[151] *Thank You for Your Service* traffics in the same standards of legibility, so that combat trauma becomes legible as such primarily through the infliction of violence in the home.[152]

Although the book focuses on the men, the women in their lives serve as screens upon which their injuries become visible in the forms of suffering borne and suffering caused. Uniformly, the women are instructed by various professionals, to "give it time"; Finkel reveals that for many of them, this is an untenable solution.[153] The book chronicles a micropolitics of anger, both of the veterans themselves and of their partners. While her husband goes to outpatient treatment, Theresa remains stranded in their apartment with the "gouged walls, and the punched-in door."[154] Sascha, contemplating a visit to see her partner at his inpatient facility, wonders, "Could she tell him that soldiers aren't the only people who have nightmares? Was he ready to hear that?"[155] Kristy keeps a secret note file on her phone where she documents all of her husband's abusive behaviors.[156] Saskia, whose husband dropped the baby, cannot quite calibrate her feelings toward him. At first, "her patience, she had decided, would be bottomless." But this commitment falls apart over time; some days she sees him as "really hurting," but other days she wants to tell him to be a man and "'Get your ass up.'"[157] She rages at him in choppy, insistent text messages when he is gone, infuriatingly, for four months of treatment in a luxe California facility.[158] In a brief stab at independence she gets a job as a caseworker for a mental health agency, which she loves until she cannot do it any longer, and quits as abruptly as she began because she realizes that "the one who needs help is her."[159] Eventually, according to Finkel's narration, she pesters her husband so much that he leaves inpatient treatment against medical advice and comes home.[160] Throughout, Finkel leverages the narrative of the victimized spouse to explore the irreparability of these emotional, cognitive, and physiological injuries, which corrupt the affection that the women might otherwise have had.

Despite all its rich detail, Finkel's account leaves no room for the elaboration of a political subjectivity. The men mostly appear as blindingly enraged: at themselves, at their wives, at the veterans' health care system. And the women are basically reactive, sometimes nobly self-sacrificing, sometimes petty, sometimes retaliatory, sometimes terrified, sometimes just muddling through.[161] This division of affective labor in the text masculinizes the categories of PTSD and TBI, although it is estimated that roughly 20 percent of reported TBI cases are women.[162] It also reduces all the characters essentially to their feelings. This is a richer view by far than the charities that put their faith in the curative powers of beef jerky, but by emphasizing the affective costs of PTSD and TBI, the book invites an essentially, if not exclusively, affective response. PTSD becomes the beginning and the end of the story, the only explanation or cause of anything negative that happens; for example, the arresting officer in a domestic violence case quietly encourages the perpetrating husband to tell the judge that he has been deployed multiple times and now has PTSD.[163] The book ends with Saskia and her baby-dropping husband trying to reconcile yet again, and leaves us, too, where we started, with a fuller sentimental experience of PTSD but not much beyond that.

The visibility that *Thank You for Your Service* affords is primarily emotional. PTSD is the operative force that hovers in every turn of the narrative. Although the book aspires to illuminate its consequences, the condition itself becomes something of a black box, knowable only through its devastation of young men who, in turn, become capable of little more than acting out toward women. The book, perhaps because it ventriloquizes the thoughts of its disaffected protagonists, evinces skepticism about any available treatment options for PTSD. Health care providers and support personnel are well-intentioned but overworked and naïve. Military bureaucracy is labyrinthine and unresponsive. Treatment programs are untested, and recovery is tortuous. Military leadership tries, conscientiously but too late, to improvise responses to veteran suicides. Given all of this, there is not much left for readers to do but feel, and feel bad.

Dust to Dust: Making TBI Visible

Finkel's narrative is one way of making PTSD and TBI visible; on the pages of *Thank You for Your Service*, these injuries look like helplessness and rage and disintegrating relationships and barely tolerable existences. But TBI can look like other things, too, and under a microscope, the scarring caused by a blast-related TBI looks like dust.[164] This dust-like pattern is not the TBI itself, but rather the widespread and multiple lesions that occur in the healing process. Nonetheless, this is a signature manifestation of this signature injury. The discovery of this distinctive dust-like pattern is one of the most provocative findings to emerge from the Center for Neuroscience and Regenerative Medicine's (CNRM) study of donated brain tissue. The CNRM is one of the many centers of the Uniformed Services University of the Health Sciences, which is ultimately overseen and mostly funded by the Department of Defense.

Since 2012, the CNRM has maintained a Brain Tissue Repository that collects donated brain tissue from veterans and active duty personnel of all military branches and civilians with TBI. The repository archives the tissue, preparing it for research use by scientists at the CNRM and at other locations around the world. The CNRM accepts donations from veterans who served at any time, whether or not they had known blast exposure or a TBI diagnosis.[165] Blast exposure, it should be noted, is a unique situation, and part of what distinguishes TBI in military personnel from that which civilians might experience after an event like a car accident or a fall.[166] The CNRM's work depends on the willingness of families to donate the brains of their deceased loved ones, which inaugurates new questions of sacrifice, gratitude, gifting, and militarization.

Dr. Daniel Perl, an internationally recognized researcher in neuropathology, is the leader of the repository at the time of this writing and its public face in most media accounts.[167] A study that Dr. Perl and coauthors published in the *Lancet Neurology* about changes in the brain after blast exposure included some of the most conclusive visual evidence yet that TBI and the PTSD that can accompany it entail a distinctive and identifiable pattern of scarring. The authors begin with this background:

> Although conventional neuroimaging for mild TBI typically shows no brain abnormalities, military personnel have reported persistent post-concussive symptoms, such as headache, sleep disturbance, concentration impairment, memory problems, depression, and anxiety,

> suggesting structural damage not detectable with routine imaging techniques. With symptoms but no biomarkers, these TBIs became colloquially termed invisible wounds.[168]

Their post-mortem research, however, revealed a "distinctive, consistent, and unique pattern" of scarring in the brain as well as insights about where such scarring typically appears—namely, at the interface of tissues with differing densities. Perl and his coauthors conclude their article with a note of optimism that future clinicians "will find ways to make these injuries not only visible, but also treatable for service members, veterans, and civilians."[169]

Shortly thereafter, *National Geographic* published an article about the findings, suggesting that they might finally have "solved" the long-standing "mystery" of shell shock by identifying a characteristic marker for blast injury.[170] In a quote published in the article, Perl notes that although the research does not yet reveal anything "obvious in terms of treatment," it will nonetheless "mean reevaluating people we've labeled as having PTSD." He elaborates that because of this new discovery, clinicians "should not think about approaching [PTSD] as a purely mental health problem." In other words, Perl recommends that clinicians consider potential biological components of PTSD diagnoses for individuals with a history of blast exposure.[171] The author of the *National Geographic* article also drew another conclusion and posited that new findings about TBI might provoke a "philosophical" dilemma for people contemplating military enlistment: "If you know that exposure to a blast event—the signature mechanism of injury in modern warfare—may well irreparably damage your brain, will you still join up?"[172] This query is somewhat puzzling in that the decision to enlist always entails a calculation about risk to one's body or one's life.[173] By conjuring a scenario of physical survival accompanied by mental scarring (both figurative and literal), the author intimates that the likelihood of blast exposure and TBI necessitates a different kind of deliberation. In this way, the author also hints that there might be a limit to the presumed generosity of military personnel and intimates further that the cost of living with brain damage is greater than the "ultimate" sacrifice of one's life that enlistees might be more freely prepared to make. As framed in the article, the matter of whether potential recruits would willingly risk this portion of their subjectivity in trade for the affordances of a military career parses whether these benefits, which include public esteem, would be enough to justify the inherent risks.

This kind of groundbreaking TBI research is deeply contingent on the absolute generosity of military personnel and their families. Most contemporary organ donation arrangements are predicated on a "gift relationship."[174] To this, the repository's work is no exception, as it depends on a very particular and dramatic generosity: the willingness of families to give what it describes as the "courageous gift" of a brain tissue donation to "future soldiers and their families." As with posthumous organ donation for transplant, the deceased veteran's next-of-kin has the ultimate authority to decide whether the donation of brain tissue will happen. Initially, the CNRM could not approach families to ask for donations and so relied on them to volunteer the bodies of their loved ones. As of the time of this writing, they are now permitted to ask for donations as well as enroll living servicemembers and veterans as future brain donors. The organization appreciates every donation they receive and recognizes that, given the magnitude of the research yet to be done on military TBI, they will still need many more donations.[175] Families' willingness to make these contributions is essential to the work of identifying the physical imprint of the previously amorphous and invisible entity of "trauma"; their generosity, a final offering of the servicemember's body, is the condition of possibility for these discoveries.

When I visited the CNRM, I inquired about what motivated families to donate brain tissue. Largely, I learned, it was the desire to help other people. Guided by this overarching aspiration, donating families do differ in their personal hopes and wishes for the process. Some understand the donation as a way to find closure in their loss and so do not want any follow-up contact with the center. Others donate in the hope that it might unlock a mystery about their loved one's life or death, though the center makes no guarantee that this information will be discovered or shared. The lab does, however, work to keep donor families and other interested members of the public informed about their research and regularly updates the "News" section of its website with links to stories about their efforts. The CNRM employees that I met expressed a deep consideration for the profoundness of the "gift" of brain tissue and accordingly follow strict protocols for honoring it.[176] In practice, the CNRM endeavors to honor these gifts by employing them in the pursuit of scientific discovery.

The donor occupies a curious place in this process, simultaneously absent (as a living subject) and essentially present (as the provider of an essential sample). Although servicemembers can make their wishes known to their families while they are still alive, they do not have final authority over the decision, as they will be dead when it is made. Once CNRM receives the donation, the tissue is thoroughly deidentified to protect the confidentiality of the donor and sent for processing in accordance with its laboratory protocols. Over the course of many months, the donated tissue will be cut and recut, washed, fixed, stained, and set on slides to facilitate a thorough evaluation of each case. At the CNRM lab, this work relies both on automation and the skill of technicians, who shepherd the tissue from machine to machine with the larger goal of maximizing the quality of the slides generated. By the end of this cycle, a single donation could be divided into hundreds of carefully cataloged specimens just a few microns thick. The slides are prepared so that they can be utilized in current studies or preserved for future research use.[177] At a technical level, the CNRM follows standard laboratory protocols for research work with donated human tissue, but they also describe their overall approach as unique in its thoroughness. I argue that these procedures also constitute another mode of figuring the veteran with TBI. Trauma is decontextualized out of the donor's individual history, social environment, and individual history and experience so that its somatic imprint might be identified and studied more precisely.

The CNRM's primary objective is to aid in the development of treatment strategies to better rehabilitate TBI patients and reintegrate them into their interpersonal and community networks. The repository's vision encapsulates its project as "Caring for America's Veterans, So No One Stands Alone." This promise appears at the top of its home page, which is replete with images of family. These visions of military domesticity are very much the opposite of those featured in *Thank You for Your Service*, with its narrative emphasis on familial dysfunction caused by the strains of injury and caregiving. The repository's red-and-white logo includes the silhouettes of two adults and a child holding hands. The adults are featureless but for subtle iconic cues that mark them as a couple. One person is slightly taller, implying maleness, while the shorter adult's body displays a slight curvature at the midriff suggesting femaleness. The male holds the hand of a gender-nonspecific child, while the arm of the female disappears next to her male companion, indeterminately positioned such that she could either have her arm draped protectively across his shoulder (intimating a new caregiving role) or laced through his arm (suggesting a more traditional gender dynamic).

Of course, this is only my interpretation. When I submitted this portion of the chapter for review and vetting by the CNRM staff, they emphasized that the logo is designed to reflect their mission, which is to "seek innovative, rational approaches for rehabilitation, therapies and prevention on long term effects of brain injury."[178] From their perspective, the logo represents family. The image is intended to speak, in part, to those who have recently lost a loved one, who might be inspired to authorize donation of their brain, and hence to support the research that might ultimately yield the insights and rehabilitation strategies necessary to secure a future of familial connection for someone else.

Toward this end, the three-image slide show that fills the page's banner features stock photos of military families that make TBI visible in a different, more hopeful, way. In one scene of this triptych, there is a smiling, uniformed woman embracing a girl in front of an American flag. In another, two parents in uniform flank a young girl who holds a small American flag beneath a blue sky, above the caption describing brain tissue donation as a "courageous gift." Sandwiched between these scenes of loving, joyful familial interaction is an affective outlier: a woman mournfully resting her head on the shoulder of a marine in uniform, her hand on his cheek and his eyes downcast, his facial expression somber but unreadable, an ineffable air of injury about him. Healing veterans here becomes a kind of repair that spirals outward to heal families and relationships, while undiagnosed or untreated TBI becomes the main obstacle to their flourishing; other military families are hailed to give a gift that will enable researchers to surmount it. Here, healing follows belatedly and obliquely from an explosive and otherwise irreversible undoing.

Unmasking TBI

Even as TBI gradually becomes more visible to select researchers, the public still generally lacks a clear representation of veterans with the condition. There are vague clusters of associations with certain phenomena, like explosions and suicide. There are resonances with the more familiar figure of the "angry vet" or the somewhat more romanticized version of the mournful, regretful one. But beyond awareness of certain symptoms, like speech impairments or memory loss, this subjectivity is difficult to imagine in part because it is a hybrid of the person before and after the injury, some things familiar, some things not.

Before the landmark discovery of astroglial scarring, another DoD initiative invited veterans with TBI to undertake their own efforts to make their injuries visible. As part of an art therapy project at Walter Reed's National Intrepid Center of Excellence, veterans with TBI created masks to give their conditions visible and tactile shape. Although participation in the program was voluntary and many of their masks reveal painstaking and thoughtful construction, the institutional context for the therapy complicates the matters of their agency and consent. The mask making was intended to give veterans a way to express how it feels to have TBI, but they are not the only ones who benefit from it. Subsequently featured in a curated series of photographs by *National Geographic*, the masks served a pedagogical function as well, explaining TBI to an audience that might otherwise be uncomprehending.[179]

With few exceptions, like the mask that playfully outfits its wearer as a mustached Viking or the one that includes the word "support" along with a cross and a bright yellow sun, the masks are unrelentingly difficult, both aesthetically and affectively. Some portray their wearers as menacing, with horns, jagged teeth, and naked skulls. Others radiate pain, with features like pins and nails driven into the head, a vise labeled "PTSD TBI," and anguished expressions to encapsulate the physical sensations of TBI. Some are political, painted in various combinations of red, white, and blue; one has "Allahu Akbar" written across a forehead in shaky cursive above a bullet hole dripping blood down the nose. Gashes, stitches, bloodshot eyes, viscera. Many are bifurcated, the injury confined to one side of the mask but tightly adjacent to the other, which looks placidly on, unharmed.

The intensity of these masks, as reflected in the attention to detail and vivid coloration, suggests that their creators felt an urgent need to explain something. Many feature depictions of brains themselves, yet none of these organs appear to be harmed: undamaged whorls of pink or gray rest above eyes, cheeks, noses, mouths, and ears maimed by combat. The visual motif of the seemingly healthy brains coheres with the idea of "invisible wounds." But all of the masks that show brains also document the gruesome processes required to uncover them, and the surrounding skulls are shattered into pieces, peeled back, cut away, or cracked open. This is an evocative reminder of the costs of exposure. They record the violence necessary to make something so intimate knowable to outsiders, the excruciating labors that recognition demands. Yet they remain masks, decorative pieces designed to hide the true face residing below them. By revealing and obscuring their subjects simultaneously, the masks also enact the paradox at the core of this figuring.

CHAPTER 5

Liberal Imaginaries of Guantánamo

In the waning moments of a 2006 interview with Donald Rumsfeld, Larry King posed a three-word question to the then secretary of defense: "Future of Guantánamo?" Rumsfeld began by framing the detention center as an "interesting problem." He cited its "bad reputation," averred that its conditions are humane, and then continued:

> We don't want to be jailer for the world. We don't. We would prefer not to have anyone. We'd like to have all of these people go back to the countries they came from and be dealt with there. Unfortunately, some of the countries we aren't allowed to give them to, because we worry that they would not be treated in a humane manner.[1]

There are good reasons to quibble with this portrayal of the United States as a reluctant power, exceptional even in its breaches, and we cannot know whether Rumsfeld was being sincere or disingenuous in his appraisal. But when I was at Guantánamo, I had an overwhelming impression of its tenuousness, reflected in its improvised facilities and practices, which lent some credence to the notion that this outcome was not part of the original plan. Nearly everyone I spoke to confirmed my sense of the place, echoing common refrains about the expense and inefficiency of the operation, the barely sustainable labor-intensiveness of the processes for confining and guarding the detainees, and the practical reality of being hamstrung by governmental refusal to either close the facility or adequately fund it so that conditions could be improved.

The origins for this chapter reside in an anecdote—of admittedly unknown veracity—that a Defense Department administrator told as we were flying back to Virginia from Guantánamo. Rumsfeld, the story goes, genuinely weary of being the "world's jailer," gave a quiet order to a group of DoD officials to generate a list of detainees who might be summarily released. Short on time and guidance, they decided to start with the oldest and began interviewing the men about what they would do if they were, hypothetically, freed. The administrator told me that they expected to hear about reunions with families and resumption of interrupted livelihoods. Instead, they were surprised when one answered that he would go home, fashion a bomb, and then detonate it in a place where it would kill as many American military personnel as possible. DoD officials scrapped the plan to release anyone on the list. Of course, this story may well be apocryphal, and I presume that the narrator had his own motivations for sharing it.

True or not, however, the story struck me because it was the first time in years of reading, thinking, and writing about Guantánamo that I had heard, albeit in a profoundly mediated way, the angry voice of a detainee. Authorized images and sound bites have been trickling out for over a decade, always under the auspices of state and military authority. But these are not the only institutions that mediate detainees' voices: so, too, do the individuals and organizations that work on their behalf. As part of their efforts to figure the detainees as recognizable and worthy objects of sentiment, this mediation filters out most traces of their political subjectivities, which are essentially unknowable to outsiders, and might even be unthinkable to them.[2] Perhaps more nakedly than in any other instance I describe in this book, the figuring of detainees plays out a well-intentioned form of objectification whereby, in Elizabeth A. Povinelli's terms, a "gap seems to open between those who reflect on and evaluate ethical substance and those who are this ethical substance."[3] In what follows, I attend to the mechanisms that fashion detainees as this "ethical substance," matter to be used as the foundation for larger political projects.

I begin with an overview of the mechanics of pity and anger in the figuring of detainees and cover the practices of imagining the enemy in warfare generally and of imagining detainees in particular. That history focuses primarily on governmental or state imaginaries of detainees, but my main concern in this chapter is the operative imaginaries of advocates for the detainees. Accordingly, I also provide a history of anti-Guantánamo activism. Much of this activism, I demonstrate, relies on an erasure of detainee political subjectivity and, more specifically, a refusal of the possibility of detainee anger, and so I offer a more sustained consideration of the politics of detainee anger before turning to my case studies. The first of these is a set of three city-council resolutions from Massachusetts and California that extended hypothetical welcomes to select detainees on the condition of their hypothetically being released. Reading the resolutions themselves, as well as the discourse surrounding them, I query the politics of this deeply conditional hospitality and, moreover, the presumptions of American exceptionalism underpinning it. American exceptionalism is central to the analysis of my next object, the Witness to Guantánamo documentary project, which collects testimonies from former detainees and a range of others connected to the facility. WTG does crucial documentary work, but in my analysis, I also attend to the ways that its structure compels the detainees to offer forgiveness and recuperative visions of American goodness. Next, I consider the politics of detainee creative production in poetry, memoir, and visual art, with a particular attention to practices of circulation and consumption and the fictive experiences of intimacy that they promise their audiences. The chapter ends with a brief reflection on a fanciful renarration of Guantánamo's past and future.

Given the unbridgeable distance at which detainees are kept from outsiders, the question of their political subjectivity remains unanswerable, even as it lurks everywhere in this chapter. I do not purport to know some special hidden "truth" about the detainees. Rather, I am interested in others' claims to have found such a thing and the means by which they circulate, support, and illustrate those claims. Here, I analyze these fabricated experiences of recognition and explore how various entities, all presumably well-intentioned, have looked for something that cannot ever appear. If we, as non-detained outsiders, are opposed to indefinite detention, it is not sufficient simply to recognize the detainees' humanity or feel compassion for their circumstances. Instead, I argue that the more urgent task is to explore how the wish to recognize the detainees instead begets patterns of non- and misrecognition that occlude detainee political subjectivity even further. In this context, the suffering of detainees is made visible only to the extent that it can be subsumed into a vision of palliative American benevolence. The objects I analyze here intimate that there is no harm perpetrated at Guantánamo that cannot be redeemed by the compassion of concerned citizens.

Constructing the Pitiable Detainee

The other five chapters of this book begin with genealogies spanning decades or even centuries of affective and imaginative sedimentation around various figures. But the figure of the detainee has a much shorter history. In her theorization of the practice of enforced disappearance, a cognate practice of indefinite detention, Banu Bargu notes that some expressions of sovereign violence "have their own historicity" rather than emerging predictably from previous formations.[4] Indeed, the category of "enemy combatant," as the U.S. government currently operationalizes it, is scarcely as old as the War on Terror itself. Although the intricacies of the detainees' legal status, their circumstances at Guantánamo, and the systems of mediation through which they appear are not entirely unprecedented, they are also peculiar.

Warfare necessarily entails imagining the enemy. Camouflage is based on a guess about how enemies perceive their environments. Psychological Operations actions leverage hunches about the enemy's beliefs and fears. Counterintelligence practice tries to anticipate enemy curiosity. Joseph Masco documents how speculative approaches to warcraft became ascendant in U.S. military strategy during the Cold War with near-obsessive efforts to "model, game, intuit, and assess" the enemy mind.[5] Contemporary counterinsurgency and counterterrorism have similar ambitions, but are also rooted in epistemological uncertainty about what the enemy knows and how the enemy thinks, insecurities that are made more pronounced by doubts about whether enemy combatants are rational actors at all.[6] Michael Barkun argues that policy makers typically do not understand terrorists as rational actors, but rather as inscrutable deviants governed by a range of dysfunctions. Consequently, as Liz Philipose notes, studies of terrorism "routinely derive unobservable motivations, belief systems, compulsions, psychopathology, self-destructive urges, disturbed emotions, and problems with authority from observable behavior."[7] These are all attempts to make terrorist interiority knowable and comprehensible within available schemas for interpreting political behavior.

In the prevailing state logic of the War on Terror, "detainee" is a functional synonym for "terrorist," and many of the same assumptions are deployed in attempts to analyze them. Fundamentally, the presumption that all detainees are terrorists or would-be terrorists serves to justify the way they are treated. But this approach also offers a roundabout acknowledgment of the possibility of detainee political subjectivity.[8] This is evidenced, for example, in an open secret among the people who work at

Guantánamo, who freely comment that if the detainees did not hate America before they were detained, they certainly—and logically—do now. Of course, this confession self-servingly perpetuates the logic of detention, but it also accounts for the transformative violence of detention in a way that reparative anti-detention discourses do not. Those discourses, which underpin most anti-Guantánamo activism, never entertain the possibility that detainees could harbor antipathy toward the United States or its citizens and presume instead that these men recognize and even appreciate America's exceptionalism.

The state does not have a monopoly on imagining detainees. While the U.S. government busies itself representing the detainees as ruthless and unrepentant, detainees' advocates traffic largely in visions of them as hapless and pitiable. These figurations, which promise to make the detainees accessible, knowable, and visible, appeal to American sympathies and a besmirched American identity and present detainees primarily in terms of their captivity. Toward that end, recently published writings by detainees themselves offer a counterpoint to assertions of their dangerousness.[9] Yet these, too, are heavily mediated by the military authorities that censor them and by the interlocutors that publish them. Overall, the claims made on behalf of the detainees often hinge on them appearing apolitical, thus allowing concerned Americans to act politically in their stead. As in the preceding chapters, pity here is contingent on innocence and a sense that someone has been victimized. Innocence, in the case of the detainees, is understood not just as the lack of wrongdoing on their parts but also, crucially, as the absence of agency.

Evacuating detainee agency enables an affective swap; because the detainees are constructed as victims, the possibility of their anger is foreclosed, allowing their advocates the pleasure of righteous, vicarious anger on their behalf. Of all the unknowable things detainees might feel, I foreground anger because this emotion poses the greatest liability for critics of U.S. detention practices. It points to something irremediable, dangerous, otherwise uncontainable, and to a problem for which profusions of sympathy by undetained outsiders are no solution. Anger lies at the core of detainee unimaginability; it is comforting, by comparison, to imagine them feeling frightened, homesick, or defeated. Sunaina Marr Maira's insights about popular distinctions between "good" and "bad" Muslims are apposite here. These perceptions, she argues, determine how civil rights are apportioned.[10] Advocating for a detainee imagined as blameless, passive, and forbearing is easier than advocating for one who actively struggles against his circumstances. Povinelli describes "red lines" of liberal recognition, perimeters that tolerance will not cross.[11] Detainee anger constitutes one such line. After all, anger implies an object, and if I consider another person's anger, I must wonder whether I might be its source or its target. And so, in anti-detention activism, detainee anger is generally absented, defanged, or radically reframed to be more palatable to American audiences, even as American indignation is a key element of anti-detention protest.

The replacement of detainee anger allows the anger of sympathetic outsiders to coexist easily with their pity. No doubt, as the American public was getting accustomed to the notion of Guantánamo early in the War on Terror, the Bush administration worked to manage popular sentiment about detention, cultivating fear of the detainees and ambivalence about their suffering.[12] But the state and military are not the only actors invested in adjusting the affective landscape around detention, and advocates for detainees seek to cultivate connections between them and a dubious American public. They do so in the hope that affective investments by outsiders might ultimately improve conditions for the detainees, but I want to query the illusion that such affective connections are possible at all.

Fictions of Transparency: Anti-Guantánamo Activism

The first War on Terror detainees arrived at Guantánamo in January 2002. In the intervening years, activist strategies protesting the facility have evolved. The earliest humanitarian and legal actions focused on increasing oversight of detention and providing services to the detainees, but these interventions were the sole purview of the professional specialists who have access to the facility. On the other hand, anyone who has the inclination can participate in cultural activism against Guantánamo. Such activism encompasses a range of practices from rhetorical opposition to public demonstrations and protests.

Perhaps the most visible form of such activism is the act of donning the iconic orange jumpsuit in public for marches, sit-ins, and performance artworks. These tableaux evoke the earliest detainee photographs released by the Department of Defense, but also risk replicating their dehumanizing visibility.[13] These images construct detainees as helpless and featureless but for the accoutrements of detention, and outfitting oneself like a detainee is a sartorial claim of solidarity that seeks, by a transitive magic, to compel recognition for their plight. Emphasizing the durational occupation of public or governmental space, these protests underscore detainee passivity as they condemn the affront to American ideals that confinement entails. They also aim to endow detention with a visible and recognizable form, reducing an array of processes and attendant harms and injuries to a single neon symbol.

What, exactly, does an outsider embody when costumed in an orange jumpsuit? After all, the instant identifiability of this outfit reveals that people do not need to be made aware of detention. Furthermore, there is no guarantee that additional recognition would cause any change in detainees' circumstances. Lauren Berlant cautions that "compassionate recognition," while valuable for particular ends, often "become[s] an experiential end in itself."[14] A similar dynamic is operative here. By rendering actual detainees hypervisible and invisible at once, these spectacles put the righteous anger of the protester on full display.[15] Clinging to this bright orange symbol fixes the image of detainees as perpetually victimized and presumes that their subject-positions can be imagined and adopted as easily as slipping into an oversized jumpsuit. In her canonical *States of Injury*, Wendy Brown queries the limits of anti-oppression discourses rooted in the oppressions of their subjects, asking, "What kind of attachments to unfreedom can be discerned in contemporary political formations ostensibly concerned with emancipation? What kinds of injuries enacted by late modern democracies are recapitulated in the very oppositional projects of its subjects?"[16] Applied to the practice of protesting on behalf of the detainees, these questions uncover how such acts depend on detainees' unfreedom as a condition for their visibility in the public sphere.

Troping of the orange jumpsuit continues alongside newer forms of activist engagement enabled by the increased transparency of operations at Guantánamo. As official restrictions on the visibility of the detention center and its inhabitants have relaxed, activist platforms have expanded to include the voices of detainees. This marginal increase in transparency has made it easier to believe that we know, or that we can know, them.[17] Along these lines, Rebecca Wanzo's consideration of political illegibility is an important counterpoint to uncritical celebrations of increased visibility. She writes, "To be politically illegible as a sufferer is to have one's story visible but obscured by historical and cultural debris, [so that] the intended audience cannot read or interpret it in a way that leads to true comprehension of the cause of suffering."[18] Thus, mere access to detainees' stories does not guarantee comprehension of them, but the *belief* that it does drives the clamor for more.

The fact of my own visit to Guantánamo is only one anecdotal confirmation that Guantánamo has become more transparent. As early as 2011, Elspeth Van Veeren noted that images of the detention center were "readily available and widely circulated," often by the U.S. government itself, eager to promote the impression that its detention practices were humane.[19] Indeed, alongside photos of the U.S. military personnel who live and work there and the facility itself, the Joint Task Force Guantanamo website offers an extensive photo gallery that includes sections on two of the camps.[20] This visibility reflects a broader tendency of counterinsurgencies to document their captives, which Susan L. Carruthers describes as an "insistent desire to make enmity apparent to the naked eye."[21] The image of a menacing enemy can serve as persuasive propaganda, particularly when accompanied by a subsequent photo of that enemy killed or subdued; hence the photographic preponderance of detainees seemingly at leisure in the official archive of Guantánamo.

Laudable in theory, transparency has costs. Claire Birchall, for example, proposes that increased transparency in some arenas might "displace certain conversations, ideals, and practices to a 'shadow field.'"[22] I understand the relative opacity of U.S. detention operations at Bagram as such a phenomenon. Moreover, the increase in transparency is not an unequivocal good for the detainees.[23] For a government at pains to prove that its detention practices are "safe, humane, legal, and transparent"—a compulsion that only intensified during the Obama administration—the work of demonstrating this falls to the detainees themselves.[24] Accordingly, they appear across the official photographic archive betraying no evidence of maltreatment or discontentment. Photographed without their consent in images that circulate beyond their control, the detainees are made visible on the state's terms. The state retaliates when the detainees endeavor to set their own; for example, following the publication of his *Guantánamo Diary*, Mohamedou Ould Slahi was stripped of his "comfort items," apparently in an effort to force him to comply with interrogators.[25]

Neither does transparency remedy the thoroughgoing disappearance wrought by indefinite detention. Bargu writes that political disappearance works physically and ontologically against the disappeared. She writes, "To the violability, torture, and destruction of [the insurgent's] body, we must now add in *erasability* from existence, as the ultimate practical proof (and fantasy) of power."[26] If disappearance makes this deletion possible, transparency is not a meaningful countermeasure, in part because it relies on a notion that our awareness and recognition are potent enough to interfere with the mechanisms of indefinite detention. In addition to understating the harms of detention and downplaying its systematic embeddedness, this faith in recognition endows Americans with exceptional powers.

Unimaginably Angry

The outward appearance of detainee anger is regulated both by the government that oversees the detention and actors that seek to contest it. For the non-detained outsider, there is the question of how to look for that anger and how to know it when we see it. The context of the War on Terror conditions American audiences to recognize Muslim anger manifested as violence, but outside of sensationalist news reports about "Death to America" brands of jihadism, representational frameworks are quite limited. In an epistemological query about the experience of seeing otherness, Kelly Oliver wonders, "How is it possible to recognize the unfamiliar and disruptive? If it is unfamiliar, how can we perceive it or know it or recognize it?"[27] More specifically, reflecting on the relative invisibility of insurgents in media representations of the War on Terror, Carruthers asks, "What combination of the practical, tactical, and political serves to make enmity so curiously disembodied?"[28] More than anger is at issue here, though anger is most glaringly absent; the blank space of detainee anger stands as a reminder of all that we cannot know about detainee political subjectivities.[29]

Practically, for people who are invested in cultivating sympathy for the detainees, their anger is a liability, particularly because Muslim men are often shadowed by suspicions of dangerousness and volatility in general. Anger is not the only thing that's missing. Media representations never, as Carruthers notes, portray insurgents as "sentient individuals," and reportage on detainees systematically avoids stories that might, in Anjali Nath's terms, "illuminate their personhood."[30] There is an epistemological violence latent in the refusal to acknowledge that detainees might have unknowable subjectivities, which is redoubled in the subsequent effort to imagine that we can know them nonetheless. Writers like Talal Asad and Ghassan Hage have described the inaccessible interiorities of so-called terrorists in methodological and philosophical terms.[31] In what follows, I explore how mediation, imagination, and affective investment expand these epistemological gaps.[32]

The unbridgeable distance between detainees and non-detained outsiders is built into the detention apparatus itself, which whipsaws detainees between miserable intimacy with the guards (this can be deeply unpleasant for them as well, as they become the immediate objects of detainee anger and protest) and total abstraction before the world outside. Reflecting on the story of the would-be bomber that I heard on my tour, it occurs to me that with the possible exception of their lawyers and humanitarian workers, the people who know the detainees and their anger best are those employed in the daily business of detention. These people also have the power to violently exploit their knowledge of detainee interiority. For example, interrogators leveraged Abu Zubaydah's purported entomophobia in an effort to coerce his compliance.[33] More generally, however, as the U.S. military has gradually pivoted away from highly technologized forms of fighting precipitated by the late twentieth-century Revolution in Military Affairs, it has turned toward what Laleh Khalili describes as an insidious form of "'emotionally intelligent'" warfare.[34] This shift reveals that affective sensitivity and militarization are not incompatible; indeed, affective recognition itself can readily be militarized.

Non-detained outsiders may never know what kinds of subject-formation are possible in the context of indefinite detention. Disappearance, as Bargu argues, is not only abduction; it is also the erasure of an existence; how might the disappeared person reconstruct a subjectivity after such deletion?[35] And if subjectivity is relational, how do practices like solitary confinement, a common condition at Guantánamo, impinge upon it? Adam Ewing argues that in solitary, a practice increasingly directed toward Muslims both inside Guantánamo and out, inmates "experience the horror of their own mental degeneration. They become invisible to themselves."[36] What allows us, then, to believe that they might be visible to us?

To be clear, my assertion about the potential unknowability of detainees is not an Orientalist claim about their inscrutability or a replication of the arguments from pro-detention quarters that they are wily or untrustworthy. It is not a reprise on prevailing media representations engineered to sustain the mystery around insurgents.[37] Those depictions aspire to silence insurgents; detention finishes the job. Yet outsiders' efforts to endow them with a voice creates different forms of muteness, so that their circumscribed visibility comes at the cost of political subjectivities left unacknowledged or invisible.[38] Moreover, the project of making detainee voices heard by American audiences either presumes that they would be intelligible or forces them into a preexisting grid of intelligibility. Ultimately, the question is how we might respond ethically to detention in the absence of such intelligibility. The answer is not to pretend unintelligibility away.

"We Believe the Men Will Be Grateful to the Community": Conditional Hospitality for Cleared Detainees

In November 2009, voters in Amherst, Massachusetts, passed a resolution to officially welcome "cleared" Guantánamo detainees who might be unable to repatriate to their home countries. Nearby Leverett passed an identical resolution in April 2010. And after a failed attempt in February 2011, in October of that year, the City Council of Berkeley, California, subsequently approved a more elaborate piece of legislation patterned on the Massachusetts models.[39] Proponents of these measures presented them as remedies for the wrongs of the War on Terror and actions that reflected each place's history of welcoming people in need of refuge. Given Congressional bans on the resettlement of cleared detainees in the United States and restrictions on government funding to pay for such relocations, the measures are essentially symbolic and nonbinding, but they are important nonetheless. Taking for granted that the detainees would *want* to relocate to the United States and to these places specifically, the resolutions ventriloquize the position of an apolitical detainee who bears no ill will toward the United States and has such unshakeable affection for the country that he would readily make it his adopted home. Emphasizing the impossibility of the detainees settling anywhere else, the ordinances are oddly resonant with Rumsfeld's assessment of the "interesting problem" posed by Guantánamo.

Reductive imaginings of the detainees, along with recuperative fantasies of American goodness and exceptionalism, animate these initiatives. These proposals channel citizens' anger over wrongs perpetrated in their names and pity for the detainees separated from families and livelihoods into a vexed and ultimately conditional hospitality. Citing the significant numbers of men who have been cleared of any wrongdoing but remain detained for lack of an alternative place for them to go, the resolutions apply to cleared detainees in general, but accompanying documents identify specific, exemplary detainees as objects of their welcome. Pioneer Valley No More Guantánamos (PVNMG), which authored the Amherst/Leverett resolution, names Ahmed Belbacha and Ravil Minzagov and includes brief biographies of each.[40] PVNMG describes Belbacha as an asylee in the United Kingdom who fled Algeria after receiving death threats and was sold for a bounty to U.S. forces while on vacation in Pakistan. Minzagov, a one-time dancer in the Russian Army's ballet corps, converted to Islam and relocated first to Afghanistan and then to Pakistan in search of religious freedom. Minzagov was arrested during a raid, driven by suspicion that one of the inhabitants knew Abu Zubaydah, on a home for Muslim refugees. He was turned over to U.S. forces, but subsequently cleared. The Berkeley supporting documents, prepared by the city's Peace and Justice Committee (PJC), also select Minzagov but pair him with Djamel Ameziane.[41] The documents do not explain the circumstances of Ameziane's capture but include a press release from the Center for Constitutional Rights deeming him an "ideal candidate for resettlement," citing his fluency in English and French, his college education, and his loves of cooking and soccer.[42] Similarly, Minzagov's biography describes him as a dutiful soldier and victim of religious persecution so fearful of being returned to Russia that he "fabricated" stories about his connections to known terrorists in the hopes that U.S. forces would take him to the relative safety of Guantánamo.[43] These biographical notes imply that the United States would be a perfect environment for these men to pursue the good lives that had otherwise been inaccessible to them.[44]

Preceded by intensive lobbying efforts and publicity campaigns, the Amherst resolution passed on the first try, but not without controversy. A local news story from Amherst noted that the issue had "divided the town."[45] The issue, apparently, teased out ideological complexities in some residents' politics, so that one avowed conservative supported it reluctantly on the grounds that such hospitality is what the American "flag represents," while some liberal voters questioned the wisdom of the plan. Things apparently proceeded much more smoothly in Leverett, and most of their news reports described the voters approving the measure "overwhelmingly" in their annual town meeting.

By comparison, the path to victory in Berkeley was more circuitous. Instead of the more directly democratic town meeting system in place in the much smaller Massachusetts communities, Berkeley is governed by a city council. Additionally, the Berkeley resolution is more elaborate and reflects sustained and serious logistical consideration from city council members. Originally scuttled by the bare fact of the Congressional ban, the revised measure references this obstacle but then sails over it. Instead, it includes lists of the community organizations that might assist resettled detainees and resources of which the men might avail themselves. In short, it displays the loving, meticulous intensity of a favorite make-believe.

Before its eventual passage, the bill encountered both ideological and practical opposition. A Sacramento-based group called Move America Forward (MAF) spearheaded early, but apparently inconsequential, opposition to the resolution. In an effort to remedy the anti-military bias it perceives in the American news media, MAF hosts patriotic rallies, provides care packages to deployed servicemembers, and coordinates "Gold Star Parent Trips" to Iraq.[46] Speaking against the first version of the resolution, MAF's communication director Danny Gonzales cited the danger that it posed to the community.[47] Rather less ingenuously, he also volunteered MAF funds to fly the city council to Guantánamo instead.[48] In the end, the failure of the first resolution was largely procedural, by a vote of 4–1, with 4 abstentions.

One opposing councilman protested that the resolution was a distraction from the business of governing the city, but the city manager's disapproval was the most persuasive. Citing the total federal prohibition of and "bipartisan opposition" to resettling detainees in the United States, he recommended that the council take no action on the matter.[49]

No doubt, the PJC knew about these practical hurdles. It even referenced them directly in its supporting documents but proceeded with its resolution nevertheless. In this way, they staged a symbolic rebuff to federal law. In its effort to usurp federal authority to resettle detainees, this move also reflects a desire that Elisabeth R. Anker identifies as a key facet of post–September 11 citizenship: an identificatory wish to act as or in place of the state. She writes, "Melodramatic political subjects envision a heroic mastery over the world performed by state action and yet also experienced by the individual. Their legitimation of state power is thus a consequence of their identification with it, in which the power symbolized by the state becomes an internalized ideal of sovereign agency."[50] In this way, the PJC's rebellious gesture against state power actually serves as a reenactment of it.

Nothing changed at the federal level between February and October, but something apparently did in Berkeley. The February and October versions of the resolution do not differ substantially. But unlike in the winter, the autumn motion moved forward. For the October meeting, in fact, the mayor added it to the council's "consent calendar," an administrative mechanism for matters that are "routine, non-controversial, easily explained and can be expected to receive Council approval without specific need for discussion."[51] During the February meeting, the resolution did not generate much attention; the matter of the detainees arose at 11:00 p.m., thirty minutes before the meeting's adjournment, with a great deal of business conducted after this item. October deliberations were also efficient. Once the council agreed on the contents of the consent calendar, it was approved quickly, and only one councilman asked that his objection be recorded. In her comments, Rita Maran of the PJC answered a few perfunctory questions about the organizations who might assist in resettlement, and another member thanked the council for the feedback on the earlier draft. All totaled, the matter occupied approximately five minutes of the meeting.[52]

Like its predecessors in Amherst and Leverett, the Berkeley resolution hails the city's history of welcoming people in exile, as emblematized in its 1971 declaration as a "City of Refuge," and notes that "private support" has been offered to the hypothetically resettled detainees. There is scarcely any consideration, even in the sixteen-single-spaced-page packet that accompanied the first resolution, of whether or not the detainees might want to land in Berkeley. During her verbal comments, Maran noted, almost as an aside, that detainees would have to agree to move to Berkeley and would be selected through a process of interviews and referrals, but no one ever entertains the possibility that they might refuse. Similarly, in Amherst, one member of the Select Board mused that they might be overstepping, reasoning that if the federal government could not decide what to do with the detainees, then Amherst might be ill-advised to wade in. Another voter said that, in light of all that the detainees have "gone through," routing them directly to Amherst rather than attending to their "physical and mental health" needs did not seem "logical."[53] Even these concessions, however, do not directly address the question of detainee agency, volition, or preference. After all, some cleared detainees have refused the terms of their imparted resettlement plans, holding out for different destinations as the Chinese Uighurs did, or deciding, for different, perhaps unknowable reasons, to stay at the facility.[54]

Who wouldn't want to live in America? In Berkeley or Amherst, no less? These questions, tacit but incredulous, underpin the resolutions' lack of consideration for detainee preference. Even when Maran concedes that detainees would have to decide for themselves whether Berkeley was the right choice, she does not wonder aloud whether they would want to come to the United States in the first place. Relatedly, Jasbir K. Puar notes that the ACLU's materials on the post–September 11 deportations of South Asian immigrants play on the emotional details of their domestic lives to "convey the desirability of U.S. residence." In this way, she writes, "There is recognition of difference, yet it is subsumed under a homogeneous image of the detainee as a typical immigrant escaping his or her politically inhospitable, culturally backward, uninhabitable, and economically deprived country of origin."[55] Thus, the melodramatic political style that Anker describes enables the seamless coexistence of idealism about the United States with outraged denunciation of its violence. The proponents of these resolutions enact a form of citizenship that derives its goodness by "suffering with others" who have also been afflicted by "sanctioned villainy."[56] The fact that the villain in this case is the United States itself complicates their position, but apparently not irredeemably.[57]

Advocates of the resolutions freely admit that they, too, derive benefits from this extension of hospitality. The leader of the Leverett campaign, for example, insisted that the voters' approval of the measure "will go a long way toward healing the wounds inflicted on prisoners and our collective conscience by U.S. torture policy and illegal indefinite detention."[58]

This may be a laudable objective, but it makes no discernible difference in the detainees' circumstances. Moreover, as Timothy Kaufman-Osborn argues against the notion that all Americans are responsible for the torture at Abu Ghraib, "such invocations of collective identity may be read as performative utterances that help call into being and constitute the foundational entity which they then claim to designate."[59] Thus, the claim of collective responsibility for Guantánamo creates a loophole through which the state can fortify its assertion that it undertakes indefinite detention on behalf of its citizens, who aspire to remake the state more humanely.[60]

Even doubts about the plan were allayed by recourse to American exceptionalism. Supporting documents circulated in Berkeley included talking points as answers to detractors' questions.[61] One such inquiry, written as if it were posed by a person dubious about this plan, asks, "Why bring them to Berkeley? Will they be happy here?" The reply begins, "We hope so," and makes reference to Berkeley's history as a place of refuge and its collective willingness to resist stereotypes, but indicates that the former detainees will have to decide for themselves whether they want to settle there and for how long. The notion of their happiness warrants further scrutiny. Khalili analyzes how "happiness" has become both a tactic and goal of counterinsurgent warfare. The happiness of an insurgent population is, she writes, "ultimately a projection of a fantasy." In this counterinsurgent imaginary, the failure of an insurgent population to be happy is read as "disaffection" that is "ultimately an act of sedition against the sovereign imperial force."[62] Thus, unhappiness in Berkeley would demonstrate a kind of political and affective backwardness, while happiness would be proof that all had been forgiven.

Another fictional skeptic challenges, "How can you be sure they won't be anti-American and angry?" The suggested response to this plausible, if heavily loaded, question is an assurance that "other released detainees have indicated that they do not blame the American people for what our government and the U.S. military has done to them. We believe the men will be grateful to the community for the opportunity to leave Guantánamo and to live freely and rebuild their lives." The first statement may be technically true, but because any such utterance would have been mediated, we cannot know what the men truly felt and believed. The second sentence, however, is entirely speculative. It imbues gratitude with the power to wash anger away. And as in Leverett, where the voters' support for the nonbinding resolution applicable to a hypothetical circumstance was a balm for their consciences, this assertion rewards the citizens of Berkeley in advance for their willingness to offer a welcome they would probably never be called upon to extend.[63]

Indeed, all of this would-be hospitality came to naught. Belbacha was released to Algeria in February 2014.[64] Minzagov was transferred to the United Arab Emirates in January 2017.[65] And the Center for Constitutional Rights reports that Ameziane was "forcibly transferred" to Algeria in 2013, "despite his fears of persecution."[66] These outcomes prevented any of the cities from having to reckon with the implications of their resolutions and the affective investments that animated them. Thus, their imaginaries of both the detainees and the exceptional American welcome they would have received remain perfectly intact and untroubled.

Uses Limited Only by the Imagination: Archiving Guantánamo

While nearly everything about the resettlement resolutions, including the detainees' voices, is imaginary, the Witness to Guantánamo (WTG) documentary project captures the actual words of actual former detainees. Undeniably, WTG offers them a unique forum to recount their experiences of detention. However, I argue that the project also creates an opportunity to affirm American goodness between the lines of those recollections, expressly distinguishing Guantánamo from the nation-state whose military oversees it. WTG is a testimonial project, and as of August 2017, it has interviewed 156 people from 20 countries, all of whom are connected to Guantánamo in some way. Peter Jan Honigsberg, a law professor at the University of San Francisco, runs the project, which operates with archival support from the University of Southern California's Shoah Foundation Institute for Visual History and Education. Describing itself as the "world's most comprehensive collection of Guantanamo stories," the goal of WTG is to "educate the public and mobilize pressure to hold U.S. government officials and private actors accountable."[67] Explicitly based on the "Shoah" model that utilized the testimonies of Holocaust survivors to counter the assertions of Holocaust-deniers (and citing Steven Spielberg's influence), WTG endeavors to "provide voice to former detainees and other witnesses so they can share their personal and poignant stories." Given this, once all of the materials are translated if or as needed, transcribed, and posted online, WTG anticipates that the "diverse potential uses of the archive will be limited only by the imagination," a claim predicated on the assumption that all of these potential uses will advance opposition to Guantánamo or indefinite detention more generally.

Reliant on a testimonial model that presumes a link between spectators' knowledge of suffering and their desire to act in opposition to it, WTG's orientation is legalistic, recording violations of human rights and international law. Their work proceeds according to a logic of transparency that endows the acts of viewing and reflecting on the testimonies with political significance. In this way, WTG participates in the type of spectatorial humanitarian display that "blur[s]," in Lilie Chouliaraki's terms, "the boundaries between watching and acting" and emphasizes self-expression in response to suffering.[68] For example, the WTG site features "Student Reactions" to its lesson plans, which take the form of two poems written from the "perspective of a detainee."[69] These poems evince their authors' keen awareness of the violence and humiliation of Guantánamo. However, this focus also illumines the limitations of such an approach, which renders detention lamentable only to the extent that it is viscerally imaginable to outsiders.

Currently, WTG archives around the edges of detention. Although WTG's coup is its collection of testimonies from former detainees, accounts from non-detainees far outnumber these. The site features roughly ninety-five such interviews with relatives of detainees, translators, attorneys, interrogators, guards, human rights activists, and government officials. As of summer 2017, the site listed fifty "Detainees" and nine "Detainee Family" members among its interviewees.[70] For each detainee, the site includes a freeze-frame from their interview, their name, nationality, and a brief biographical caption. Only about a quarter of the former detainees have videos available. When recordings are not yet available, WTG invariably promises a "clip coming soon"; when they are available, they are excerpted vignettes, one to five minutes in length. Some feature voiceover translations, while other former detainees give their interviews in English.

These videos are as revelatory as they are opaque. They provide, on the one hand, rare personal accounts of the experience of detention. But they are, on the other, intensely mediated: framed, edited, excerpted, with no clear explanation of the logic by which they were selected. The project offers an important corrective to the absence of detainee perspectives from discussions of Guantánamo, but its other consequences are more ambiguous and include a certain fetishization of the detainee's voice.[71] Eliciting individual stories, WTG ushers its interviewees into a liberal register of comprehensibility. Asma Abbas queries the value of the "voice" itself in liberal regimes; against the prevailing notion that "the mere unleashings of voice [are forms of] 'resistance' or . . . 'counterhegemonic,'" she "finds the representational, expressive imperative to be thoroughly hegemonic itself" because it facilitates entry into liberal economies of sentiment.[72] Oliver frames the conundrum differently, arguing that the "experience of testifying to one's oppression repeats . . . objectification even while it restores subjectivity."[73] These former detainees appear and disappear as subjects simultaneously, as the documentary apparatus makes them both personable and consumable.

Sharing their stories, the ex-detainees become Guantánamo pedagogues. One man, Haj Boudella, characterized his experience as empowering in a quote featured on the website:

> [My] participation in your project is motivated by—let us not allow anybody to commit that kind of deed in the future. . . . It is a kind of contribution of [mine] to the world's culture, the world's civilization. . . . Our religion says that if the judgment day comes, and you have in your hand a tree you want to plant, then plant it. Because even the judgment day is not a reason to stop working the good deeds.[74]

I cannot know the complex decision-making processes that led the detainees to participate; here again, I confront my epistemological limits. But the power differentials inherent in the documentary process compound a burden common to racial, ethnic, and religious minorities, whereby they are compelled to teach the dominant culture about their beliefs and customs. For example, Puar describes how after September 11, Sikhs were encouraged, often by their community leaders, to "patiently educate" nervous airport security screeners about their turbans.[75] Here, the portrayal of the detainee as an enthusiastic participant conforms to WTG's palliative assertion that they harbor no ill will toward the United States or its citizens, to whom all of the detainees' comments are implicitly addressed.

In their interviews, government officials and military personnel provide the larger narrative context for the detainees' experience; in turn, the detainees offer personalized stories about the harms of detention. These accounts document the micropolitics of Guantánamo, but rarely scale up to a criticism of U.S. policy. One of the few detainees to reference geopolitics is Abubakir Qasem, a Uighur, who says that he was only mistreated during a 2002 visit from the Chinese delegation, when the U.S. government turned over information about Uighur detainees' families to the Chinese government. Qasem ruefully notes that this betrayal undermined his belief in the virtue and integrity of the United States and so reframes this international political event as individual grief. Of course, WTG's goal is to capture personal experiences of detention, but it does so at the cost of decontextualizing its violence, recruiting former detainees to be its apolitical chroniclers. Relying on the palatability and of the "good" and rights-worthy Muslim that Maira describes, WTG offers viewers the gratification of sympathizing with a real (former) detainee. Humanizing the detainees in this way is an expediency mandated by prevailing representations of them as adherents to inhuman ideologies.[76] But terms of those representations are such that the only viable alternative is the apparent absence of any ideology at all.

WTG largely avoids controversial acts of detainee resistance like hunger strikes. As signifiers, such acts are supersaturated with meaning, and many parties clamor to ascribe their preferred interpretations onto them. Official governmental explanations of detainee suicides, for example, cited them as evidence of the detainees' lack of regard for human life or deemed them, preposterously, acts of terrorism or asymmetrical warfare targeting the United States.[77] On the other hand, detainees' lawyers invoke bodily harm, whether self-injury or punishment from the guards, as irrefutable evidence of "guard brutality and prisoner desperation."[78] But beyond the practical consequences of these acts, Lauren Wilcox argues that "by harming their own bodies," the detainees "attempt to exercise power over meaning." She continues that, in trying to martyr themselves, they deny the presence of the sovereign and assert their own sovereignty over their bodies, even as this effort to "enact subjectivity comes at the cost of the very materiality of their bodies."[79] Asad makes a compatible claim about perceptions of suicide bombing as uniquely horrifying because they embody "the limitless pursuit of freedom, the illusion of an uncoerced interiority that can withstand the force of institutional disciplines."[80] One of the few WTG references to hunger comes in the testimony of Ruhal Ahmed, who describes being overwhelmingly hungry on the rendition flight to Guantánamo, his mouth watering for the peanut butter sandwich that he could smell but could not eat because he was masked, shackled, and hooded. By leaving drastic forms of resistance generally unremarked, WTG keeps its detainee subjects intelligible and safely resident within the pale of the governable.

In these interviews, former detainees describe resistance in terms of solidarity with one another more than deliberate opposition to U.S. authorities. They recall encouraging each other to be strong or to engage in acts that are nonviolent. For example, Khaled Ben Mustafa explained that detainees do not have "many options to show that we're angry." He then recounted, with a faint smile, how the detainees would collectively refuse to return the plates from their meals after eating, a peaceful act of defiance that tired and exasperated the guards. Another detainee, Sami al-Hajj, makes a passing mention of refusing to break a hunger strike and then finding out he was being released, a dramatic turn of events that quickly overshadows his account of striking. And in a second, longer, clip, he reflects on all the good Americans he met while detained, acquaintances who proved that "not all the people in the U.S. are bad."[81] Abdurahman Khadr recalls attending an Al Qaeda training camp as an adolescent, wryly describing himself as such a troublemaker that Osama bin Laden's bodyguards had to discipline him, but emphasizes that he does not believe in suicide bombing. The only detainee who explicitly describes self-harm is Moazzam Begg, who speaks of banging his head on the wall for want of stimulation and psychiatric help while he was in isolation. Importantly, this act, which Begg himself describes as evidence that he had "lost [his] mind," is portrayed not as a willful display of resistance but rather as a sign of desperation and an evacuation of the self.[82]

Conversely, stories of intentional brutality by the guards figure prominently in the archive. Mosa Zemmouri (who is something of an outlier in the collection; more about that later) describes someone being force-fed, but as a secondhand account. It is not clear from his testimony to whom he is referring, and the details overall are sketchy. Other accounts are much more explicit in their descriptions of maltreatment.[83] Murat Kurnaz describes seeing a man strung up and left hanging, dead, for twenty-four hours.[84] Bisher al Rawi recounts a story of guards taunting a detainee afflicted with kidney problems by withholding his blanket.[85] Ben Mustafa describes guards responding with firehoses to the peaceful refusal to return plates. Khalil Mamut recalls having the wrong tooth—whether through negligence or deliberate malfeasance is unclear—pulled by a Guantánamo dentist.[86] Brahim Yadel tells of undergoing surgery only partially anesthetized so he could be interrogated during the procedure.[87] Doubtless, these accounts are crucial documents of detainee experiences. However, even as the videos perform this function, the details of these specific cases aggregate into an impression of the detainees as passive victims as opposed to agentic political subjects and hence ideal repositories for sentiment.

WTG offers an ideological and affective offset for these discomfiting stories, featuring personal accounts from detainees who view their captivity philosophically and their captors generously. Feroz Ali Abbas explains that although his experience of Guantánamo stripped him of any naïveté about the world and taught him that life is suffering, he also knows that "truth will prevail" and that his trials in Guantánamo were decreed by Allah in divine, if humanly unfathomable, "wisdom and mercy."[88] Al-Hajj reasons that all things have positive and negative dimensions and, because he is a journalist, wagers that there was something lucky about ending up at Guantánamo and having firsthand access to this otherwise restricted story. Ben Mustafa portrays Guantánamo in surprisingly gentle terms, describing peaceful mornings, food that improved over time, and the opportunities to learn about other cultures; the "whole world," he muses, "is in Guantánamo."

Relatedly, some clips offer a recuperative vision of the United States. Qasem characterizes himself as disillusioned as he explains that many Uighurs idolized the United States and its freedoms, a roundabout compliment. But the last words attributed to al-Hajj, in which he acknowledged that not all Americans are "bad," affirm his renewed faith in American goodness. And Abdurahman Khadr emphasizes the need for intercultural understanding, saying that he would like to show Al Qaeda members videos from people who lost loved ones in the World Trade Center and videos of everyday life in the Middle East to Americans, so that these groups might come to understand one another. Abdul Rahim Janko, after asserting that not all Muslims are terrorists, hints that he might want to visit the United States someday, saying that if he does, it will be "as a friend," because he "love[s] the American people."[89]

Thus it falls to the former detainees to recuperate American identity. Donald E. Pease theorizes contemporary American exceptionalism as the "fantasy that permitted U.S. citizens to achieve their national identity through the disavowal of U.S. imperialism."[90] Pease wagers that in the years after the September 11 attacks, U.S. citizens essentially traded their civil liberties for the opportunity to consume spectacles of enemy dehumanization.[91] The WTG testimonies permit Americans to trade them back while being educated about this dehumanization within a larger narrative that reassures them about American virtues. Hugh Gusterson notes that "insurgents play a key role in securing American identity" as its foil.[92] Insurgents do that symbolic work by appearing as irredeemably evil. In WTG, however, the former detainees, who are licensed to speak because they have been processed by the American legal system, serve this purpose because they are its truest, most intimate witnesses, who nonetheless insist that it can be repaired.

Zemmouri deviates most sharply from this pattern. He is not the only detainee to cast aspersions on the United States; Al Rawi, for example, notes that U.S. interrogators are smarter than their "Third World" counterparts and know better than to leave visible injuries. But he does not look, speak, or reconcile like the other interviewees. He addresses the cam-

Both visually and narratively, Mousa Zemmouri is an outlier in the *Witness to Guantánamo* documentary project. Compared to the other former detainees interviewed on the site, Zemmouri never shows his face to the camera, and in his speech, he is much more pessimistic and far less willing to exonerate the United States for its detention practices.

era in English, but somewhat haltingly, and the interviewer interjects more frequently than in any other conversation. In response to Zemmouri's description of force-feeding, the interviewer asks why the detainee in question was "so angry." Zemmouri begins his answer by enumerating some of the violence and privation he witnessed and then scoffs, "What do you think?" The clip ends abruptly. In the next segment, Zemmouri describes being held in isolation and then taunts his interlocutor: "You should try it." Finally, he conjures a dark global conspiracy of "worshipers" of America who collude to deprive former detainees and others of their rights; he depicts the United States as bloodthirsty and always desperate for an enemy. The United States chose Muslims, he reasons, because it had already exterminated the Native Americans. The interviewer leaves this unchallenged, but Zemmouri's appearance and comportment undermine the credibility of his claims. Visually, Zemmouri is an outlier; like a handful of other detainees, he appears on camera with his face obscured, but he also seems to be wearing a garment with a pointed hood. His silhouette is vaguely menacing, alien, and also dehumanizingly reminiscent of the hooded detainee made iconic in the photos from Abu Ghraib.[93] Formless, faceless, and irate, Zemmouri enacts a political subjectivity that does not court recognition from his American audience, which makes his words easier to disregard.

Alternatively, Al Rawi offers an account that is much easier to hear. Speaking neutrally, a partial smile intimating a tepid hopefulness, he makes no accusation and talks around the violence that undid him. Narrating his halting process toward reassembling himself in an excerpt entitled, "Getting Back to Normal," Al Rawi says that it is taking a long time. The interviewer prompts him encouragingly, saying, "But you're working on it, as you said." Al Rawi replies:

> I definitely. . . . I mean it's funny we meet today. But really, really a few things happened the last two or three weeks, and very strange. And nothing . . . it's not like acting for this interview. But about a . . . you know a short time ago, really days, weeks, a couple of weeks, I started looking at myself and I'm thinking, Well, maybe I am closer to where I want to see myself than I think I am. Like I'm feeling, I'm closer to normal, I'm getting there. I could actually sense it. I wrote a letter to somebody and I wanted to tell them that. I skipped it. But I was telling them, like—they're going through a hard time—and I said, "Look it's been over four years and I'm just now, I am feeling, I can sense that I am getting normal. I can actually sense it." Before I could sense definitely I'm not normal. I could . . . and that's when I told my wife this on many, many occasions. I said this, "I am not normal." But now I can sort of, very slightly, I can sort of dare say, "I am getting there."[94]

As he struggles to measure and articulate his recovery, Al Rawi reveals the destructive violence of detention and the difficulty of returning fully to oneself. His optimism is compelling, and I too wish that he would "get there." Yet in the context of the WTG archive and the still-larger one of embattled and belligerent American exceptionalism, his progress toward normal is readily recast as evidence that we might all, with enough perseverance, simply put Guantánamo behind us.

Art Therapy for Indefinite Detention

Even cleared detainees are barred reentry to the United States, and even initiatives that allow them to address American publics modulate their appearances. Cultural productions by current detainees circulate further than their creators ever could and so provide their audiences with a tantalizing, but ultimately fictive experience of intimacy with the artists. Heavily promoted as windows on detainee interiority, these creative works are typically displayed with minimal explanation from the detainees themselves, a silence that leaves our own interpretations unchallenged. W. J. T. Mitchell, in a meditation on the vexed status of evidence in the Combatant Status Review Tribunals that determine the fate of Guantánamo detainees, describes the sensory regime in which "the apocalyptic indifference of readability and unreadability installs revealability without revelation."[95] That is, detainees are fundamentally available for exposure that still leaves them unknown and unknowable.

In addition to whatever forms of self-censorship the detainees apply, their artistic productions are always filtered by other institutions—military, governmental, journalistic, editorial—before they reach their audiences. Although the genre is radically different, Amira Jarmakani's analysis of the popularity of romance novels about sheikhs during the War on Terror is instructive here. Jarmakani offers a range of explanations for the counterintuitive spike in interest in these stories after September 11, including an abiding imperialist desire for the Other and the novels' fetishizing of security. But Jarmakani also notes that these novels uniformly portray the sheikh as relatively progressive and inclined toward Western visions of modernity so that the stories become lusty apologias for liberalism.[96] Because detainee creative work always appears with such heavy mediation and such minimal interpretive guidance from the creators themselves, it is readily appropriated to perform similar functions.

Published detainee poems are essentially the remainder of what Elisabeth Weber describes as a "massive censorship" operation by the U.S. government to regulate information about detainees.[97] Moreover, she notes that to read the poems as we do is to "read them torn out of their cultural and especially their linguistic and poetological context." She argues further that we encounter the poems rewritten in the "language of the jailor," because they were translated by people whose credentials were adequate security clearance rather than translation expertise in Arabic poetry.[98] Released with tremendous fanfare by the University of Iowa Press, *Poems from Guantánamo* includes twenty-two such poems written by seventeen detainees, each profiled in a brief biography preceding their work. An extensive editorial apparatus scaffolds the project, which includes essays about the collection, its significance, and the genre of Muslim prison poetry. Editorial contributions reiterate the labor-intensiveness of the book's creation, and the acknowledgments thank "hundreds" of "volunteer lawyers, professors, paralegals, law students, and human rights activists" and many improvised translators. The editor also references the tremendous effort expended by the detainees to compose, circulate, and protect the poems prior to their publication, which were hidden, smuggled, etched on Styrofoam cups, and often confiscated, censored, or destroyed despite their authors' efforts to preserve them.

The anthology was widely and divergently reviewed; some critics heralded it unreservedly, while others questioned its aesthetic merits.[99] Despite these disagreements, no one doubted the significance of the collection itself. Perhaps the most definitive academic response to the poems came from Judith Butler, who identifies them as containers and conveyances of affect, writing that they "interrogate the kinds of utterance possible at the limits of grief, humiliation, longing, and rage."[100] This characterization implies a responsive interlocutor whose affective and political task is made explicit in the volume's closing essay by Ariel Dorfman: "Think that perhaps someday, perhaps soon, if we care enough, if we are troubled enough, it will not be just the verses that are set free to roam the world but the hands and lips and lungs that composed them."[101] But how "troubled" is enough? And, moreover, by what mechanism does our being "troubled" amount to "enough" for the detainees?

Despite the heavy editorial hand operative throughout the book, the detainees' role in the process is never fully explained. Although the poems were apparently published with their consent, it is not clear whether they imagined an audience beyond Guantánamo as they were initially writing.[102] Weber observes that many of the poems rely on motifs of tears and being unable to uphold the Quranic responsibility to "car[e] for the elderly, widows, and orphans." These devices, she wagers, "may make it possible for the poems to reach English-speaking readers in spite of the obstacles of structural censorship."[103] But the passive-voice construction of her argument obscures the question of authorial intent; did they intend to reach English-speaking audiences by making these universal appeals, or was that simply a fortunate by-product? From a different vantage, Neel Ahuja critiques the tendency to frame the poems as expressive of a universal human voice, arguing that such a move "elides geographic and political specificities underlying these complex writings."[104] The matter of authorial intent is confused within the volume's supporting essays as well, which claim both that the detainees wrote poetry in the hopes that outsiders might read it and that they wrote it in the despair of isolation, composing only for themselves and God.[105] In the end, the matter is settled by the editor's assertion that the poems were published so that they might "prick the conscience of a nation," and the urgency of this pricking apparently overrides any other consideration.[106]

Preoccupation with the national conscience may explain the popularity of one poem in particular, Juham al Dossari's "Death Poem," which is excerpted frequently in discussions of the collection and reprinted on the inside flap of its hardcover jacket. "Death Poem" is a sustained apostrophe that can be read as addressing the wider world. Al Dossari implores his readers as "people of conscience," allowing or inviting us to imagine ourselves as such. The poem begins, "Take my blood / Take my death shroud and / The remnants of my body. / Take photographs of my corpse at the grave, lonely. / Send them to the world, / To the judges and / To the people of conscience. . . ." In addition to the affirmative hailing of the audience, it also affords them, through a radical gesture of self-abnegation, the ultimate authority over his body, precisely the sort of responsibility that many outsiders have already claimed.

Compared to the poems, Slahi's expressly autobiographical *Guantánamo Diary* presents a more straightforward interpretive task. The printed book runs to nearly four hundred pages. Slahi originally composed the manuscript longhand during the summer and fall of 2005, and at first the U.S. government classified the entire draft. Subsequently redacted for publication, the book retains blocks of blacked-out text, sometimes a few words (like the names of interrogators), a few sentences, or many consecutive pages. It also includes reproductions of Slahi's handwritten manuscript, covered by impenetrable black stripes.[107] Even with these details unavailable, the book provides a clear picture of Slahi's experiences in the custody of the U.S. military and its secret allies, including multiple renditions, interrogations, and torture.[108]

Larry Siems, a writer and human rights activist, edited the *Diary*. In the prefatory front matter, he makes explicit that Slahi himself was not involved either in the government's redaction of the diary or in his editing of it, which he describes as "relatively minimal."[109] Subsequently, in the introduction, he admits that "other than sending [Slahi] a letter introducing myself when I was asked if I would help bring his manuscript to print—a letter I do not know if he received—I have not communicated with him in any way."[110] As of the *Diary*'s publication, his closest encounter with Slahi was a brief sighting on a Pentagon video feed while he watched Slahi's Periodic Review Board hearing. Siems asserts that Slahi "explicitly authorized publication" of the diary and lauds the narrative content ("both damning and redeeming") and the beauty, irony, and humor of Slahi's writerly aesthetic.[111] He notes that Slahi wrote in English, his fourth language, and demonstrates a lexical variety comparable to that of a Homeric epic.[112]

In the text, Slahi's voice comes through as conversational and frank in vivid accounts of the conditions of detention and his mental, bodily, and emotional responses to them, tracking the gradual deterioration of his psychological and physical health. He tracks the situations of other detainees when possible and offers perceptive, often sympathetic, sketches of his guards and interrogators. Relaying his knowledge of their backgrounds, politics, religious beliefs, and feelings about their work, Slahi archives the unexpected intimacies begotten by these confined circumstances. Yet the visible redactions stand as stark reminders of the distance between Guantánamo and the outside world and ways that state power both forces and corrupts that intimacy.

Thus, for all Slahi's candor, to read his diary as an unfiltered confession might underestimate him. Given his deep knowledge of the mechanics of detention, even if Slahi wrote with an eye toward publication, he almost certainly would have anticipated that his work would be censored prior to release, in much the same way that the detainees I saw when I was at Guantánamo would almost surely have known that they could be observed by people behind the one-way mirror. To read him as entirely guileless is to presume, comfortingly, that he would have absolute faith in his readers and no political sensibility of his situation. The challenge then is to give Slahi credit for being smart enough to intuit the layers of mediation that would be inserted between him and his readers without becoming so skeptical as to dismiss his account. To clarify, I am not suggesting that Slahi was disingenuous or deceitful; rather, I am observing that the apparatuses of detention refract truth so profoundly that any kind of certainty is impossible. This distortion undermines even the earnestness of the "author's note" at the end, which is written in the third person and avers that Slahi asked his lawyers to tell his readers that he does not hold a "grudge" against anyone described in the book and looks forward to someday sharing a cup of tea with them under different circumstances. Depending on the reader's own beliefs about U.S. detention practices, this might appear either as a remarkable gesture of forgiveness or a calculated and cynical feint for clemency. At a minimum, however, this ambiguity is a reminder of the epistemological chasm that detention opens up.

In October 2016, Slahi was released from Guantánamo back to his native country, Mauritania. To mark the event, the American Civil Liberties Union released a statement that quoted Slahi as follows: "'I feel grateful and indebted to the people who have stood by me. I have come to learn that goodness is transnational, transcultural, and trans-ethnic. I'm thrilled to reunite with my family.'"[113] We cannot know whether or not this is a full or faithful summary of Slahi's feelings about detention. But in the larger context I have described in this chapter, Slahi's words read easily like a benediction to Americans perturbed about Guantánamo, a hint that they can be part of this larger "goodness" that Slahi describes. No doubt, this is an encouraging thought, and my critique here is not aimed at Slahi. Rather, I highlight the ACLU's repetition of his comments to caution against reading them as verification of an essential and ineluctable American goodness. The ACLU points out that Slahi was one of two "Special Projects," detainees whose abuse was directly approved by Donald Rumsfeld, which perversely endows Slahi's words with the status of special evidence for American benevolence should one be inclined to look for such a thing.

Unlike Slahi and the poets, who wrote clandestinely, detainees who produce visual art do so under the supervision and auspices of the command at Guantánamo. For a variety of reasons—including international law mandating that prisoners of war have access to enrichment and educational activities and a faith in the powers of art therapy to calm restive detainees—JTF-GTMO offers art classes to detainees that it deems compliant. Indeed, when I saw detainees at Guantánamo, they were just finishing their art class. Beyond the legal mandate to provide such opportunities, the art classes reflect Khalili's contention that liberal counterinsurgency operations are defined by a "will to 'improve' and reeducate detainees."[114] Given this coercion, we cannot assume that the detainees' artwork reflects their "authentic" selves or visions. Although the pretense of art therapy lends an air of truthfulness to the pieces, they stand in complex relation (or nonrelation) to the subjectivities and interiorities of the men who created them.

Art has played an important role in liberal efforts to refute "clash of civilizations" rhetoric in the aftermath of the September 11 attacks. Jessica Winegar observes that producers of Islamic art are forced into a "humanity game," wherein they are recognized as human precisely to the extent that Western elites recognize their creative works as art. She notes that after September 11, art was often held up as a "bridge of understanding," a way to reconcile the liberal dilemmas provoked by the attacks. Good intentions notwithstanding, the countless exhibitions of Islamic art were underpinned by "universalist assumptions" about the meaning of both process and product: "that art is a uniquely valuable and uncompromised agent of cross-cultural understanding" and "constitutes the supreme evidence of a people's humanity, thereby bringing us all together."[115]

Presumably, some detainees derive a benefit from painting or drawing. But they are not the only ones. The public circulation of their work amounts to a command performance, by proxy, whereby the detainees demonstrate that they are being treated well and are in reasonably good spirits. "Comfort items"—those objects like toiletries, Korans, and flip-flop sandals afforded to detainees—ostensibly to make their detention more tolerable, serve a similar function. Yet the very notion of "comfort items" echoes hauntingly in light of Marita Sturken's argument about the primacy of comfort in American popular culture during the War on Terror and her contention that making Americans comfortable with extralegal practices is a "primary mode through which the U.S. practice of torture is mediated."[116] Detainee artwork, which circulates far beyond the reach of its creators, can readily become a comfort item for American spectators made uneasy by the notion of Guantánamo, particularly because they are so much easier on the eyes than other visual documents of detention, like the photographs of shackled detainees.

The paintings and drawings were originally displayed for internal viewing only at the Guantánamo library, but the command there lifted its photography ban on detainee artwork prior to the military commission of Omar Khadr.[117] With that prohibition removed, BBC World publicized them first, and they have since appeared in a range of other outlets including *Slate*, *Huffington Post*, and *Business Insider*.[118] None of these venues offer substantive explanations of the art, however, instead presenting it as the manifestation of an amorphous, essential, and apolitical creative impulse. This impression is heightened by the seeming tranquility of the images, which appear devoid of any animus. The circumstances of their creation matter; they were produced under the supervision and at the behest of Guantánamo's staff, and the commonality in subject matter across this archive suggests that the detainees received instruction on what to draw. Of course, we cannot assume that context fully determines their meaning; it is possible, as Andreja Zevnik theorizes, that creative practice in detention is a mark of the kind of ecstatic "excess" that escapes sovereign control, a small practice of freedom.[119] But this expression of freedom is short-circuited when the art is reproduced in spaces beyond the artists' control. The apparent neutrality of the visual art makes it possible to imagine its creators as calm or quiescent, even as the ostensible lack of a message invites us to look and imagine more deeply on the hunch that surely there must be *something* unsaid or encoded here. The news outlets that recirculated the artworks directly invite such speculation with their vagueness. Indeed, the paintings and drawings appear without any attribution to a named artist, devoid of any biographical details that might obstruct viewers' imaginative identification with them.

The pieces evince varying degrees of training and aptitude but uniformly bear the traces of artistic effort, displaying attention to detail, texture, light, shadow, and shape. Perhaps in accordance with a Koranic prohibition against the depiction of living beings, no such creatures appear in the landscapes both natural and human-made, seascapes, and still lifes that comprise this corpus.[120] Some of the pieces are quite lovely; all are tantalizingly opaque. A few confront the conditions of detention directly, while others suggest it only elliptically. Some seem surreally decontextualized. It is tempting to take the darker and more foreboding images as manifestations of anguish or to read the pieces for subtext, whether political or otherwise. But speculation about their content presumes that we have the capacity to discern what the detainees really meant, thought, or felt when they put brush, charcoal, or pastel to paper and assumes—more problematically—that they would have wanted us to know.

Given our radically restricted access to the galleries and our speculative ties to the circumstances of their creation, any interpretations we might concoct would be entirely conjectural.[121] Some detainees painted impressions of their cells; one drawing focuses tightly on the details of a lock and hasp. This content is perhaps unsurprising, but how can we know what the detainee-artists intended to communicate? Do these cell-scapes indicate a lack of creativity and imagination or the capacity to find beauty even in austere circumstances? Is this kind of creating a way to commit an image to memory or an effort to expiate it? Conversely, many detainees conjured landscapes far removed from their immediate environs: beaches, cityscapes evocative of Islamic architecture, meadows, mountains, and woods. Here again, what can we know? Do these reveal artists unfazed by detention or desperate to escape it? What is the specific significance of each of these places, these buildings, boats, trees, and rocks? The still lifes are even more inscrutable. Uncannily resonant with canonical forms of visual art and elementary art training, what can be said with any certainty about these table settings, fruits, and vases? Was the detainee just practicing? Trying to communicate something? Or trying not to communicate anything, enjoying a respite from being cajoled or forced to speak to interrogators and guards?

When we look at the work from a vantage far removed from Guantánamo, the superficial transparency of the artifacts folds in on itself. Regarding the art as the expression of a basic human creative impulse sentimentalizes it. Beyond the aesthetic pleasures of spectatorship—some of the art is quite lovely—this humanist approach understates the violence of detention, presumes a limit to its power, and removes a barrier to hoping that it can be forgiven.[122] Compared to the poems and the diary, which offer clear political and emotional cues to their readers, the visual art stubbornly refuses to betray ideological or affective content. If viewers are so inclined (and there is a great deal of evidence to suggest that we are), even those images that represent detention, those locks and closures and cells, can be readily assimilated into a simplistic, commonsense understanding of incarceration, a reframing that strips indefinite detention of its extraordinary character.

Simultaneously, idealizations of the creativity underpinning these works make it possible to believe that the destructive power of indefinite detention can be mitigated or offset. Leaning too heavily on detainee artwork as a sign that all is not lost in detention exonerates the United States. Such an interpretation replicates what Berlant describes as the "relationship between the normative affect of liberal optimism and ongoing structural violence."[123] The urgent but unanswerable question is how to balance acknowledgment of the transformative force of detention without imaginatively reducing the detainees to forms of bare life emptied of all but the most utilitarian impulses.

The thing that is always filtered out, mediated, here is detainee anger, whether at their captors or at the undetained people who freely consume their work. The thing that is always accommodated is spectators' pity. In the *Poems*, detainee anger is rationalized as a reasonable reaction to detention, its wildest manifestations domesticated and channeled toward the easy comforts of the appropriately indignant reader's conscience. In the diary, Slahi's postscript assuages anxieties about the lingering consequences of America's actions. If anger buzzes behind any of the visual art, it operates at a register below that which we can perceive. Paradoxically, the outsiders who might be most attuned to the possibility of detainee anger are those that support their detention, whether ideologically or practically. Otherwise, neither the form of activism that imagines the detainees as mute captives nor the one that mythologizes them as writers and artists captures the structural dimensions of their subject-formation. The failure to acknowledge the potential for their anger minimizes the destructive power of detention and invites us to fill in the space where their anger might otherwise be with the pleasure of our own outrage, uncomplicated by any obligation to its supposed beneficiary.

"A Passed Future": Commemorating Guantánamo

It seems I am too late. Paid annual memberships to the *Guantanamo Bay Museum of Art and History* have sold out for this exhibition season. So I signed up to be a "network member" instead. The welcome email I received, signed Guern N. Ka, says that the museum appreciates my support nonetheless, invites me to contact them with any ideas I might have for collaboration, and hopes I will find their resources useful. While it does not have the panache of a paid annual membership, my reduced status as a mere network member actually entails a level of commitment that is comfortable for me. Because after all, the *Guantanamo Bay Museum of Art and History* does not exist. Or at least, not exactly.

On its website, the museum envisions a world in which Guantánamo was decommissioned in 2012 (complete with a photograph of President Obama signing the order to do so) and the facilities subsequently converted into an art museum and education center.[124] While the organizers acknowledge that getting there can still be tricky, they welcome queries from school groups and invite artists to apply for residencies, though for now they will have to be virtual. The website features photos of visitors in its exhibition spaces, all apparently partaking in its vision: "Collectively remembering a passed future." Immaterial though it is, the museum is real enough to showcase actual artworks on its website; more fancifully, it promises that its galleries are staffed by trained docents, and daily tours are available. It is real enough too to have hosted a 2013 "satellite exhibition" in Berkeley, which featured a screening of *Performing the Torture Playlist*, a collection of karaoke renditions of the songs used as soundtracks for interrogations and torture.[125] I find myself wondering what would happen if Berkeley someday got its detainee after all, whether he would be interested in attending such a performance. Perhaps. It is real enough, too, for the *Atlantic*, which profiled the Museum in 2012, calling it "brilliant" and endorsing its logic: "If Gitmo exists because of one fiction, perhaps it can be closed by another."[126]

Perhaps.

But there is a curious omission here: the detainees. Only one is mentioned by name. For his troubles, Al Dossari, author of the death poem, earned an eponymous "Center for Critical Studies." The texts listed on the "Resources" page include writings by Butler and Derek Gregory, but nothing by detainees.[127] Neither is there mention of the detainees' artwork, which does not seem to have garnered a place in this archive. Instead, all of the art featured online is by citizens or residents of the United States.[128]

President Obama signs the order, officially closing the Guantanamo Bay Prison. (2008)

Visitors celebrating during the opening gala of the museum. (2012)

Obama signs order to close Guantanamo

Closure halted by congress

International campaign begins to close Guantanamo immediately

The direct-action group "No Pasaran" coordinates unprecedented day of action and blockades

Curators organize series of exhibitions to galavanize action against Guatnanamo

Congress bows to international pressure and finalizes closure. The camp is soon decommissioned

Construction begins on the museum

"No Pasaran" organizes blockades around the globe after congress halts the closure of the detention camp. (2010)

The Guantanamo Bay Museum of Art and History conjures a future in which the site, having been decommissioned by President Obama in 2012, is converted into a space of learning and cultural enrichment. These photographs illustrate key moments in its story.

One layers audio from Omar Khadr's interrogation—prominently featuring the moment in which he appears to be entreating someone, anyone, to "Kill me, kill me, kill me"—over images of the so-called Torture Memos that provided the Bush administration with a legal justification for its interrogation practices. Another is a guerrilla project that stencils arrows pointing toward Mecca in unlikely places.[129] If the fetishization of detainee creativity enacts one kind of erasure, the exclusion of it from even this imaginary archive reflects another kind of violence. Despite the promise of a collective remembering, this is a different colonization of the imaginary, which claims that the only people who have the authority to fantasize a different future for Guantánamo are citizens of the state that created it in the first place. This space, now imaginatively, magically devoid of detainees, sanitized and repurposed, becomes at long last a site of total expiation.

CHAPTER 6

Feeling for Dogs in the War on Terror

If a military working dog (MWD) seems clingy, restless, confused, exhausted, or is yawning, panting, or scratching excessively, she might, according to some canine behavioral health experts, have canine post-traumatic stress disorder (C-PTSD).[1] The diagnosis is not uncontroversial, and there is no direct way to verify the presence of the condition. Unlike the process of evaluating humans for PTSD, which depends heavily on a patient's ability to self-report symptoms, C-PTSD can only be observed and inferred by experts, who must be able to vouch for a sufficiently traumatic event in the dog's past and assess the dog's subsequent behavior as abnormal. Veterinarians and animal behaviorists argue about whether the label is accurate, but no one asks whether the dog might be faking it to get out of performing his duties, while human patients who present with PTSD symptoms are often greeted with suspicion. Even if they quibble about the nomenclature, everyone agrees that the symptoms constitute a problem, because MWDs need to be able to work, and work reliably under stressful conditions. Treatment options for a distressed animal include rest, extra gentle care, reconditioning, and pharmaceuticals; if these fail, the dog will be retired.

The notion of C-PTSD assumes that canine interiority—a dog's feelings, preferences, memories, and decisions—is organized like human interiority and that canine trauma would manifest itself through maladaptive behaviors as it does in humans. C-PTSD skeptics, on the other hand, reason that dogs do not process and register phenomena like humans do. At its core, the dispute about C-PTSD is an argument about the nature of animal subjectivity, and each side affords it only partial recognition. C-PTSD advocates recognize trauma in dogs not on its own terms, but only where it manifests in a manner akin to human trauma. Conversely, C-PTSD doubters suggest that dogs need a different diagnostic lexicon, but they may fail to recognize some of the MWDs' distress. If nothing else, this disagreement reveals the essential impenetrability of animal subjectivity, particularly around the issue of suffering. This dilemma resides at the core of this chapter, which focuses on the figuring of war dogs.

My inquiry in this chapter was animated by a single event. In March 2008, a cell phone video depicting a U.S. Marine cooing over a live puppy and then throwing it off a cliff surfaced on YouTube and promptly went viral. The video, viewed hundreds of thousands of times, seemed a bewildering spectacle of cruelty; public and media responses circled around vexing questions about the character of American military personnel, what circumstances might inspire human beings to such depravity, and how the culprit and his cameraman ought to be punished. These inquiries, primarily concerned with facts and procedures and establishing tidy narratives, generally ignored the larger and more unsettling questions provoked by the collision of violence, sympathy, and militarized American identity as it unfolded in sixteen seconds of shaky mobile-phone footage. Short, untitled, and low resolution, the video provides no guidance on how to answer them and explains nothing; multiple viewings only amplify this indeterminacy, posing the unanswerable questions again and again, reminding us of the grisly stakes of getting them wrong and the impossibility of getting them right.

Despite (or because of) the video's ethical, political, and affective inscrutability, it became the object of vigorous discursive proliferation. There were angry, often inarticulate reaction videos posted back to YouTube; rejoinders from the American Society for the Prevention of Cruelty to Animals (ASPCA) and other anti-cruelty organizations; news stories about the perpetrators or the video itself; and apologies, disavowals, and disciplinary action from the Marine Corps. There was something about the video that I, too, found haunting, as it provoked in me a choking kind of sadness and a blinding, scarcely expressible form of anger that did not diminish appreciably over time or get defused by my theorizing about it. It was from a place of curiosity about the intensity of my own feeling that I began this inquiry. The video is simultaneously garish in its clarity and stultifying in its indecipherability; from the thicket of this contradiction, I use it to refract a broader inquiry about mediated suffering, sympathy, and American identity. In this chapter, I ask how the luckless puppy managed to reconfigure, albeit at mortal cost to itself, the usual relationships between those phenomena.

By throwing two long-established sympathies into crisis, the video illuminates the mechanics of figuring and its centrality to the public culture of contemporary American militarism. More than this, however, I argue that the puppy itself has the potential to orient us toward a different and more substantive way of responding to the suffering of other beings in wartime. I begin the chapter with a detailed description of the video, attending to its key visual, audio, and narrative elements. Then, I consider the role of anger and imagination in responses to it. To provide historical context for the outrage that the culprits Lance Corporal David Motari and Sergeant Crismarvin Banez Encarnacion provoked, I trace the evolution of anti-cruelty discourse and practice in the West. I then consider the politics of humane care for animals and the significance of this orientation for American identity, which intensifies the crisis of sympathies that the video provoked. Images of military personnel and dogs working together are the focus of my next section, in which I explore the varied affective and ideological appeals of these partnerships. Returning to the Motari-Encarnacion video after that, I reflect on the intensity of the reaction it provoked and the crucial role of the puppy in engineering it, despite the multiple layers of mediation through which audiences encountered its death. In the chapter's conclusion, I begin to sketch the contours of a different response to this kind of animal suffering and perhaps the suffering begotten by militarization more generally.

"So Cute Little Puppy": The Motari-Encarnacion Video

Visually, the video displays the grainy texture and relatively poor resolution of a cell-phone camera from the mid-2000s and so oscillates jarringly between ordinary and extraordinary. On the one hand, the fact of its being recorded signals that the perpetrators believed something momentous was occurring. On the other, the choice of this visual medium imbues it with the sense of quotidian spontaneity that suffuses nearly everything we document on our always accessible phones.[2] The panorama is desolate: a cloudless pale blue sky and a rocky brown landscape, with some scrubby green vegetation visible in the background. The forbidding landscape offers a stark contrast, however, to the charm of the puppy, captured in a brief close-up. It is black and white, with a sad-looking face visible enough to be wrenching despite the large pixels of the low-quality video. With the exception of the few seconds where he positions himself in front of the phone, awkwardly inserting himself into the frame, the film basically follows the gaze of the cameraman: at his fellow marine, at the puppy, along the terrible arc of its flight, and then back to the man who threw it.

The audio quality is also relatively poor. There is the constant, muffling, ambient noise of the wind and a similarly pneumatic sound, likely the men's breathing. Though they might not be readily audible on a casual listen, there are also acoustic details like boots crunching the gravel, the material of the perpetrator's uniform rubbing against itself, and another dog barking in the distance. At times, their dialogue is hard to hear, though their laughter comes through clearly. By far the most distinctive element of the audio is the yelping of the puppy as it flies through the air, which ends abruptly with the sound of it hitting the dusty ground below.

At sixteen seconds long, the video has a very compressed plot. It begins with the perpetrator posing for the camera, holding the dog by the scruff of its neck. He then turns to face the dog and turns the dog to face him. They regard each other for a brief moment, and he asks, apparently rhetorically, "Cute little puppy, huh?" Then there is a brief shot of the empty horizon before the cameraman ducks into the frame, smiling, and responds with a falsetto "Ohhh, so cute, so cute little puppy." He then moves back behind the camera and follows his comrade with it, lingering on the dog for a moment while the other man cocks his arm back. And then, as he is preparing to throw, he laughingly offers a disingenuous explanation for what comes next: "Oh, oh, oh—I tripped," as he launches the puppy over the cliff. The camera tracks the animal on its parabolic flight toward the ground and pauses again in the dirt, where the creature landed and bounced and finally came to rest. Almost immediately thereafter, this time in a normal voice but with an inscrutable affect, the cameraman narrates a drawn-out "That's mean" and "That was mean, Motari," during which he pans the camera back rapidly to him. Motari shrugs and briefly tilts his head toward the camera. He smiles. Somebody laughs goofily (probably the cameraman), as Motari offers a fatalist assessment of his actions, "I do what I do," and he turns half away from the camera, as the video ends abruptly with him in quarter-profile.

Although initially, much about the video was unknown, including whether it was real, where it was shot, and what became of the puppy, most of the facts have since been confirmed.[3] These include authenticity, location (Haditha, Iraq), and the identities of both the culprit, Lance Corporal Motari, and the cameraman, Sergeant Encarnacion. The fate of the puppy seems fairly obvious. Contemporary spectators often greet digital images with some measure of skepticism, but these relatively superficial concerns about the video masked the much more profound and irresolvable questions that it posed. If the video seemed unbelievable, this was not simply because the details were murky. Rather, I argue that the video's incomprehensibility arose from the crisis it provoked when it put two affective commitments, both professed to be central to American identity, into conflict: support for American military personnel and humane compassion for domestic companion animals like dogs.

Adriana Cavarero argues that the "casualness" of casualties in contemporary militarized violence is precisely what makes them horrifying.[4] The insouciantly flat affect that Motari and Encarnacion exchange reveals that this casualness can stretch beyond the boundaries of species. Relatedly, Jacques Derrida, in his discussion of the beast and the sovereign, unspools the notion of "*bêtise*," a term that translates loosely to "beastliness." Importantly, *bêtise* is the purview not of beasts but is instead "proper to man" and manifests as a deliberate and knowing infliction of harm with an eye toward securing one's position and power.[5] Comprehending the video requires accepting both the humanness of *bêtise* and the reality that Americanness offers no exemption from it. While American spectators of the War on Terror might have become accustomed, if not inured, to the mediated sight of human casualties, images of animal casualties from this conflict are rare by comparison, and images of animal casualties created by American troops scarcer still.[6] Beyond the casualness of Motari's actions, his "cute little puppy" mockery of sentimentality multiplied the video's affront to such sympathies.

Even after YouTube removed the video, it continued to circulate widely and provocatively. There were video responses in the form of angry, often inarticulate monologues posted back to YouTube and "reaction videos" that filmed unsuspecting witnesses watching the original for the first time.[7] The "Comments" sections beneath these videos became sites of textual exchange, much of it hostile and profane. The ASPCA and other anti-cruelty organizations issued more staid condemnations. News stories, prefaced with warnings about graphic content, replayed parts of the video or developed sketchy profiles of the perpetrators. The Marine

Corps proffered apologies, disavowals, and—eventually—punitive action. Three months after the video first appeared, Motari was "separated" from the Marine Corps and Encarnacion disciplined by it. Even prior to this verdict, however, the corps distanced itself from Motari and his actions, publicly describing the video as "shocking and deplorable" and inconsistent with the "high standards" of behavior for marines.[8]

For my part, I have watched the video a hundred times at least, and there is so much I do not or cannot know. I find myself wondering about the puppy, Motari, Encarnacion, all the happenstances that conspired to bring them together. I wonder if the dog had any pleasures in its short life—naps under a shady tree with its mother and littermates, a gentle human touch, scraps from a table. I wonder if it had an instinctual reaction to Motari, if it ran to greet him, or cowered, or maybe, already accustomed to human neglect, showed no interest at all. I wonder what slipped and came loose inside Motari and when. Was it long before deployment, at some chaotic juncture on an ill-defined battlefield, or maybe just right there, at the edge of the cliff? And how did he feel with the warm, soft, furry weight of the puppy in his hand? I wonder if Encarnacion knew what was going to happen there, why he turned on his camera to record, if he had that quicksand anxiety-dream feeling in the instant when he realized what was unfolding before him, how he settled on meanness as his diagnosis for what Motari did, how he understood his photographic role in this little death. I wonder about all the people who watched what he recorded and began to wonder themselves, began to doubt, began to speculate. I wonder about how it would feel to be so staggeringly angry that threats of homicidal violence against these marines and their families are the only words that come out.

Although responses came from a variety of quarters—official and otherwise, and nearly everyone who saw the video (it seemed) had something to say about it—they displayed a remarkable uniformity. Only a minority dismissed the video as insignificant or found the conduct of the marines morally neutral or even amusing, and they were loudly and significantly outnumbered—likewise those who tendered half-hearted exonerations of the marines' behavior, citing combat stress. For the most part, the response was unequivocally reproving and often even vengeful; after the video was posted, there were death threats against Motari, whom the U.S. Marine Corps then took into protective custody, and his family. Particularly in comparison to the mixed, ambivalent, tepidly critical popular reaction to wartime atrocities like the torture at Abu Ghraib, response to this video was uniformly and intensely critical. Colleen Glenney Boggs, in her analysis of the use of dogs at Abu Ghraib, set forth the possibility that Marco, the large black dog pictured lunging at a detainee, might have "interpellated" that man into the U.S. "disciplinary regime" operative at the prison.[9] The yelping and unnamed puppy in this video, it seems, hailed spectators into a morass of ideologies, discourses, and sympathies, where they struggled to find any kind of purchase.

In this way, the animal accomplished something remarkable. The dog was powerless in the hands of its captors and anonymous to its audience. Despite this, and the mediated way we encountered its death, and even the countervailing force of popular loyalty to "the troops" that has exonerated them from countless other transgressions, the puppy provoked a reaction. And this provocation points to the possibility of a response to suffering that goes beyond the empty and alienating gesture of figuring.

The Little Dog Becomes a Conduit: Anger and Imagination

On the face of it, my claim that anger is the dominant affect in the figuring of this dog may seem counterintuitive in light of the depth and complexity of feeling that so many people have for their own dogs and of the intensity of the reaction that the video provoked. But the object of my analysis is not individual responses to the video; rather, I am interested in the larger patterns of affective investment the video elicited and the histories that inform them. The apprehension I described in Chapter 1 implies the presence of a childish political subjectivity at risk of corruption; this does not apply in the case of the puppy. And while people surely feel affection for their own beloved animals, this kind of "love" does not factor into the responses to this video in a meaningful way (nor is love a key element in most anti-cruelty discourse, about which more later). It is unconventional to speak seriously of admiring a dog in general, and because the admiration I have described throughout the book requires agency on the part of the object of that admiration, it does not adhere in this case, as the puppy is clearly a victim. While we might have abstract inklings of gratitude for dogs that perform military service in wartime, it is not the same as that for the humans that we thank for fighting on our behalf or in our stead. And anyway, this is not one of those dogs. Pity in this instance is both too little and too much. Too little, in that it does not fully scratch the affective itch of the video, which seems to demand more than feeling bad for the puppy. Too much, in that it is so incurable, so futile, and ultimately so unbearable that it curdles into something else.

Anger. The little dog becomes a conduit for it, channeling it onto Motari and Encarnacion. This kind of figuring is unique among all those I have analyzed in previous chapters, because it is directed *around* or *through* its object rather than directly at, onto, or into it. In this way, figuring renders the dog both essential and incidental to this affective work, constructing the animal not so much as a repository for affective investment but rather as a relay for it. In the process, the animal is expended: already killed by Motari and Encarnacion, it is subsequently overrun by this surge of feeling. Because the creature in question is nonhuman, the work of figuring the puppy depends, more than any other I have described thus far, on imagination. Of course, imagination of animal interiority or suffering is the necessary foundation for any kind of ethical claim on behalf of an animal, because they cannot communicate that kind of information to humans in a reliable way. This does not mean that communication between animals and humans is impossible. But it is intensely mediated, and mediation in this case is a function of species or, more precisely, species difference. This is a massive dilemma. It is important, on the one hand, to acknowledge what Derrida describes as the absolute, perhaps ultimately unknowable otherness of the animal's point of view.[10] On the other, there is an urgent need to make that animal suffering humanly legible, so that humans might act to prevent or ameliorate that suffering.

Typically, and for the variety of historical and ideological reasons I will outline, we tend to default to sentimentality when trying to make sense of our connections to suffering dogs. This pathos is gratifying, but problematic. Jean Baudrillard claimed that our "sentimentality" about animals is precisely "proportional to" our "disdain" for them, our inability to see them on their own terms even as we focus relentlessly upon them.[11] As an alternative to this, Alice Kuzniar conjures a vision of affective reciprocity between humans and animals, dogs in particular, that eschews sentimentality and all the oppressive hierarchies that come with it, emphasizing instead what she describes as more complex ties borne of "intimacy, compassion, propinquity, and mourning over their death."[12] All of these feelings, with their underpinnings of love and closeness, are part of a much richer vocabulary for understanding relationships between humans and their pets. But this is difficult, if not impossible, to scale up to feelings for animals in general or animals with whom one does not share an intimate bond.

Even in the case of a connection with an animal we know or believe we know, how can we sync ourselves emotionally with them? How would we determine if we were successful? These questions become even more urgent and unanswerable when an animal is in distress. To sympathize at a distance with a suffering animal, to woefully recognize and care about the animal's suffering, is patronizingly incomplete because sympathy alone amounts to nothing for the animal. Even to empathize with the animal, to partake actively and profoundly of their feelings, would require surmising about them in our own terms and interpreting them through our own structures for managing and processing feelings and sensations.[13] Something essential will always be lost in these attempts.

Offering a trenchant assessment of the situation of animals in contemporary Western societies, Anat Pick writes, "When it comes to animals, power operates with the fewest of obstacles."[14] Animals do not seem to resist, or do not resist like people do, and so this operation of power seems both smooth and justifiable. When Motari delivers a patently absurd explanation for throwing the puppy—"I tripped"—he reveals how easily that power is justified, but also how superfluous the justifications are. As embodied in the marines, U.S. military power in this video appears both absolute and craven; in her analysis of the torture of Iraqi detainees at Abu Ghraib, Anne McClintock detected a similar combination of omnipotence and paranoid weakness.[15] At its most naked, power looks like abuse or utter disregard for the sentience or suffering of another being, and that form of power is clearly on display for these sixteen seconds of grainy video. But power can also manifest as the opposite of that, an overflow of care or concern, and the imagining of another being's suffering in ways that gratify our senses of righteousness or feelings of empowerment to act in response; this form of power courses through the responses to the video. Anti-cruelty movements in the West have run on these feelings for nearly two centuries, consistently leveraging the near-absolute power of humans over animals to make their claims.

The Anti-Cruelty Movement and the Origins of This Crisis of Sympathies

In the self-evident gratuitousness of Motari and Encarnacion's choice of victim and the unequivocal cruelty of their actions, the video is terribly lucid. But this very clarity was what made it so unsettlingly difficult to parse. This obvious viciousness cornered spectators into a crisis of sympathies, forcing them to choose between two types of beings—dogs and U.S. military personnel—that usually seem uncomplicatedly worthy of affective investment. Heeding the ubiquitous imperative to "support the troops" in this case would necessitate disregarding the suffering of the dog. But disregarding the suffering of the dog would require contravening a long history that valorizes sentimental regard and humane care for animals, a history that is tied intimately to the process of American self-definition as compassionate and civilized.[16] Any response to the video thus entails making a choice, and making that choice is also a way of making a political claim about oneself by determining who counts as an agent, a perpetrator, and a victim.[17] Lauren Berlant wagers that "when we are taught . . . to measure the scale of pain and attachment, to feel *appropriately* compassionate, we are [also] being trained in stinginess, in not caring."[18] Typically, these choices are made easy by the systems that instruct us how to apportion our sentiments. The primary function of "sentimental politics," as Rebecca Wanzo argues, is to dictate the "logic that determines who counts as proper victims" in American culture.[19] Sentimental politics is a metric for determining how to distribute our sympathies, and obeying this guidance is often as easy and immediate as yielding to a reflex. Here, however, the choice is anguished because Motari and Encarnacion have thrown those systems into irresolvable conflict. In the process, they imperiled some of the most salient markers of contemporary American identity as it hinges on particular sympathetic identifications with others.

The American anti-cruelty movement found its first expressions in the mid-nineteenth century. It was roughly coeval with myriad other efforts to extend rights to the previously disenfranchised, platforms including abolitionism, the early suffrage movement, and advocacy for children's rights.[20] Champions of those ideas frequently found common cause with anti-dogfighting activists and others concerned with various forms of animal maltreatment.[21] In the 1860s, the anti-cruelty movement galvanized through a range of public welfare organizations that were largely the purview of middle- and upper-class men and women.[22] As a discourse and a practice, "humane" treatment of animals had emotional, legal, and ideological elements. The first American anti-cruelty statutes were passed in 1828.[23] However, the idea of humane care for animals—as an attunement, feeling, and set of practices—did not gain widespread traction until later in the century. Over the following decades, anti-cruelty activists extended their campaigns beyond obvious types of harm to animals, such as blood sports like dogfighting, to subtler forms of maltreatment and neglect. This period was marked by what Katherine Grier describes as an "increased self-consciousness" about human-animal relationships.[24] This self-consciousness was enacted as a "domestic ethic of kindness" toward animals, whereby one demonstrated one's identity, status, and sensitivity in compassionate care for nonhuman beings, particularly those associated with one's household.[25] Accordingly, pet-keeping became much more common in the 1800s. In all of this, the emphasis was on the helplessness of animals, who had the capacity neither to speak nor to significantly improve their circumstances.[26]

During the nineteenth century, American animals became visible as beings whose suffering and deaths were lamentable, and good citizens demonstrated their emotional refinement through that lamentation.[27] The origin story for the ASPCA is illustrative in this regard. The ASPCA's promotional materials describe its founder, Henry Bergh, intervening on a New York City street as the driver of a coal cart whipped his exhausted horse.[28] Bergh, already an activist for animals abroad, was then inspired to build on the example of the British Royal Society for the Prevention of Cruelty to Animals and work to institutionalize a comparable movement in the United States. Bergh asserted that the protection of animals was "'purely a matter of conscience,'" not politics, and he founded the ASPCA according to this precept in 1866. Empowered by the New York state legislature to enforce new statutes against cruelty to animals, the ASPCA made its earliest efforts on behalf of horses and livestock, but also extended its care to cats and dogs, with the latter figuring especially prominently in its story.

Importantly, the anti-cruelty movement was not rooted in a recognition of animals' subjectivity. Instead, it advocated for them as being worthy of protection *despite* their lack of meaningful subjectivities. A long-standing emphasis on animal voicelessness and helplessness generated a movement that saw its mission as speaking for those who could not ever hope to speak for themselves. As the protections afforded to human citizens were stretched to encompass domestic animals, governments increasingly acted as their defenders.[29] At the individual level, those humans who cared responsibly and lovingly for their animals were able to define themselves through these relationships. This afforded a different mechanism for subject-formation than those that emphasized the differences between animals and humans, a dynamic that is especially powerful in the case of human relationships to pet dogs.[30]

The American anti-cruelty movement was one facet of a broader attempt to extend the protections of civil society to those who had previously been excluded from the realm of the human and to reenvision the United States as a society that cared for the powerless. Deliberations about what constituted ethical—and lawful—treatment of animals unfolded alongside other debates about which humans qualified for full membership in the nation-state and how their claims to citizenship would be adjudicated and enforced.

But as Maneesha Deckha notes, "Human problematizations about nonhuman beings are rarely ever just about the nonhuman, but mediated by other circuits of difference."[31] Thus, in the nineteenth century, displays of humane concern for animals became marks of refinement and class status. Over time, the seams between sympathy for animals and sympathies for other disenfranchised groups began to pull apart. Concern for children—or certain children, as I describe in chapters 1 and 2—became paramount, and the practice of humane care for animals often took the form of disciplining the working classes and people of color for their apparently inhumane relationships to the animals in their world.

The primary emotional modes of anti-cruelty discourse were pity or compassion for animal victims and outrage at human perpetrators. Anti-cruelty discourse, often steeped in the rhetoric of civilizedness, aligned neatly with American imperialism, and sympathy for animals often displaced sympathy for people of color or was rearticulated as vitriol against them. Janet Davis, for example, traces the linkages between American opposition to cockfighting and U.S. imperial agendas for the Philippines, Cuba, and Puerto Rico in the nineteenth and twentieth centuries.[32] More generally, as Deckha notes, "Upper-class Victorians . . . began to associate superiority along class and racial lines, at home and abroad, with the inability to tolerate certain forms of suffering."[33] Deckha argues further that anti-cruelty policy interventions were not simply or even primarily about violence against animals but rather were aimed at "[reinforcing] civilizing missions with respect to both domestic and colonial populations though a legislative purpose that can be properly understood only by reference to race, religious, class, and gender dynamics."[34] All of this was accompanied by a suite of prescriptions for appropriate ways to think and feel about animals. Domestically, anti-cruelty platforms coexisted comfortably with hierarchies of race and class.[35]

By the mid-nineteenth century, the poor and working classes, who often had different relationships to their animals by necessity, became targets of anti-cruelty campaigners. Sympathy for the plights of animals and for the poor and people of color began to appear incompatible. These patterns continue to repeat into the present and played out vividly, for example, in the aftermath of Hurricane Katrina. Dogs who were separated from or abandoned by their families became the objects of intense rescue efforts. This outpouring of care, in Kelly Oliver's estimation, surpassed that which was directed at the human victims—largely poor and African American—of the storm.[36] The constellation of race, dogs, and sympathy shifted again roughly two years later with the indictment of Philadelphia

Eagles quarterback Michael Vick on dogfighting charges. Condemnations of Vick's actions slid easily into racialized discourses about the violence or ferocity of black masculinity that were rooted in long histories of dehumanizing black men.[37] In Vick's case, sympathy for animals took the place of sympathy for humans, a trade readily facilitated by entrenched discourses about race, class, and subjectivity. In making this claim, I am certainly not suggesting that the public should have sided with Vick instead, but am rather trying to locate this swell of outrage in a larger historical context.[38]

For the last century and a half, humane concern for animals has been an affective mainstay of American public culture, while support for the troops, as I noted in Chapter 4, has had a more inconsistent history. From the outset, anti-cruelty discourse, sentiment, and practice have emphasized the helplessness of animals. By contrast, affective and ideological constructs of American troops emphasize their volition, expressed as willing sacrifice for the rest of us. Both opposition to animal cruelty and profession of support for the troops are defensive postures. Anti-cruelty activists speak as human protectors of vulnerable animals. Similarly, the mandate to "support the troops" generally presumes a position from outside their ranks. To rhetorically support the troops is to confess that, for whatever reason, we have chosen not to locate ourselves among them. Civilians can thus prove their allegiance to them, and to the state, by defending them against their detractors.[39]

"You're Gonna Make Me Cry!": Dogs in Wartime

In many ways, the Motari-Encarnacion video is an outlier in the visual and popular cultures of the War on Terror, which are flush with stories that harmonize sentimental visions of dogs and military personnel. Just as supporting the troops regardless of one's feelings on the war in question has become something of a transpolitical imperative in recent decades, humane care for animals like dogs has long been portrayed as apolitical, simply and unequivocally moral. In both cases, however, adherence to these affective positions has become the marker of a properly American identity. Chris Hedges argues that in wartime, dogs often serve contradictory purposes for soldiers, becoming objects of love and violence, sometimes simultaneously.[40] In a context where both military personnel and dogs are so magnetic to affective investment, extreme examples of military personnel displaying either love for or violence against dogs draw public attention. Stories of the special bonds that arise between military personnel and their animals, of dogs that are victimized by America's enemies, and of coura-

Gracie the dog became an Internet celebrity for her effusive greeting of her human, a member of the National Guard, when he returned from Afghanistan. Videos like these offer the doubled pleasure of seeing two preferred objects of sentiment (dogs and military personnel) in rapturous connection.

geous MWDs reinforce these sympathies for military personnel and for dogs by binding them together. The resulting amalgamation is deeply and pleasantly compatible with militarization. But when these sympathies are caught between Motari and Encarnacion on one side and the puppy on the other, they become irredeemably complicated and irreconcilable.

Some dogs seem to display a perfected kind of wartime citizenship, an affective pedagogy of how to feel about military personnel. For example, dogs like Gracie, who became famous in a 2008 YouTube video for effusively greeting her human upon his return from Afghanistan, enact the scene of a fairy-tale homecoming at the conclusion of deployment.[41] Gracie was given to high-pitched whines, a wagging tail, and an adorable habit of throwing herself into her human's lap, hopping out, and doing it again. In so doing, Gracie cued her people (a woman off-camera coos repeatedly and says, "Aww . . . you're gonna make me cry!") and more than 21,000,000 other YouTube viewers how to respond to returning military personnel, an exuberant and forgiving form of wartime citizenship. Gracie's performance was just the beginning of a whole subgenre of user-generated homecoming videos in which heedlessly, wildly happy dogs provide the kind of welcome that all stateside citizens are supposed to offer. Moreover, because they seem only to be acting on instinct, driven by a rush of love, joy, and appreciation, they naturalize these emotions, making them seem straightforward and automatic.

Similarly, there are the dogs who lose their humans in combat and seem to know just how to mourn. For example, it was widely reported that at the funeral of Jon Tumilson—a Navy SEAL killed when his Chinook helicopter was shot down in Afghanistan—his dog spent the entire service lying beneath his master's flag-draped casket. Photos and videos circulated widely online, documenting the durability of this bond.[42] Ideologically and emotionally, this is an easy combination, even if (or perhaps because) it is also overwhelming. When we respond appropriately to these videos—with watery eyes, warmth in the chest, lumps in the throat, a wash of indescribable feeling—the circuit is complete: the troops are supported, the dog is an object of humane concern, and we have vouched for our status as good subjects of this affective order.

Stories about soldiers rescuing dogs, usually strays from "savage" places like Iraq or Afghanistan, offer explicitly nationalized gratifications.[43] They work, according to Purnima Bose, as a feint, highlighting U.S. humanitarianism while obscuring Iraqi civilian casualties.[44] Whether the bond forms through happenstance (as when a dog "adopts" a serviceman or group of them) or intention (if a soldier deliberately sets out to rescue a particular animal) the result is a feel-good story about human-dog relationships that secure each party's role in established hierarchies of species while valorizing military personnel.[45] When a servicemember befriends a dog even as he is in mortal danger, the act makes the servicemember appear all the more heroic and deserving of gratitude, as he has extended the American humanitarian mission beyond even the bounds of the human. In these narratives, the dogs often seem helpless or even pathetic, but spectators can identify with their vulnerability and admire their instinctively good taste in humans, falling for the same men in uniform that we do.

This parallel explains the popularity of stories about soldiers finding healing through relationships with dogs. In such accounts, dogs serve varied functions, working as comfort animals for military personnel who have PTSD or veterans who have lost limbs or functionality in combat or as pupils for veterans who find purpose in training them to help others.[46] In these roles, the dogs perform more direct services for military personnel than the humans who simply profess their rhetorical support for the troops; I would argue further that their willingness to do the work that humans have outsourced to them is part of their allure. The human-dog relationships here are a bit more complicated than those of soldiers and dogs working together or soldiers rescuing dogs, because in these instances, the hierarchies are scrambled. These service dogs are still objects for human use, even if they seem in some ways to be fuller, more functional, and capable subjects than their humans. But this is often represented as a temporary condition, a transitional stage on the way to the proper order of things, as the soldiers are restoring their humanity through their interactions with the dogs. In turn, the animals guide them toward an ideal subjectivity: disciplined, autonomous, healed.[47]

In cases where dogs seem especially defenseless or victimized, human spectators get the righteous pleasures of feeling achingly sorry for them and appalled at the behavior of their human aggressors. While I was searching for YouTube content related to the Motari-Encarnacion video, the platform's algorithm suggested I might also want to watch reports about the cull of wild dogs in Baghdad. Prefaced by a warning about graphic content, CNN's story begins with a shot of brown puppies lying on the dusty ground, one absently chewing on the other's tail, as the narrator observes that "if these stray puppies were born in America, they might have a better chance of finding a loving home."[48] Instead, because they had the misfortune to be born in Iraq, the municipal government deemed them pests and attempted to eradicate them as such. Dispatching veterinarians and other civil servants to kill them, whether with guns or strychnine-laced meat, Iraqi officials presented these strategies as the necessary solutions to problems with rabies, attacks, and feral dogs feeding on human corpses. The subsequent montage shows dogs being shot, expiring, lying dead on the sidewalk, followed by the commentary of a woman from SPCA International who suggests that the far more humane option would be to tranquilize and then medically euthanize the animals. The camera lingers intently on the dying dogs, while the visual and narrative elements of the piece emphasize the violence of their deaths, juxtaposed with the seeming unconcern of the officials in charge of meting them out.

As a representation, the video is located squarely within a long Orientalist legacy of depicting Arab men as wantonly violent through discourses that associated cruelty to animals with racialized "savagery." Indeed, some of the earliest anti-cruelty activists explicitly couched their campaigns in the racist frame of a struggle against "barbarism."[49] Here, they are defined directly against the civilized compassion of the woman from SPCA International and implicitly against the example of U.S. military personnel tasked with saving this place from itself. My own responses to the video—sadness, outrage, revulsion—were exactly those that its creators intended to elicit. With its soundtrack of pained animal cries and white-coated veterinarians whistling to call the dogs to the poisoned meat, I found the viewing experience agonizing. My intellectual understanding of the politics of the video did nothing to ameliorate the discomfort of watching it. Nor, of course, did my discomfort make any difference for the dogs. It is difficult, if not impossible, to disentangle my reaction from that Orientalist history; part of me wants to insist that my reactions are automatic, a sign of my compassionate nature, but the apportioning of compassion is never a matter of pure reflex, regardless of how immediate it feels. After all, I felt virtually no compassion for the officials dispatched to do this work. Although I tried to consider the situation from their perspectives, this thought experiment changed virtually nothing about how I felt watching the story.

Accounts of intense, companionate, and even collaborative bonds between dogs and military personnel haunt the Motari-Encarnacion video, while the footage of the Baghdad cull signifies ambivalently in juxtaposition with it. Of course, I cannot know what cued the YouTube algorithm to suggest this video repeatedly after my viewings of the Motari-Encarnacion clip; perhaps it was some combination of tags like "dogs," "killed," and "Iraq." But with the pairing established, the video of the cull can serve multiple contradictory functions. For example, it might stand as an equivalency, suggesting that what the Iraqi government did to those stray dogs was the same as what Motari and Encarnacion did to the puppy. Along these lines, it might also work to intensify viewers' horror at the behavior of the marines, in part by intimating that they are no better than the "savages." On the other hand, these depictions of inhumane treatment of dogs—state-sanctioned, premeditated, and on a large scale—could serve to minimize the significance of the single death Motari and Encarnacion caused, making it seem trivial, even merciful, by comparison.

A more recent example of enemy action against dogs traffics in tamer affects. In February 2014, in an apparently unprecedented propaganda move, the Taliban boastfully tweeted that it had captured a NATO MWD and illustrated its victory with a video of its furry prisoner.[50] In this video, it is daylight. The dog is surrounded by five bearded, armed men in traditional clothing, in an enclosure with dirt and stone walls—a pit or a trench, or some kind of cellar. The dog still seems dressed for his work, wearing a protective vest, and is leashed with a chain that attaches to one man's waist. The camerawork is inexpert, as the cameraman seems to have trouble training his viewfinder on the dog, who sometimes wanders in and out of the frame. The filming is also a little tentative, and there is a long shot where the camera just lingers on the torso of one of the captors and the large gun he's holding and then drops artlessly down to the animal.

Uncertainty structures the scene. The video opens with the dog pacing aimlessly from person to person. While American news stories about the video provide snippet translations of the men rejoicing over their victory, the video was released without English subtitles, and for those members of the audience who do not speak Pashto, the dog's apparent noncomprehension mirrors their own. Occasionally, he perks up his ears or wags his tail slowly, warily, or furrows his brow. Late in the clip, the dog seems to get a notion to leave the enclosure, moving suddenly toward the exit, pulling the man to whom he is leashed along behind him. In this act, the dog's body language conveys intention, ears pinned back and tail up. But once outside, he seems to get muddled or disheartened and just walks in circles. The camera always shoots high-angle, so we see the dog from above. Perhaps the cameraman adopted this vantage to demonstrate his hierarchical position over the animal in a culture where animals do not invite the same emotional and physical closeness that they do in America. Or perhaps he did not really know what he was doing. Either way, for American spectators, this orientation locates the animal in the lower and beloved position of a pet.

But compared to the incensed coverage of the Baghdad cull or Motari and Encarnacion, the news coverage of the dog-napping seems to downplay its significance, typically describing the animal's behavior as a manifestation of uncertainty rather than fear.[51] For example, the *Washington Post* offers a reassuring interpretation of the dog's condition. It describes the dog as "slightly befuddled-looking" and "more confused than terrified."[52] This depiction underplays, perhaps for its potentially sensitive readership, the possibility that the dog might be traumatized by its captivity. Even as the *Post* includes quotes from an interview with a former Air Force dog handler who emphasizes that the captive animal's human partner must be "'devastated'" by the separation, the tone of the story—and many others—remains somewhat mirthful, maybe a little bemused, emphasizing the ridiculousness of the Taliban crowing over their canine hostage.[53] This serves also to minimize the Taliban's victory and, by extension, the significance of the resurgent Taliban more generally. Overall, the subdued tenor of this coverage reveals that the figure of the suffering dog can be deployed selectively and flexibly, depending on strategic or political exigency.

Within the array of dogs populating the affective landscape of contemporary American militarism, MWDs occupy a unique place. Given

In 2014, the Taliban captured a NATO Military Working Dog and tweeted a video of its hostage. American news stories about the video emphasized the loyalty and stoicism of the dog while minimizing the significance of this stunt by the Taliban.

their vaunted status, spectators are likely to attribute a more complex canine interiority to this kind of dog than perhaps any other.[54] It is easy to see the animal, who betrays no outward signs of panic, as noble, brave, and stoic even under intense stress. At the same time, the dog offers subtle cues of his unease in facial expressions and body language. In this way, he demonstrates his sentience by registering the danger of his situation, while appearing to remain composed in the face of that danger, and so enacts an ideal militarized comportment. The poignancy of his situation is made more acute by his clear but futile desire (as exemplified in his seemingly purposeful departure from the enclosure and subsequent flagging of his spirits) to do the job he is still uniformed to do.

Visions of soldiers and dogs working together are perfect gratifications for a militarized imaginary, and the presence of a dog in the frame multiplies the affective pull created by scenes of military life. Witness, for example, the popular fascination with the Belgian Malinois dogs involved in the raid on Osama bin Laden's compound in 2011. Their story became a key impetus for the 2012 introduction of the Canine Members of the Armed Forces Act, legislation stipulating that MWDs were entitled to recognition for their service and certain benefits as a reward for it. Our imaginings about the loyalty and dedication of MWDs tinges their role with a singular pathos. Indeed, the title alone combines three prized things in American popular culture: military, work, and dogs. Thus, collaborations between dogs and soldiers are easy, ideologically and affectively: two favored objects of sentiment are working together, with each occupying its proper place in a clearly delineated hierarchy of master and helper.[55] At the same time, representations of MWDs often endow them with some kind of agency. It might be unsavory to think about the animals being conscripted into service; better to imagine them "choosing" to serve, not because they understood their mission, necessarily, but because they were so faithful to their masters that they followed obediently and unfalteringly. On the use of military animals, for example, the ASPCA notes that it "recognizes the value" of these creatures for the U.S. military, but adds the important caveat that they should not be "unnecessarily put at risk or sacrificed" for the mission, a reminder that they operate firmly within a hierarchy of species, that they are always under human control.[56] In this way, it is not so much dogs themselves but their apparent desire for connection with humans that elicits our emotional responses; we care about dogs because they seem to care so much about us. The example of the MWD infuses this emotional connection with a militarized intensity.[57]

Just as the demands on humans change in contexts of militarization, so too do those placed on dogs. Megan Glick argues that, in general, for dogs to be legible as creatures worthy of our affiliation and care, they "must be docile subjects," suited and used for companionship. She continues that "when dogs are used for other purposes—as food, religious sacrifice, or sport—they cease to become dogs" as we know them.[58] The construct of the MWD deviates slightly from this pattern because the identity of the MWD is dependent on its performance of military *work*. They become legible as subjects through their activity, obedience, and contribution to the mission. Their bonds of affection or comradeship with their human handlers are touching and important, but secondary.[59]

Of course, the work of an MWD is not always so romantic or even palatable. The dogs involved in the torture at Abu Ghraib are much more difficult to accept, affectively and ideologically, than any of the other creatures I have described so far. Some of the photos taken at the prison showed large MWDs snarling at terrified prisoners. This apparently easy and fearsome collaboration with the guards bespeaks what Boggs describes as a "symbiotic relationship" between the two parties.[60] The two species work together—though not evenly—to achieve, in Boggs's terms, a bestialization of the detainees.[61] Of course, dogs were not the only animals used to terrorize prisoners during the War on Terror; insects, in particular, figured significantly in torture.[62] But these creatures do not have the same cultural significance as dogs in the West, rarely if ever eliciting the same kind of sympathies or identifications from humans.

The Abu Ghraib dogs do not seem so straightforwardly noble as other MWDs. Their agency is complicated. They did not engineer the torture in which they participated and are presumably incapable of reasoning out the implications of their actions. They are merely following orders, but also seem to demonstrate a volition in doing so that indicates a kind of subjectivity. Typically, as Oliver notes,

> Animals occupy either pole of the limits of *the moral community*: they are absolutely innocent because they act on instinct and therefore do not control their behavior and cannot be morally blameworthy; they are absolutely monstrous because they cannot control their violent instincts and are therefore beyond the pale of the moral community.[63]

But at Abu Ghraib, the dogs interacted with humans who were their masters and their prisoners. This pair of dynamics muddies the question of their participation in torture and our imaginings of them more generally.

One such dog, Marco, appeared in a photo lunging at a prisoner. The man, with his hands shackled behind his back, is on his knees, eye-level with Marco. His face and Marco's face mirror one another; it is significant that the dog's name is mentioned in the news accounts, but the man's is not. Both of their mouths are open, both of their teeth exposed. But the man looks terrified while Marco looks ferocious. His handler, Sgt. Michael J. Smith, is astride him, restraining him with both hands. Smith's expression is indecipherable, and he appears to be either grimacing with the effort or smiling in delight, or both. Eventually, Smith was court-martialed, charged, and sentenced for using the dog to intimidate prisoners, but adjudicating Marco's actions is much more difficult.[64] Elisabeth de Fontenay notes that the "animal . . . appears to be the only being in the world unable to be treated as either a subject or an object."[65] The animal's dual position as neither/nor and both/and has philosophical, legal, and ethical ramifications and, moreover, affective consequences for those of us who simply bear witness to its behavior.

The work of figuring all of these dogs entails calculations of their lovability and their agency. In certain instances, we find dogs more loveable if they seem highly agentic and display a willingness and ability to negotiate the environments that humans construct for them. Consequently, MWDs, like service dogs for veterans, are regarded as heroes for their shows of skill and agency that benefit both their human companions and, by extension, the nation-state as a whole. They doggily embody the fantasy of a militarized subject who is both fearless and sensitive. In other cases, it is the relative absence of agency, as in the case of the Baghdad strays or Motari and Encarnacion's puppy, that generates the affective draw; they do not invite love, exactly, but the connection is still profound. Alternatively, the Abu Ghraib MWDs appear as highly agentic but less than lovable. In the photos, they seem disinterested in human connection, scarcely cognizant of the people who are apparently struggling to control them.[66] In those instances, the agency of the dog is obvious, but affectively troubling.

The positions of the state and military vis-à-vis dogs are incoherent, a snare that traps even venerable MWDs. According to Boggs, MWDs occupy a generally precarious position, which dramatizes the disjuncture between the professed admiration for these animals and their actual treatment.[67] For example, the Canine Members of the Armed Forces Act was predicated on a fairly straightforward claim—namely, that MWDs ought to be reclassified as active "members" of the military rather than as "equipment" for its use, a position that elicited vocal support from advocacy groups.[68] The bill died in committee. But the Defense Authorization for the 2013 fiscal year did designate MWDs as members of the armed forces and outlined procedures for their retirement and adoption. Where applicable, the document offered guidance on the matters of recognition or decoration for dogs that are "killed in action or perform an exceptionally meritorious or courageous act in service to the United States." The 2018 Defense Authorization mentions MWDs in two separate sections, one that orders the secretary of defense to prepare an annual report about military dogs (their numbers, status, cost) and another that establishes a policy for them to receive awards and commendations as appropriate.[69] It does not, however, recognize them explicitly as sentient members of the military.

The American system of military justice provides some protections to other types of animals that military personnel might encounter. In its massive *Manual for Courts Martial*, animals appear in two main places: among the list of possible victims of sodomy listed under Article 125, and—in the case of so-called "public animals" who are the property of the state or government—as potential objects of abuse in Article 134. In both instances, however, the offenses are framed as transgressions against the military itself, affronts to "good order and discipline" or a way of defaming the service, an orientation reflected in the marines' sanction of Motari and Encarnacion. That is, the animals who are victimized by such behaviors are somewhat ancillary, or incidental, to the actionable offenses themselves.

Military regulations also restrict companionate interactions between military personnel and animals. A series of "General Orders"—1-A (issued 2000), 1-B (2006), and 1-C (2013)—enumerates "Prohibited Activities" for U.S. military personnel fighting on various fronts. These all forbid personnel in the theater of operations from establishing relationships with animals that would elsewhere seem commonplace, even salubrious for both parties.[70] These prohibitions are couched within the larger framework of preventing behaviors "prejudicial to the maintenance of good order and discipline."[71] The documents also cite sensitivity to customs in countries "where local laws and customs prohibit or restrict certain activities which are generally permissible in western societies." Accordingly, the General Orders list among their prohibited activities "adopting as pets or mascots, caring for, or feeding any type of domestic or wild animal."[72] Despite the hardship that it causes individual members of the military and the way it might grate against our sensibilities about soldiers and animals, the prohibition stands.[73]

In all of this, the case of the puppy in the Motari-Encarnacion video is singularly complex.[74] The animal is clearly helpless and adorable in a very typical puppy way, but wrenches human categories into crisis, forcing the American military personnel out of their presumed cultural roles as heroes, civilizers, and gentle stewards.[75] This backs viewers into a sentimental corner in multiple ways. It essentially forces them to sympathize with the puppy, but doing so comes at the cost of dislocating the marines from their privileged affective and ideological positions. The animal does not seem to have any volition or the capacity to be anything other or more than a victim, doomed by the simple misfortune of encountering these two men: the puppy is a good, but very unlucky, dog. The familiar hierarchy of human and animal was on clear display, but only as a scaffold for cruelty. There was no apparent utility in what Motari did, no reason for Encarnacion to film it, no persuasive way to exonerate either of them: their cruelty was wanton.[76] But to simply call them "uncivilized" would be unsatisfactory, because any such attribution runs up against discourses about the nobility of American military personnel and about the essential compassion and humaneness that characterizes America itself. Typically, our inherited sentimental histories help us organize, manage, and hierarchize our feelings, but in the case of Motari, Encarnacion, and the puppy, they could not do that work. The little black and white dog provoked a crisis, revealing not only the limits of these sentiments but also their incommensurability. The dog compels spectators to make a painful choice. Consequently, I wager, the anger that the video elicited was not so much at the marines themselves or even at their unthinkable, unbelievable actions but rather at the untenable decisions that they required of spectators as a result.

Despite Mediation, Beyond Figuring: Reconsidering Anger

In regimes of sentiment, the recognition of another's suffering begins with meticulous examination of it: its causes, its magnitude, whether or not it is deserved. This process, as many scholars have noted, often compounds the suffering it is ostensibly meant to eliminate. Asma Abbas argues, for example, that the dispensation of justice in liberal societies requires a rehearsal or "performance of suffering" as a ritual that is more procedural than it is meaningful.[77] Berlant suggests that these displays amount to command performances, staged at the behest of the privileged, who earn a "sense of general moral propriety" by enacting "empathic identification" with the less fortunate others who suffer before them.[78] The impulse to watch and rewatch the video is a legacy of these practices; although the dog itself cannot perform its suffering on command, the scene can be revisited easily with just a click of the replay button. This is not an uncomplicated form of spectatorship. Brett Mills makes a provocative claim for animals' rights to privacy during crucial corporeal moments of their lives, like mating, giving birth, and dying.[79] If we take this claim seriously, it points to an added violation in what Motari and Encarnacion staged and leaves us, as spectators, in awkward, after-the-fact complicity with that transgression.

The visibility and perceptibility of suffering are always circumstantial and contingent, even more in the case of an animal. It is only by chance that the suffering of this animal came to be known at all, and it is unclear how exactly this video made its way to YouTube. One news story about the incident suggests that Motari and Encarnacion themselves uploaded the video, but most accounts describe the video "appearing" or "being posted" online.[80] Perhaps either Motari or Encarnacion wanted to get Internet-famous, or perhaps someone else sought to expose them, or perhaps whoever it was simply posted the video with no real intention at all. Still, videos of cruelty to animals—whether they are meant as exposés or entertainment—are hardly scarce. As with any viral media phenomenon, there was surely a measure of randomness in the global ignition of this one. But once the video became known, it rocketed around U.S. media outlets because of who the aggressors were. The puppy became visible because it fell into, and subsequently flew out of, the hands of U.S. Marines, a bizarre dividend of the popular fascination with and esteem for military personnel.

The swift and vociferous response to the video was characterized by anger, which found expression in shaming vituperations, pleas for justice, and a desire for vengeance. Offline, the death threats against Motari and his family are perhaps the most dramatic evidence of this. Online, many—but not all—of the expressions are less overtly menacing, but equally affectively intense. Some are comedic fantasies of the dog getting revenge, as by throwing Motari and Encarnacion off a cliff, whereas others are rather more earnest expressions of a wish for justice and punishment. Sianne Ngai argues that "ugly feelings" often arise at the site of "suspended agency,"[81] and here, with the puppy long since dead and Motari and Encarnacion beyond the reach of any civilian spectator's power to punish them, unwieldy anger is the only available recourse for viewers upset by the video.

The black and white puppy that Motari threw off a cliff as Encarnacion filmed on a cell phone became a relay for spectatorial anger at the crisis of sympathies that the marines provoked. But it also nudges us toward the possibility of new and different ways of responding to wartime suffering.

Among the most obvious examples of this are the spluttering reaction videos posted back to YouTube. The speakers are often inarticulate—sometimes with rage, other times with what seems to be a shortage of words or comprehension. Something about the video makes it hard to think; horror often leaves us dumbfounded.[82] But although we might find ourselves thunderstruck or speechless, at least momentarily, the puppy itself was apparently able to verbalize until virtually the moment of its death. The struggle to articulate our feelings is not simply a struggle to find the right words; rather, it marks the effort required to get one's emotional bearings in the gap between two irreconcilable sympathies. We find the words only when we find our footing in one sentiment or the other.

Comments posted about these videos seem vitriolic in all directions. Certainly, this is characteristic of much online conversation, but I argue that the outrage here is a reflection of that affective bewilderment. Some viewers expressed direct hostility toward Motari and Encarnacion; others chimed in to support these condemnations, rather than crafting their own. The vast and vocal majority of respondents to the video criticized the marines, but a minority did seek to exculpate or defend them, often with rhetoric that was aggressively defamatory of their detractors. Some appealed to spectators' militarized sympathies, suggesting that they did not or could not understand the trauma of war and what it drove humans to do. Others were more defiant, insulting the sense and the patriotism of those who would criticize Motari and Encarnacion, insisting that concern for the animal—which is, after all, just a dog—instead of the marines was insultingly misplaced. Predictably, none of these exchanges resolved anything, but instead further entrenched the irreconcilability of the sympathetic loyalties that motivated them.

Superficially, the majority of spectatorial anger was directed at the marines: whether on behalf of the puppy, on behalf of the nation besmirched by their conduct, or both. However, I suggest that the actual, if often unnamed, target of this anger was the crisis of sympathies that they provoked. Anger at the marines was actually somewhat incidental. Consternation over the human consequences of cruelty to animals is as old as anti-cruelty legislation; in early law on the matter, as Deckha notes, "the offense was understood as an assault on public morals and not a direct harm to the animal," so that any prosecution was, first and foremost, "motivated by anthropocentric interests."[83] In callously tossing the puppy off the cliff, the marines forced a failure in available discourses that instruct spectators and citizens how to allocate their sympathies, as well as those that justify American militarism. In effect, Motari and Encarnacion threw American identity itself into question. Sympathizing with the dog recuperates it partially, eases some of the awkwardness of disapproving of the actions of military personnel. It serves also to reaffirm a vision of American identity that includes humane care for animals. But as Wanzo points out, stories that "touch the heart" or play on the emotions very often reinforce the status quo.[84] Here, what is reinforced is a particular and deeply political belief about the quality of American compassion.

Importantly, the near-universal condemnation of the marines' behavior contrasts starkly with the minimizing, waffling, and equivocation that predominated in official and public responses to other atrocities like the torture at Abu Ghraib. I do not mean to suggest that we ought to compare the two violences or that one should be lamented more than the other.[85] But the differential allocation of sympathy in response to these mediated scenes of cruelty is revelatory. For a nation accustomed to discounting the suffering of people of color in general and people from Middle Eastern countries in particular, the human victims at Abu Ghraib did not provoke a conundrum of sympathies, and so the reaction was hesitant by comparison. With nothing to mitigate the ideological difficulty of criticizing American military personnel, there was neither an alternate affective pleasure like feeling compassion for a dog nor a sustained affective crisis, and so the scandal dissipated. However, there was a similarity in the ways that the two acts were punished: disciplinary action for those caught on camera but few if any penalties for the higher-ranking officials who might condone or overlook such behavior.

For viewers seeking the emotional release of sympathizing with someone in the video, the marines, by their callousness and lack of remorse, made it relatively easy to opt for the puppy. Consequently, expressing angry sympathy for the puppy is not a courageous ethical act. Instead, it might best be understood as obedience to a long-programmed emotional default, a sentimental drift toward an easy target. And no matter the depth or intensity of our empathy, there can be no meaningful redress for the animal's mortal suffering, and any response is, necessarily, too late to be of any real use. Even if the puppy had survived the fall, there remains an unbridgeable gap between it and even the most compassionate of spectators: how could we apologize to it? What could any one of us, really, hope to communicate? Yet despite all this essential futility, I still want to insist that attending to this kind of cruelty, caring about this kind of suffering, is an ethically important thing to do. I suggest the yelping puppy might nudge us toward a different way of responding to suffering. This requires finding a form of attunement to suffering that operates outside the discourses that seek to confine or channel it.

The rush to anger over the video obscured not only the suffering of the animal, but also its ability to articulate it. This is the essence of the harm engendered in any act of figuring. A too-narrow focus on human reactions to the video risks eclipsing those of the puppy and its capacity to communicate suffering in a compelling way. This capacity is especially remarkable, given that American audiences are by now thoroughly accustomed to bearing mediated witness to the suffering and dying of humans in wartime, to the extent that doing so has become, for most spectators, banal.[86] Somehow, the puppy overrides the uncertainty so often begotten by mediated representations of others' suffering. Even mediated by the cell phone camera, then mediated again by the computer screen, and in some cases mediated a third time by rebroadcast of the video on television, the puppy made something happen. In this way, the puppy's capacity to elicit a reaction is more remarkable than the human responses themselves.

Sentimental culture trains us to marvel, frequently and unabashedly, at the quality of our feelings for others; this is just one of the ways that it recenters the more privileged party. A preoccupation with our own feelings about the video renders the suffering of the puppy incidental and overshadows the capacity of the puppy itself to provoke them. When a spectator responds to its frightened yelps with his or her own scarcely articulate expressions of anger, this indicates that the puppy has accomplished something, an efficacy that might go unrecognized if we focus only on the intensity of our own reactions. Of course, the connection that it has generated cannot be intersubjective, as the puppy will never be able to respond, but nonetheless, the puppy has communicated a version of its experience despite multiple layers of mediation. Watching, it would have been difficult, if not impossible, to claim that the animal did not suffer. Even those who defended Motari and Encarnacion did not make this argument; they could say only that the puppy's suffering did not matter. To the extent that its suffering was undeniable, the need to imagine it was obviated. Marie-José Mondzain's description of a violent image is instructive here. She writes, "When we say that an image is violent we suggest that it can directly act on a subject without any linguistic mediation."[87] And so the puppy ought to prompt us to explore the possibility of an extradiscursive response to the suffering of another being, a response that exceeds the blunting force of mediation and slips past the constraints of sentimentality.[88] With its nonlinguistic but profoundly communicative utterance, the puppy forces us to recognize it as a sentient being, proving itself to be piercingly eloquent and surprisingly dexterous in its utilization of human channels of communication.[89]

What Lingers, Inconsolable

Our reactions to suffering often seem automatic or involuntary. The very ideologies and institutions that condition us to feel certain things in response to certain stimuli also train us to ignore that conditioning and encourage us to think of ourselves as naturally sensitive. Thus, our affective responses start to seem, in Berlant's words, like "expressions of [a] true capacity for attachment . . . rather than effects of pedagogy."[90] It is gratifying to experience emotions that we understand as rightful and appropriate, even, or especially, if they are sharp or unpleasant. Many animal rights platforms institutionalize these avenues for sentimental gratifications. Such initiatives afford important protections to animals but do not necessarily undermine the hierarchy of human dominion over them. Instead, they merely soften it.[91]

Historically, sympathy for suffering animals has been patterned on a particular structure that emphasizes their helplessness, their voicelessness, and their ultimate lack of control over their fates. In this model, the job of the compassionate human is to defend the animal against other humans. This ignites a complex of feelings for the creature, accompanied by animosity toward those who would harm it and perhaps a feeling of charitable satisfaction or having protected it or wanting to. By reacting to suffering animals in all of these ways at once, we seem to demonstrate the quality of our character, our ethics, our sensitivity. Yet the puppy demands something different. The animal does this without being an agent in the human sense, without intending to demand it, or really intending to do anything at all beyond vocalizing its own experience in the moment. Simultaneously, as our own discourses fail, we are rendered incoherent or even mute.[92]

Thinking through this commonality requires dwelling on the puppy's yelps and attending to the puppy's appearance before the camera, but moving beyond its "cuteness," which apparently was insufficient protection against cruelty. John Berger writes eloquently about the power of the zoo animal's gaze to make us discomfitingly aware of our own spectatorship, arguing that in such encounters, humans become attuned to the reciprocity of gazes between themselves and the captive animals and the obligations that might imply.[93] Yet there can be no such reciprocity here; the scene is mediated, past, and the puppy is already dead. But the significance of the video lies partly in its foreclosure of this reciprocity. The puppy underscores the radical limitations of our ability to set this wrong right, the fundamental inadequacy of our various human mechanisms for redressing animal suffering, even as it leaves us more certain of the need to do so.

About a year after the Motari-Encarnacion video was released, another instance of cruelty toward a dog was caught on camera in Baltimore, where I live and work. The case reveals a vexing confluence of cruelty, anger, and mediation. In late May 2009, Baltimore city police rescued a pit bull that had been doused with lighter fluid and set on fire, suffering serious burns on 98 percent of her body. A local shelter named the dog "Phoenix" and labored for a few days to save her, but ultimately decided to euthanize. At the time of her death, her attackers were unknown. With the help of one of Baltimore's street surveillance cameras—ubiquitous in neighborhoods where the residents are generally poor or of color—they were later identified as seventeen-year-old twin brothers, Travers and Tremayne Johnson. Though much of the public and the news media—and even, it seemed, the mayor—had already convicted them, the first case ended in a mistrial because one of the twelve jurors held that the surveillance footage was not clear enough to definitively incriminate the Johnsons.[94] Mediation, in that case, fostered skepticism. And on retrial, the brothers were found not guilty.[95]

Many Baltimore residents, along with the national anti-cruelty community, responded to the news of the mistrial and acquittal with dismay, and some with outrage. Their frustration highlights the insufficiency of extant, juridical forms of redress for animal pain.[96] This failure haunts the cases of the Johnson brothers and Motari and Encarnacion in different ways. In the trial of the Johnson brothers, advocates for animals felt like the lack of a conviction meant that the system had failed, though in the strictest sense, it had not, as a legally empaneled jury found that the prosecution's case did not clear the threshold of reasonable doubt. Motari and Encarnacion, on the other hand, were punished, but the initial empathic response to the suffering of the puppy has no place in official or legal adjudications of the crime; the hovering residual anger cannot be sufficiently redressed. This little dog is dead and cannot be resurrected. There is no restitution to be made.[97] So something lingers, inconsolable.[98]

There are ways to reduce that unsettled feeling back to manageable intensity and partake of righteous affective pleasures. We could return, for example, to sanctioned defenses of animal rights. One of the ASPCA's signature television commercials is a silent montage of pitiable domestic animals who have presumably been abandoned, neglected, or abused, overlaid with a nondiegetic song, most famously Sarah McLachlan's "Angel."[99] The logic behind the commercials is presumably that the sight of the animal combined with the poignancy of the song will move people to donate. These commercials hover on the edge of unbearable. They are evocative, even discomfiting, but not totally overwhelming, in part because they dictate a clear course of action in response to the eruption of feeling. The act of making a donation, in turn, serves to modulate the emotions that the commercials are designed to generate. The advertisements license us to tap into this wellspring of feeling and wallow for a while; perhaps this bruising pleasure is part of what we are paying for when we donate in answer to the commercial's entreaty. In these commercials, the animals are silent. Generous in its provision of the comforts of pathos, the ASPCA's appeal is never complicated by questions of the animals' agency and its sentimental imperative is obvious and easy to heed.

In 2005, the ASPCA launched a new campaign under the slogan "We Are Their Voice." The claim that "We Are Their Voice" suggests that the organization speaks for those creatures who cannot speak for themselves. Because the ASPCA has persisted in this mantra for over a decade so far, it stands to reason that it is effective in cultivating support and raising funds. When the ASPCA asserts that "We Are Their Voice," it domesticates the articulations of the suffering animal, silencing the creature, institutionalizing the potentially unwieldy human reaction to it, and inserting a mediation or a translator where none might have been required. So what would happen if it let the animals "speak," or whimper, or howl, or whine for themselves?[100] The puppy's cry hints piercingly at the answer.

During the same period, the ASPCA also mailed out fundraising brochures describing itself as moving "Beyond Emotion to Action: Saving Animals' Lives." The message here is that the unruliness of emotion is insufficient or impractical for their urgent work. But as I think about Motari and Encarnacion's puppy, or what little must remain of it, I wonder if this kind of transcendence is necessary. I appreciate that the ASPCA notes the insufficiency of feeling bad about animal suffering. But the problem with this model is that it emphasizes the transformation or evolution of the beings that are already more powerful while overlooking the capacity of animals themselves to unmake us.[101] The thing about the Motari-Encarnacion video that staggers me is the fact that I cannot move "beyond" it.[102]

There might be a value in the way the video mires its spectators.[103] After all, the historically established modes of sympathizing with animals have not succeeded in eradicating animal cruelty and clearly failed to prevent it in this case. I am not claiming that there is a source somewhere of a true or authentic sympathy that is somehow untouched or uninflected by ideologies, something magically pure and apolitical, just waiting to be tapped. Nor am I implying that I have a solution to the so far insoluble problem of cruelty to animals. But the puppy loudly reminds us of its continued urgency and of the inadequacy of the policies and mechanisms we have in place to address it. In a matter of seconds, Motari and Encarnacion provoked the crisis of these long-established sympathies. The puppy demands that we look beyond those sympathies altogether.

The black and white puppy was, presumably, a stray. Strays occupy a liminal space; they wander close to spaces of domestication and might even graft some sustenance from them but retain their fundamental wildness. Feral creatures refuse the comforts and safety afforded by captivity and serve as reminders that there are other possibilities for making one's way in the world, even if they entail more risks. Whether the dog sought out Motari and Encarnacion or they collared it, Motari and Encarnacion brought the animal into a human environment and then threw it out. The dog's status as an apparently ownerless and autonomous creature affixes another affective complexity onto it. If we imagine the animal as deprived or bereft for lack of a loving human home, it becomes more pitiable. But if we look to the animal itself, rather than to its connection with humans, a new affective potential comes into view. Feral creatures live in uneasy, partial, shifting symbiosis with their human cohabitants, neither totally dependent upon them nor entirely free of their influence. We might try to cultivate our own feral sentiments in response to the suffering, animal and otherwise, that militarization engenders. Such affects would bear the traces of their tamer and more regulated counterparts but refuse the enchainments that come with them.

CONCLUSION

A Radical and Unsentimental Attention

None of the figures I consider in this book are permitted to be angry. An angry civilian child belies the construct of the helpless, innocent, and apolitical young person. An angry military child poses an even greater threat to that construct and testifies to the extraordinary demands that militarization places on young people. An angry military spouse flouts gendered and national expectations about sacrifices willingly and cheerfully borne. An angry injured veteran reveals the inadequacy, or even the absurdity, of most civilian displays of gratitude and impugns the state for its miserly protections. An angry detainee grates against fantasies of enduring American exceptionalism. An angry dog insists that its suffering matters, even when that suffering is inflicted by the type of American otherwise assumed to be ideal. This prohibition on anger is similar to that imposed on marginalized groups more generally as a way of preempting opposition or minimizing their claims of grievance. Ultimately, that prohibition sustains militarization by making it easier to imagine that no one really objects.

Figuring always proceeds under the mantle of deep concern for the suffering of its objects. But in practice, figuring occludes that suffering, largely by obscuring the subjectivities of the beings afflicted by it and the specificities of that affliction. Figuring posits imagination as an alternative to such potentially unbearable knowledge. Under imaginative control, these figures become repositories for affective investment. These investments are often framed as anti-militarist positions. In practice, however, they work paradoxically but inevitably to render militarism more tolerable, even pleasant, for those who do not encounter its consequences directly. Figuring fabricates affective connections where they do not exist while smoothing over vexed dynamics of accountability, power, and disenfranchisement. Feelings like apprehension, affection, admiration, pity, gratitude, and sentimental or vicarious anger tidy these dynamics into familiar and comforting patterns of sentimentality. Superficially, these feelings indicate an awareness of militarization's consequences, but as I have argued in this book, they actually enable an evasion of the profound ethical questions that militarism raises. In these final pages, I want to advance the possibility that certain forms of anger, distinct from that provoked by the crisis of sympathies I described in the previous chapter, might point us toward more meaningful ethical responses to the forms of suffering that militarization begets. In this context, such anger calls out the insufficiency of extant systems designed to organize and route our feelings. It is an allergy to sentimental balms for uncertainty, agitation, and discomfiture.

Ethically (and personally), to advocate for anger feels strange and risky, a little unsettling. After all, anger could get it so wrong. But anger also seems right for a conflict in which so much remains unresolved—despite avowals that it is ending, has ended, will end. And anger also seems reasonable for a conflict in which so much suffering goes unremarked or, alternatively, gets mediated and then recycled into displays of ostentatious compassion by those who can only imagine it.

Figuring, as one such display, masquerades as a substantive response to the suffering of others. It offers an imagined and ultimately imaginary transcendence of the distance between those who suffer under militarism and those who feel for them. In actuality, it preserves and expands that distance by declining to acknowledge the political subjectivity of those beings and presuming to know and understand their suffering. At the same time, figuring smudges over questions of accountability for that suffering and neatly diverts the blame from settling on militarized violence or the state. Figuring provides easy answers to hard questions about our responsibilities to one another and replaces a substantive accounting of those responsibilities with unilateral experiences of emotional connection. To all this, certain forms of anger might offer an ethical alternative. Such anger would be relatively selfless, not motivated by or even concerned with the ways that the suffering of another might be an impingement or an inconvenience. Instead, it would entail a radical and unsentimental attention to the suffering of others, not to consume or exploit it for the pleasures of sympathizing with them, but as an expression of commitment to finding creative and meaningful responses to their predicaments.

In such an undertaking, the ethical operates as a framework for calibrating our responsibilities to one another in a materially grounded rather than imaginative way. Judith Butler defines the essence of the "ethical" more as an attunement than an action. She writes, "The ethical does not primarily describe conduct or disposition but characterizes a way of understanding the relational framework in which sense, action, and speech become possible."[1] The work of ethics in the context of militarization is to consider which forms of sense, action, and speech it enables, and for whom, and then reflect on the consequences—affective, ideological, material—of these distributions. Whose senses are left vulnerable? Who can exploit them? Who experiences militarization as increased freedom of action or greater latitude, and who experiences it primarily as constraint, restriction, or obligation? What forms of speech are encouraged or valorized, and what kinds of silence are mandated? And how can we answer these questions accurately and precisely without turning to the reassurances so reliably provided by the state, the media, or the imagination? Anger, with its stubborn refusal to be placated, could create the habitat for the emergence of more meaningful ethical responses to that suffering and to all the losses that precipitated it.

Of course, not all forms of anger can provide the necessary foundation for a more meaningful response to suffering. The sentimental anger I wrote about in the previous chapter, which was essentially an aggravation at having to make an unpleasant choice, cannot serve this purpose. And clearly, the anger of the powerful arising in response to a perceived loss of privilege or the capricious, indiscriminate anger of the sovereign are antithetical to the work of crafting those responses. Nor am I endorsing the anger that underpins hatred or inspires meanness or desires an expression of force or the anger that operates blindly, irrationally, and races down the path of least resistance, which almost invariably tracks in the direction of the less powerful.

If the limited ethical utility of these forms of anger is obvious, that of anger that seems virtuous is less so. I am not advocating for anger on behalf of less powerful others or anger felt vicariously for them. Anger-on-behalf-of is central to most of the figuring practices I have described here. It presumes that others are emotionally accessible while insisting that they feel nothing that cannot be resolved by an infusion of sympathy from outsiders. Across the preceding chapters, I have written about anger on behalf of civilian children stripped of their putative innocence, military children deprived of a parent, military wives separated from their husbands, veterans with TBI and PTSD in need of substantive care, detainees held with impunity beyond the reach of international law, a dog victimized wantonly. Even if we might agree with the stated ideological underpinnings of these forms of anger—doubtless, there are good reasons to oppose the militarization of children, for example—they are problematic. Vicarious anger is a variation on anger-on-behalf-of. As a sympathetic reflection of what we believe someone else ought to be feeling, vicarious anger actually amounts to an affective prescription specifying the proper response to a particular situation. Both forms of secondhand feeling short-circuit the possibility of the other being's anger. They presume that the other being is not angry or is not emotionally sophisticated enough to recognize that something is wrong. They also refuse to entertain the possibility that the person feeling vicarious anger might somehow be responsible for it or the circumstances that provoked it. Any type of anger experienced *for* someone else denies their subjectivity and seeks to predetermine the nature of the relationship between the person feeling that anger and the other meant to be its beneficiary.

The anger that I am endorsing here is different. It is a way of registering the suffering of others without making any kind of epistemological or affective claim on it. This kind of anger notes what is unknowable about the suffering of others, and it is a frustration not with that unknowability, an annoyance rooted in a greed for experience and sensation. Instead, it is an objection to the efforts to make the experiences of others falsely intelligible.[2] It is anger directed at the promises of sentimentality not because they leave us bereft—as in the crisis that Lance Corporal David Motari and Sergeant Crismarvin Banez Encarnacion provoked—but because they only compound the harms to others that they are meant to ameliorate. It is an anger nourished by griefs that leave us, or other beings, inconsolable.

With this line of argument, I intend to think through the complexities of this anger, but I do not prescribe the forms it should take. Such prescription is antithetical to any ethical project, because ethical action must be grounded in a choice and cannot be externally imposed. Obedience—even if it may be right, necessary, or wise, in certain circumstances—is not a substitute for ethical deliberation. Moreover, the ethical, as a mode of attunement to others, cannot be reduced to a set of behaviors. Depending on our own political subjectivities, the positions we occupy vis-à-vis the state, and the roles we play in its militarized violence, the expression and objects of an ethical anger will vary. In my case, for example, the best thing I could think to do was write this book.

The crux of this ethical anger is its rejection of the affective pleasures of militarism.[3] Militarization, as an expression of the state's anger, has a way of taking the oxygen out of everyone else's, becoming all-consuming. In this way, countless criticisms of state practice during wartime are defused by reassertions of military necessity or simply subordinated to the state's demands. In contrast, anger that acknowledges but fails to comply with efforts to make militarization palatable constitutes a mode of affective resistance and objection.[4] The ethical anger I envision here does not make an entitled demand for more direct affective contact with a suffering other, which is basically a wish to encounter the affective pleasures of militarism that much more intensely. Instead, this anger operates in the distance between those who suffer because of militarization and those who do not and insistently marks that distance as the foundation of its ethical practice. It takes measure of their losses while recognizing them as immeasurable.

Still, to end with anger feels risky. Anger is unpredictable and often exceeds our control, no matter our intentions.[5] Crucially, anger has a peculiar relationship to mediation. Other strong feelings can be captured in mediated form far more readily, and the act of mediating those feelings can be deeply enjoyable (how good does it feel to write a love letter?). But anger, like pain or suffering, often exceeds the words or images with which we might try to describe it, virtually guaranteeing that any attempt at translation will be unsatisfying. Anger emerges precisely at the site of failed transmissions and thwarted connections.

As the foundation of an ethical response, this kind of anger must be chosen, even if our options are already circumscribed by the exigencies of militarization. As Butler writes, the ethical "enter[s] . . . precisely in that encounter that confronts me with a world I never chose, occasioning that affirmation of involuntary exposure to otherness as the condition of relationality, human and nonhuman."[6] The necessity of ethical practice, in other words, is often foisted upon us by circumstances that we cannot control or might have preferred to avoid altogether. In such situations, we find ourselves confronted with other beings who make demands upon us, whether tacit or explicit. Figuring seeks to dictate the terms of that exposure to otherness. It is an attempt to fabricate a comfortable kind of relationality in which the other is present, but only as a mute and distant fiction, and so incapable of making any demands at all. Intersubjectivity, which is necessary for any affective exchange, is also a condition of mutual exposure shared with another. Figuring operates precisely in the absence of intersubjectivity. As a consequence, it will always undermine the possibility of ethical engagements with others.

Anger resists the enticements of other emotional possibilities. It accepts the insuperability of the distances separating us from those whose political subjectivities are unfathomable and the difficult necessity of attending to them nonetheless. Other affective positions cannot do this work. Apprehension, affection, admiration, gratitude, and pity all demand the other's conformity with a particular vision for how they should sense, endure, feel, grieve, delight, and suffer. By contrast, the anger I am sketching here is unconditional in its response to suffering.

Necessarily, this anger resists the temptations of self-aggrandizement and the solipsism that insists that feeling—bad, sorry, or otherwise—is enough. To end with anger is to concede that there is a scarcity of affective alternatives and that the available options are inadequate at best, while committing to scavenge and capture what they cannot contain. This anger marks their limits and then moves past them, leading us to otherwise unimaginable places—affective, ideological, political—that are wild, loud, and vertiginously full of potential.

ACKNOWLEDGMENTS

True gratitude is radical and transformative. Of all the things I learned in the process of writing this book, that lesson may be the most abiding. I learned gratitude intellectually as I hunted through the literature that theorized and deconstructed it, seeking to understand the problematics of its militarization. But I also learned it personally, a thousand times over, from the people who sustained me through these years of work. Those who know me well know that I wrote this book through the most painful and dispiriting season of my life to date. I hope that they will hear, in what follows, the bottomlessness of my appreciation for their steadfast care. Of course, many of the people I mention here did not know me or my circumstances well at all, but the completion of this book would have been impossible without their expertise, guidance, and resources. Everything in me wants to resist the notion that I am dependent on others, but the depth and intensity of my gratitude remake me and remind me otherwise, daily.

My first passes at the questions that would become foundational for this project were papers at the 2012 and 2013 meetings of the Society for Cinema and Media Studies, gracious forums for works very much in progress. Subsequently, I organized a panel—"Beyond Blood and Treasure: Rethinking War Debt"—for the 2013 meeting of the American Studies Association that emboldened me to begin thinking about these issues in earnest. Thanks to Benjamin Cooper, Irene Garza, and Ji-Young Um for their stellar contributions, and especially to Meredith Lair for her provocative challenge to rethink the imperative to say "thank you" to the troops. The opportunity to present my work at "Affect Theory: Worldings, Tensions, Futures" in 2015 was critical to this project, and feedback from Sara Cefai, Ben Anderson, and Michael Richardson greatly improved it. Stacy Takacs's invitation to join her panel for Console-ing Passions 2017 was ideally timed for the refinement of my thinking about media production for military families. For their participation in a spirited roundtable at the 2017 meeting of the American Studies Association, thanks to Wendy Kozol, Kara Thompson, Elisabeth R. Anker, and Neve Gordon.

I am fortunate to have had the opportunity to publish earlier versions or smaller sections of the book elsewhere. My thanks to those editors for their keen commentary and permission to reprint versions of that work here. The inaugural issue of *Unlikely: Journal of Creative Arts* provided a welcoming venue for my article "'That Was Mean, Motari': Spectatorship, Sympathy, and Animal Suffering in Wartime." Thanks especially to Alexis Harley and Norie Neumark. My analysis of children's encounters with wartime atrocity images began as an essay entitled, "'Coffins After Coffins': Screening Wartime Atrocity in the Classroom," in *The War of My Generation: Youth Culture in the War on Terror*, edited by David Kieran and published by Rutgers University Press. Finally, my consideration of the affects and politics of detainee creative production benefited tremendously from Sarah Cefai's readership, and I am grateful that she included that essay in her special issue of *Cultural Studies* on the theme "Mediating Affect," where it appeared as "Fictive Intimacies of Detention: Affect, Imagination, and Anger in Art from Guantánamo Bay."

So many people helped bring this book, as a material thing, into being. I couldn't quite believe it when Richard Morrison's name popped up in my inbox with an invitation to talk more about the proposal I'd sent; I still can't quite believe it now. But I am so grateful for his patience, creativity, and commitment to this project, as well as the investment and endorsement of Fordham University Press. John Garza provided essential information and reassurance every time I asked. Eric Newman was a pleasure to work with. Richard also coordinated a fantastic review experience for its first draft. As part of that process, Donald Pease's praise clarified the stakes of my argument and helped me see this as a book worth writing. Carrie Rentschler absolutely deserves her reputation as a reader who is clear-eyed, constructive, and generous in equal measure. Thanks also to Leslie Mitchner and Daniel Bernardi at Rutgers University Press. And to Lisa Parks. I remain grateful to Brian Halley for giving my first book a home at the University of Massachusetts Press.

My access to key primary source material for this book was contingent on the generosity of many people. Ann Hoog at the Library of Congress's American Folklife Center made available the children's drawings of September 11 on very short notice and provided me with a wealth of backstory on the collection. Thanks to Jake at the National Center for Telehealth and Technology for background details on *Military Kids Connect.* In part because their online archive is so user-friendly, I never had to contact the people who maintain the website for the American Presidency Project at the University of California, Santa Barbara, but my research on presidential proclamations would have been virtually impossible without it. Tara Jones, then with the Office of the Secretary of Defense, facilitated my trip to Guantánamo Bay, the experience that catalyzed this research; I am very grateful to her and everyone else who made time to speak with me while I was there, for their hospitality to this skeptical guest. For enabling me to visit the lab at the Center for Neuroscience and Regenerative Medicine, thanks to Stacey Gentile and also to Sarah Marshall at the Uniformed Services University of the Health Sciences, and to Dr. Daniel Perl, for working with me over many months to make sure I got their story right. For their speedy assistance tracking down necessary secondary sources, thanks to the staff at the UMBC library and to Kristian Pollock for sending me a copy of her groundbreaking work on social stress.

I also want to thank, quite belatedly, Benjamin Thompson. I met Ben when I was working on my Ph.D. in Comparative Studies at Ohio State. He was an undergraduate in the program and a fierce anti-militarist voice in Iraq Veterans Against the War. Ben asked if I would supervise an independent study in which he could reflect on his experiences as an MP at Abu Ghraib and his role in the documentary *The Prisoner, Or: How I Planned to Kill Tony Blair*. Patient, humble, and whip-smart, Ben taught me to look beyond the most obvious and spectacular forms of state violence to document more precisely the mechanisms that drive American militarism. This has inspired virtually all of the research I've done since.

Philip Armstrong was instrumental in my graduate education. Although our paths don't cross often these days, I remain grateful for his example, his work, and his mentorship.

My research assistant, Ibrahim Er, worked diligently to get my many citations into shape. He is a model of conscientiousness, industry, and enthusiasm for the academic enterprise.

The University of Maryland, Baltimore County has provided material support for my research from the outset and has underwritten *Figuring Violence* in many crucial ways. The College of Arts, Humanities, and Social Science's Dean's Research Fund offset travel expenses, research assistance, graphic design consultation, and a necessary shopping spree for books; thank you to Scott Casper and Eva Dominguez. The Dresher Center for the Humanities has funded my work multiple times, enabling me to travel for conferences and invited presentations, and so enriching my thinking immeasurably. The center is an unwavering advocate for the humanities on campus and in the world; thank you to the board, Natalia Panfile, Courtney Hobson, and to Jessica Berman for her indefatigable leadership.

When I learned, in 2016, that I had won a Regents' Award for Excellence in Research from the University System of Maryland, I was stunned. It simply would not have happened if not for the Nominations Committee, for UMBC leadership, for Dina Glazer, who handled the logistics of my nomination, and for the letters of support from Wendy Kozol, Deepa Kumar, Bonnie Miller, and Carrie Rentschler. The award was especially meaningful given the skepticism and occasional derision with which interdisciplinary humanities scholarship is often viewed; I am privileged to work in an institutional context where this kind of work is not only accepted but encouraged and celebrated. The award, with its implied endorsement of my past work and future potential, was a perfectly timed infusion of confidence when my need for it was acute.

The Department of Media and Communication Studies provides a lively home base from which to work. Our students challenge me daily to think differently and better. Jason Loviglio is a leader flush with compassion and good cheer, and we all benefit from the climate he has created. For administrative support, thanks to Samirah Hassan and to Abigail Granger for that and so much more. I have wonderful colleagues in Elizabeth Patton, Donald Snyder, Bryce Peake, Kristen Anchor, Bill Shewbridge, and Fan Yang. I am especially grateful to Fan for being such an astute reader of my work, and I feel fortunate to count her and Marc Abram as friends. The department of Gender + Women's Studies has become a second professional home. Thanks also to Visual Arts for letting me wander into their territory from time to time. It's been a pleasure serving on MFA thesis committees for Parastoo Aslanbeik, Chinen Aimi Bouillon, Melissa Penley Cormier, Chanan Delivuk, Jeffrey Gangwisch, and Jaclin Paul.

I am immeasurably grateful to everyone who invited me to share my work with them. Thanks to the Baltimore chapter of the American Institute for Graphic Arts (AIGA) and Joseph Anthony Brown for inviting me to talk about the "art of oppression," and to the audience for their smarts. Aimi Bouillon curated a stellar show, Feminism Fights Patriarchal Power, and talking about the militarization of gratitude there was deeply energizing. Thanks to Meg Handler and Michael Sharp at *Reading the Pictures* for allowing me to nerd it up, publicly, about visual culture on their site.

Two speaking invitations in the spring of 2016 compelled me to think anew on key assumptions about the relationships between affective investment and militarism. Thanks to Scott Laderman for bringing me to Hong Kong to join a conversation about American imperial benevolence, and to Andrea Noble for inviting me to Italy for "Photography and Its Publics." I feel lucky to have met Andrea while she was still with us in this world and think often and with astonishment about the kind words that she and Thy Phu had for my first book. Even though the work I presented in those places does not appear in *Figuring Violence*, my thinking was indelibly shaped by the exchanges I had there.

Enthusiasm for my work came as happy surprises from Mehdi Semati, Piotr Szpunar, and Sarah Sentilles; the inspiring coincidence of Sarah's praise for my first book as I was racing to finish this one was exceedingly fortuitous.

Erin Thompson's curatorial work on "Ode to the Sea: Art from Guantánamo Bay" was, in a word, amazing. After I saw the show, I was flummoxed by the unfamiliar experience of not having anything critical to say, and since then, I have been continually astonished by her humility and generosity. She connected me with Alka Pradhan, a lawyer for Ammar Al-Baluchi, the artist whose work appears on the cover. I am overwhelmingly grateful to Alka for taking on the extra work that my request to reproduce Ammar's work entailed, and to Ammar for considering and granting it.

My workdays are enlivened by colleagues like Nicole King, Ana Oskoz, Nicoleta Bazgan, Amy Froide, Kate Drabinski, and Mark Durant. Thanks to Carole McCann for her mentorship and her vision. The most rewarding service work I do is cochairing the Women's Faculty Network for our college; thanks to Bev Bickel for her help rebooting it and for all the ways she brightens this world. Mejdulene Shomali's company is a wry Midwestern delight. Tamara Bhalla was the first person who befriended me when I came to UMBC, and I'm so grateful for her leadership of the Women's Faculty Network, her sharpness, her energy, and the closeness we've cultivated.

I have a fantastic family. Aunties, uncles, cousins: lucky, lucky me. I'm grateful to Carol Baumhardt for the love she showed my dad for years, and for keeping me close now that he's gone. My mom, Karen, remains the most important teacher I've ever had, and somehow finds the balance between encouraging me to work hard and urging me to find rest. She remains the force that anchors me to home, wherever I am. Two of the people who would have been most excited for this book did not live to see it in print. My nana, Virginia Zawistowski, was inexhaustibly proud of her grandchildren and kept all the books we published prominently displayed in her living room; I'm sorry this book won't find a home on that table. Then there's my dad, Gene. I signed the advance contract on this book three days before he got sick for the last time, two months before he was gone for good, and his relatively sudden departure knocked a whole lot of my life upside down. Grief taught me that some feelings simply cannot be intellectualized away, but over time I have found them softened a bit by gratitude for almost thirty-five years with a father who was miles beyond quirky, a little impossible, and so, so loving.

Beyond the good fortune of my family and the home that they provided, so many other people have since helped me make my own way in this world. Thanks to Monica Ott for running the coolest yoga studio around. Donna Morgan brightens our neighborhood and my day every time I see her. I am grateful to Patrick Kelly for making a faraway place feel familiar. The promise of a trip to visit with Dominique Farag and Nikolaj Svorin in their respective home countries was serious motivation to tie up this project's loose ends. I want to thank Rob Pawloski and his family for all their kindness and care. Rita Trimble restores my faith in humanity by the mere fact of her existence. Becky Eager is the smartest, steadiest running buddy I could ask for. After all these years, Dave Kieran remains for me a paragon of collegiality, dedication, and unparalleled excellence at ordering *dim sum*. I am still surprised, pretty much every day, that Wendy Kozol would see fit to count me as a friend and collaborator. It's as much fun to work with her as it is to shop with her, and I hope that if I watch her carefully enough, I'll eventually become a fraction as brilliant and good as she. The toughest girl I know, Rebecca Skidmore Biggio, teaches me constantly about persevering and is as warm and wise and creative as they come.

And Meghan Marx. The work she's done for this book is relatively easy to quantify: skillful design assistance, delivery of amazing meals when I was too busy or weary to make them, unshakable faith in the project, and me, every time I came up short on both. It would take many more millions of words, many more thousands of pages to account for everything else she has given me over the course of our friendship. If the truest form of gratitude is the staggering acknowledgment of a debt that I could never repay, then she has mine, absolutely. Somewhere along the way, at her insistence, I stopped trying to tally it all, and have come to understand and accept that, for having her in my world, I am simply lucky beyond all measure.

NOTES

ON THE COVER IMAGE: "VERTIGO AT GUANTANAMO"

1. The Rendition Project, "Ammar Al-Baluchi," 2018, https://www.therenditionproject.org.uk/prisoners/ammar-albaluchi.html.

2. "AE448 (AAA) Mr. al Baluchi's Motion to Remove Designation as 'High Value Detainee,'" in *United States of America V. Khalid Shaikh Mohammad, Walid Muhammad Salih Mubarak Bin 'Attash, Ramzi Bin Al Shibh, Ali Abdul-Aziz Ali, Mustafa Ahmed Adam Al Hawsawi*, September 7, 2016. In their motion requesting that the military commission remove Al-Baluchi's status as a "High Value Detainee," his defense suggested that the category of HVD was essentially a "post hoc" justification by the CIA for its use of "Enhanced Interrogation Techniques" (5). The defense also noted that HVDs have no mechanism—e.g., good behavior—to improve the terms of their confinement and detailed the ways that Al-Baluchi's circumstances will "over time degrade his capacity to participate in his own defense" (8). The motion was denied. See also Helen Klein Murillo, "This Week at the Military Commissions, 3/20 Session: Medical Records, High-Value Detainee Designations, and Classification Guidance," *Lawfare Blog*, March 23, 2017, https://lawfareblog.com/week-military-commissions-320-session-medical-records-high-value-detainee-designations-and.

3. "The Guantánamo Docket: Abd Al Aziz Ali," *New York Times*, 2018, https://www.nytimes.com/interactive/projects/guantanamo/detainees/10018-abd-al-aziz-ali.

4. Thomas H. Kean et al., *The 9/11 Commission Report: Final Report of the National Commission on Terrorist Attacks upon the United States*, U.S. Government Printing Office, 2004, https://www.gpo.gov/fdsys/pkg/GPO-911REPORT/pdf/GPO-911REPORT.pdf, 434.

5. Andrew Buncombe, "Alleged 9/11 Plotter Held at Guantanamo Illegally Should Be Released Immediately, says UN," *Independent*, February 28, 2018, https://www.independent.co.uk/news/world/americas/ammar-al-baluchi-9-11-plotter-un-release-guantanamo-gitmo-illegal-detention-a8233451.html.

INTRODUCTION: FABRICATED CONNECTIONS, DEEPLY FELT

1. Rebecca A. Adelman, *Beyond the Checkpoint: Visual Practices in America's Global War on Terror* (Boston: University of Massachusetts Press, 2014), 63–65.

2. Amira Jarmakani argues for the centrality of desire in the War on Terror, particularly in the imperial ambitions underpinning it, despite its purported grounding in fear; Jarmakani, *An Imperialist Love Story: Desert Romances in the War on Terror* (New York: New York University Press, 2015).

3. Thanks to Dave Kieran for helping me think through possible periodizations of the War on Terror.

4. Jon Simons and John Louis Lucaites note a paradox of wartime visibility in the United States—namely that war is normalized because it is everywhere but rendered invisible by that same ubiquity. They suggest that one "register" through which war becomes invisible is that of "displacement," a cultural pattern of looking away (or being diverted) from the truths of wartime suffering; Simons and Lucaites, "Introduction," in *In/Visible War: The Culture of War in Twenty-First-Century America*, ed. Jon Simons and John Louis Lucaites (New Brunswick, N.J.: Rutgers University Press, 2017), 2.

5. Cynthia Enloe, "Ticonderoga, Gettysburg, and Hiroshima: Feminist Reflections on Becoming a Militarized Tourist," *American Quarterly* 68, no. 3 (September 2016): 532.

6. Reflecting on contemporary wartime rhetoric, Barbara Biesecker argues that the "signature stylistic gesture of the War on Terror was to have put other bodies and materialities—from our adversaries to our infrastructure—*under erasure*"; Biesecker, "No Time for Mourning: The Rhetorical Production of the Melancholic Citizen-Subject in the War on Terror," *Philosophy and Rhetoric* 40, no. 1 (2007): 157.

7. Asma Abbas, "Voice Lessons: Suffering and the Liberal Sensorium," *Theory & Event* 13, no. 2 (2010).

8. Avery F. Gordon, *Ghostly Matters: Haunting and the Sociological Imagination* (Minneapolis: University of Minnesota Press, 1997), 4. Complex personhood, in Gordon's formulation, "means that the stories people tell about themselves, about their troubles, about their social worlds, and about their society's problems are entangled and weave between what is immediately available as a story and what their imaginations are reaching toward."

9. Critical attention to affect need not, and should not, displace ideological analysis; Ruth Leys, "The Turn to Affect: A Critique," *Critical Inquiry* 37, no. 3 (Spring 2011): 434–72.

10. Mary A. Favret, *War at a Distance: Romanticism and the Making of Modern Wartime* (Princeton: Princeton University Press, 2010), 11.

11. Judith Butler encapsulates the paradoxical dynamic linking power and subjectivity as follows: "No subject comes into being without power, but . . . its coming into being involves the dissimulation of power, a metaleptic reversal in which the subject produced by power becomes heralded as the subject who founds power"; Butler, *The Psychic Life of Power: Theories in Subjection* (Stanford, Calif.: Stanford University Press, 1997), 15–16.

12. Ben Anderson, "Modulating the Excess of Affect: Morale in a State of 'Total War,'" in *The Affect Theory Reader*, ed. Melissa Gregg and Gregory J. Seigworth (Durham: Duke University Press, 2010), 162.

13. Conventional accounts of political subjectivity privilege speech, but Abbas argues that "political subjectivity need not be limited to the sayable *qua* logos; it can include other forms of enunciation and demonstration as well"; Abbas, "In Terror, in Love, out of Time," in *At the Limits of Justice: Women of Colour on Terror*, ed. Suvendrini Perera and Sherene H. Razack (Toronto: University of Toronto Press, 2014), 507.

14. Noëlle McAfee writes that citizenship is "integral to one's self-understanding as a being worth heeding, as having subjectivity, having a place in the world"; McAfee, *Democracy and the Political Unconscious* (New York: Columbia University Press, 2008), 7.

15. Marita Sturken, "Feeling the Nation, Mining the Archive: Reflections on Lauren Berlant's *Queen of America*," *Communication and Critical/Cultural Studies* 9, no. 4 (December 2012): 355.

16. My understanding of "political subjectivity" is influenced by John Protevi's definition of "bodies politic." He describes "bodies politic" as "cognitive agents that actively make sense of situations: they constitute significations by establishing value for themselves, and they adopt and orientation or direction of action"; Protevi, *Political Affect: Connecting the Social and the Somatic* (Minneapolis: University of Minnesota Press, 2009), xiv. In my reading, Protevi's analysis can apply to individuals or political collectivities.

17. Kelly Oliver, *Witnessing: Beyond Recognition* (Minneapolis: University of Minnesota Press, 2001), 206.

18. Beginning in the nineteenth century, the capacity to feel deeply and publicly express strong emotions became privileged indicators of one's subjectivity and worth; Brenton J. Malin, *Feeling Mediated: A History of Media Technology and Emotion in America* (New York: New York University Press, 2014), 45–46.

19. Teresa Brennan, for example, described affect as literally, physically contagious; Brennan, *The Transmission of Affect* (Ithaca: Cornell University Press, 2004).

20. Rei Terada, *Feeling in Theory: Emotion after the "Death of the Subject"* (Cambridge, Mass.: Harvard University Press, 2001), 15.

21. Sara Ahmed, "Affective Economies," *Social Text* 22, no. 2 (Summer 2004): 117.

22. Terada elegantly disambiguates these concepts (Terada, *Feeling in Theory*, 4–6). Most important for my purposes is her definition of *pathos*, which she describes as "the explicitly representational, vicarious, and supplementary dimension of emotion" (5).

23. Kathleen Stewart, *Ordinary Affects* (Durham: Duke University Press, 2007), 59.

24. Ann Cvetkovich employs the term "affect" "in a generic sense" to refer to affects, feelings, emotions, impulses, and desires; Cvetkovich, *Depression: A Public Feeling* (Durham: Duke University Press, 2012), 4.

25. Barbara H. Rosenwein, "Worrying About Emotions in History," *American Historical Review* 107, no. 3 (June 2002): 842.

26. Protevi, *Political Affect*, 50.

27. Ibid., 9.

28. Andrew A. G. Ross describes affective contagion as a process of "harmonization" rather than direct "mirroring"; Ross, *Mixed Emotions: Beyond Fear & Hatred in International Conflict* (Chicago: University of Chicago Press, 2014), 2.

Relatedly, affects can instantiate divisions even as they bring certain collectivities into being. Gregory Seigworth and Melissa Gregg argue that "affect marks a body's *belonging* to a world of encounters or; a world's belonging to a body of encounters but also, in *non-belonging*, through all those for sadder (de)compositions of mutual in-compossibilities"; Seigworth and Gregg, "An Inventory of Shimmers," in Gregg and Seigworth, *Affect Theory Reader*, 2.

29. Lauren Berlant, "The Subject of True Feeling: Pain, Privacy, and Politics," in *Cultural Studies & Political Theory*, ed. Jodi Dean (Ithaca: Cornell University Press, 2000), 45. In her account of the ascendance of melodramatic political discourse in the United States, Elisabeth R. Anker defines it as a form that "traffics in *intensified affect* that relies on certain kinds of *identifications* with the suffering of others"; Anker, *Orgies of Feeling: Melodrama and the Politics of Freedom* (Durham: Duke University Press, 2014), 34.

30. Brian Massumi writes, "What is not actually real can be felt into being"; Massumi, "The Future Birth of the Affective Fact: The Political Ontology of Threat," in Gregg and Seigworth, *Affect Theory Reader*, 54. Here, Massumi is referring to the way that imagining a threat can make a subject believe and feel that she is threatened, which is a different line of analysis than the one I am pursuing.

31. For two perspectives on this debate, see Susie Linfield, *The Cruel Radiance: Photography and Political Violence* (Chicago: University of Chicago

Press, 2012), and Susan D. Moeller, *Compassion Fatigue: How the Media Sell Famine, War, and Death* (New York: Routledge, 1999).

32. Lilie Chouliaraki, "The Mediation of Suffering and the Vision of a Cosmopolitan Public," *Television & New Media* 9, no. 5 (2008): 372.

33. On feeling nothing, see Adelman and Wendy Kozol, "Discordant Affects: Ambivalence, Banality, and the Ethics of Spectatorship," *Theory & Event* 17, no. 3 (2014), http://muse.jhu.edu/journals/theory_and_event/vol17/17.3.adelman.html.

34. Kozol, *Distant Wars Visible: The Ambivalence of Witnessing* (Minneapolis: University of Minnesota Press, 2014), 200.

35. Wendy Brown describes the composition of the state as follows: "The domain we call the state is not a thing, system, or subject, but a significantly unbounded terrain of powers and techniques, an ensemble of discourses, rules, and practices, cohabiting in limited, tension-ridden, often contradictory relation to one another"; Brown, *States of Injury: Power and Freedom in Late Modernity* (Princeton: Princeton University Press, 1995), 174.

36. Jacqueline Rose, *States of Fantasy* (Oxford: Clarendon, 1996), 8.

37. Raymond Geuss, *Politics and the Imagination* (Princeton: Princeton University Press, 2010), 69, 68.

38. Anderson, *Imagined Communities: Reflections on the Origins and Spread of Nationalism*, 2nd ed. (New York: Verso, 1991). Anderson contends that mass media and the uniformity and simultaneity it engenders are central to the formation and maintenance of imagined national communities; his analysis emphasizes print media, specifically newspapers, a form of mediation that is less pertinent for my purposes here.

39. Scholars from a range of disciplines, including geography, history, political science, and visual culture studies, have documented the functions of imagination in contemporary American militarism. Among them: W. J. T. Mitchell, who argues that terrorism is "largely an affair of the imagination"; Michael Barkun, who analyzes the discourse of invisibility in policy approaches to terrorism; and John Mueller, who contends that we have imaginatively "overblown" the threat of terrorism, so that "terror continues to boom even as terrorism struggles"; Mitchell, *Cloning Terror: The War of Images, 9/11 to the Present* (Chicago: University of Chicago Press, 2011), 14; Barkun, *Chasing Phantoms: Reality, Imagination, and Homeland Security Since 9/11* (Chapel Hill: University of North Carolina Press, 2011), ix, 1; Mueller, *Overblown: How Politicians and the Terrorism Industry Inflate National Security Threats, and Why We Believe Them* (New York: Free Press, 2006), 173. Relatedly, Marc Redfield contends that the rhetoric of a War on Terror clears imaginative space: "The declaration of a War on Terror is undecidably and incalculably performative and constative, real and fictional, literal and

rhetorical, consequence and nugatory, radically singular and endlessly iterable and generalizable"; Redfield, *The Rhetoric of Terror: Reflections on 9/11 and the War on Terror* (New York: Fordham University Press, 2009), 58–59.

40. Joseph Masco, *The Theater of Operations: National Security Affect from the Cold War to the War on Terror* (Durham: Duke University Press, 2014), 1.

41. Ibid., 18.

42. Louise Amoore, *The Politics of Possibility: Risk and Security beyond Probability* (Durham: Duke University Press, 2013), 159.

43. Donald E. Pease, *The New American Exceptionalism* (Minneapolis: University of Minnesota Press, 2009). Pease builds extensively on Rose's understanding of the relationship between states and fantasies. In his analysis, Pease distinguishes fantasy from both mystification and delusion, calling it instead a "dominant structure of desire" and a form of imagination (1). He argues that "over and above its monopoly on legitimate violence, the modern state relied on fantasy for the authority that it could neither secure nor ultimately justify" (2).

44. Eva Cherniavsky, *Neocitizenship: Political Culture after Democracy* (New York: New York University Press, 2017), 144.

45. Ibid., 142.

46. Many of the melodramatic eruptions of feeling that Anker describes in *Orgies of Feeling* devolve to legitimating state action (111). She identifies empathy for fellow citizens in particular as an "express demand to legitimate state power" (4).

47. Kevin McSorley, "Doing Military Fitness: Physical Culture, Civilian Leisure, and Militarism," *Critical Military Studies* 2, no. 1–2 (2016): 114.

48. Enloe, *Maneuvers: The International Politics of Militarizing Women's Lives* (Berkeley: University of California Press, 2000), 2.

49. Enloe, *Bananas, Beaches, and Bases: Making Feminist Sense of International Politics*, updated ed. (Berkeley: University of California Press, 2000), 147.

50. Simon Keller describes patriotism as an intensely possessive form of love for a country understood as the patriot's very own (beyond simple appreciation for it); Keller, *The Limits of Loyalty* (Cambridge: Cambridge University Press, 2007), 60.

51. Andrew J. Bacevich, *Breach of Trust: How Americans Failed Their Soldiers and Their Country* (New York: Metropolitan, 2013).

52. Liberal counterinsurgencies, as Laleh Khalili argues, also endeavor to make their operations seem agreeable even to target populations, deploying happiness as strategy, tactic, and objective. She writes, "On a battlefield, the uses of happiness work both to coerce the population and to veil this coercion"; Khalili, "The Uses of Happiness in Counterinsurgencies," *Social Text* 32, no. 1 (Spring 2014): 38.

53. Citing individual creativity in the uptake of various emotions, Ross, *Mixed Emotions*, says that it "ensures that emotional entrepreneurs do not fully control the emotions they deploy" (57).

54. Ibid., 81.

55. Anker, *Orgies of Feeling*, 7.

56. In Jenny Edkins's terms, "The political is that which enjoins us not to forget the traumatic real but rather to acknowledge the constituted and provisional nature of what we call social reality"; Edkins, *Trauma and the Memory of Politics* (Cambridge: Cambridge University Press, 2003), 12.

57. Berlant, *Cruel Optimism* (Durham: Duke University Press, 2011), 23.

58. Jasbir K. Puar and Amit Rai's analysis of the composition of the "monster-terrorist-fag" icon is an obvious conceptual precedent for the work I do here. But my focus differs in two fundamental ways. First, the imaginaries I am concerned with are all appreciative rather than hostile, seeking to draw figures closer rather than repel or exclude them. Second, while Puar and Rai argue that the development of the monster-terrorist-fag coproduces "docile patriots," I focus on the active participation of Americans in these processes of figuring; Puar and Rai, "Monster, Terrorist, Fag: The War on Terrorism and the Production of Docile Patriots," *Social Text* 20, no. 3 (2002): 117–48.

59. I originally envisioned this project as being located in affect *theory* but understand it now as an undertaking in affect *studies*.

60. Puar's *Terrorist Assemblages* maps the coproduction of an ideally homonormative and nationalist queer subject alongside the figure of the terrorist marked as sexually perverse. This analysis provides a key methodological precedent for analyzing how figures are constructed in wartime as my book maps out a larger network of them; Puar, *Terrorist Assemblages: Homonationalism in Queer Times* (Durham: Duke University Press, 2007). Additionally, Sturken, in *Tourists of History*, has queried how particular, often otherwise valueless commodities, are imbued with emotional and political significance in the aftermath of trauma; *Tourists of History: Memory, Kitsch, and Consumerism from Oklahoma City to Ground Zero* (Durham: Duke University Press, 2007). My project here is to consider how particular living beings are conscripted to similar work.

61. Christine Sylvester, *War as Experience: Contributions from International Relations and Feminist Analysis* (New York: Routledge, 2013), 125–26. Sylvester's advocacy of a "collage" method for doing this work coheres with my own approach.

62. Anker, *Orgies of Feeling*, 3.

63. Berlant, *Cruel Optimism*, 182.

64. Butler, *Frames of War: When Is Life Grievable?* (London: Verso, 2009). Butler argues for a political and ethical agenda that works "to shift the very terms of recognizability in order to produce more radically democratic results" (6).

65. Berlant, "Subject of True Feeling," explores how mourning for others can be a form of social death making, noting that "even progressives do it . . . when 'others' can be ghosted for a good cause" (43).

66. Oliver, *Witnessing*, 193, 213.

67. Terada, *Feeling in Theory*, 19.

68. Enloe, *Maneuvers*, 293.

69. Sianne Ngai, *Ugly Feelings* (Cambridge, Mass.: Harvard University Press, 2007), 1, 27.

70. Ethnographies of militarization, like those by Catherine Lutz, Kenneth T. MacLeish, and Zoë H. Wool, provided important empirical points of reference for my work; Lutz, *Homefront: A Military City and the American Twentieth Century* (Boston: Beacon Press, 2001); MacLeish, *Making War at Fort Hood: Life and Uncertainty in a Military Community* (Princeton: Princeton University Press, 2013); Wool, *After War: The Weight of Life at Walter Reed* (Durham: Duke University Press, 2015).

71. Ross, *Mixed Emotions*, describes the methodological challenge of retaining the critical capacity to "account for [affect's] elusive, ephemeral, and unpredictable effects" (39). For my part, I exercised a preference for unique objects, previously unanalyzed, but avoided things that were culturally obscure.

72. Berlant, "Introduction: Compassion (and Withholding)," in *Compassion: The Culture and Politics of an Emotion*, ed. Lauren Berlant (New York: Routledge, 2004), 5–6.

73. Masco, *Theater of Operations*, 3.

74. Bacevich, *Breach of Trust*, 110, 31–32. Mary L. Dudziak makes a comparable claim: "As war goes on, Americans have lapsed into a new kind of peacetime. It is not a time without war, but instead a time in which war does not bother everyday Americans"; Dudziak, *War Time: An Idea, Its History, and Its Consequences* (New York: Oxford University Press, 2012), 135.

75. Enloe's definition of militarized complicity is instructive here; she defines complicity as "lending one's moral or behavioral support to any institution's legitimacy, effectiveness, and credibility, without speaking out directly or taking explicit actions to contribute to that support"; Enloe, "Ticonderoga, Gettysburg, and Hiroshima," 530.

76. Eddie Harmon-Jones, Cindy Harmon-Jones, David M. Amodio, and Philip A. Gable, "Attitudes toward Emotions," *Journal of Personality and Social Psychology* 101, no. 6 (2011): 1332–50.

77. Abbas, "In Terror, in Love," argues that love and its attendant pangs "index our political locations" with respect to the "bindings of state, society, and ideology" (503).

78. Berlant admonishes, "Your objects are not objective, but things and scenes that you have converted into propping up your world, and so what seems objective and autonomous in them is partly what your desire has created and therefore is a mirage, a shaky anchor"; Berlant, *Desire/Love* (Brooklyn, N.Y.: Punctum, 2012), 6.

79. Ahmed, "Affective Economies."

80. Danielle Celermajer offers an elegant and provocative challenge to consider how we might cultivate attunement to suffering "not as a performance of an ethical precept, but as a response to the phenomenon itself"; Celermajer, "Unsettling Memories and the Irredeemable," *Theory & Event* 19, no. 2 (April 2016), under "The Messianic Rubble."

81. Stewart, "Afterword: Worlding Refrains," in Gregg and Seigworth, *Affect Theory Reader*, 340.

82. Ross, *Mixed Emotions*, 44.

83. Berlant, "Thinking About Feeling Historical," in *Political Emotions: New Agendas in Communication*, ed. Janet Staiger, Ann Cvetkovich, and Ann Reynolds (New York: Routledge, 2010), 229.

84. Butler, *Senses of the Subject* (New York: Fordham University Press, 2015), 5.

85. In *Depression: A Public Feeling*, Cvetkovich calls for and models a different and more expansive scholarly approach to affect. Her book is part memoir and part scholarly analysis, and Cvetkovich positions them as two equally revelatory ways of knowing the same thing.

1. ENVISIONING CIVILIAN CHILDHOOD

1. N. Wayne Bell, *We Shall Never Forget 9/11: The Kids' Book of Freedom/A Graphic Coloring Novel on the Events of September 11, 2001* (St. Louis: Really Big Coloring Books, 2011), 31.

2. Ibid., 19.

3. Ibid., 24.

4. Patricia Pace, "All Our Lost Children: Trauma and Testimony in the Performance of Childhood," *Text and Performance Quarterly* 18 (1998): 238. Anneke Meyer makes a similar claim, arguing that children are "*valued exclusively* in emotional terms"; Meyer, "The Moral Rhetoric of Childhood," *Childhood* 14, no. 1 (2007): 96.

5. Steven Mintz, *Huck's Raft: A History of American Childhood* (Cambridge, Mass.: Belknap Press of Harvard University Press, 2004), 199. Mintz argues that the emphases on both types of vulnerability circumscribed

children's interactions with adults and functionally delimited the spheres in which they could acquire and demonstrate their own competence.

6. James Schmidt, "Children and the State," in *The Routledge History of Childhood in the Western World*, ed. Paula S. Fass (London: Routledge, 2013), 174.

7. Ibid., 174–75.

8. Sara Ahmed, *Willful Subjects* (Durham: Duke University Press, 2014), 75. This emphasis on the child's will arose, she suggests, with a reordering of sources of authority in the home and family as paternal authority replaced that of the church (63).

9. Peter N. Stearns, "Childhood Emotions in Modern Western History," in *The Routledge History of Childhood in the Western World*, 163. Stearns argues that this was an offset for the tandem expectation of sexual restraint.

10. Meyer, "Moral Rhetoric of Childhood," 87.

11. Ibid., 88.

12. Schmidt, "Children and the State," 185.

13. Stearns, "Childhood Emotions," 166. Stearns specifically cites the archetype of the happy salesman. Sunaina Marr Maira contends that, in general, "the notion of youth-as-transition is necessary to the division of labor and the hierarchy of material relations specific to various formations of the capitalist state"; Maira, *Missing: Youth, Citizenship, and Empire After 9/11* (Durham: Duke University Press, 2009), 15.

14. The human rights paradigm of childhood vulnerability is enshrined in the United Nations Convention on the Rights of the Child, November 1989, http://www.ohchr.org/en/professionalinterest/pages/crc.aspx. The combination of the definite article and singular noun here—the child—implies that all children are similar to the point of interchangeability, universal in their needs and entitlements.

15. Meyer, in her research on popular beliefs about pedophilia, describes the transformation of childhood into an unassailable "moral rhetoric" through various processes of "sacralization"; Meyer, "Moral Rhetoric of Childhood," 85.

16. Robin Bernstein, *Racial Innocence: Performing American Childhood from Slavery to Civil Rights* (New York: New York University Press, 2011), 2.

17. Elisabeth R. Anker, *Orgies of Feeling: Melodrama and the Politics of Freedom* (Durham: Duke University Press, 2014), 5.

18. See, for example, ChildHelp, "Child Abuse Statistics and Facts," n.d., accessed January 26, 2016, https://www.childhelp.org/child-abuse-statistics/.

19. Anker, *Orgies of Feeling*, 15.

20. As a counter to pervasive worries of this sort, Allen Meek argues that traumatizing images can, potentially, become the grounds for transforma-

tive social and political action and intimates that this might apply to young people as well; Meek, "Media Traumatization, Symbolic Wounds, and Digital Culture," *Communication and Media* 11, no. 38 (2016): 91–110.

21. Maira, *The 9/11 Generation: Youth, Rights, and Solidarity in the War on Terror* (New York: New York University Press, 2016), 194–233.

22. Asma Abbas, "In Terror, In Love, Out of Time," in *At the Limits of Justice: Women of Colour on Terror*, ed. Suvendrini Perera and Sherene H. Razack (Toronto: University of Toronto Press, 2014), 503.

23. Lauren Berlant, "The Theory of Infantile Citizenship," *Public Culture* 5, no. 3 (1993): 399.

24. Ibid., 407.

25. Marita Sturken argues that infantile citizenship reveals the mechanics of "identification with the state at one's own expense" and the processes by which "modes of sentiment that might have been perceived as weakening its stature become the terrain through which it is recuperated"; Sturken, "Feeling the Nation, Mining the Archive: Reflections on Berlant's *Queen of America*," *Communication and Critical/Cultural Studies* 9, no. 4 (December 2012): 360, 357.

26. Maira, *Missing*, 14.

27. Stearns, "Childhood Emotions," 161.

28. Bernstein, *Racial Innocence*, 4. See also Stearns, "Childhood Emotions," 160.

29. Stearns describes guilt as a "'new weapon of emotional discipline'"; Stearns, "Childhood Emotions," 165.

30. Ibid., 169.

31. Kenneth B. Kidd, "'A' is for Auschwitz: Psychoanalysis, Trauma Theory, and the 'Children's Literature of Atrocity,'" *Children's Literature* 33 (2005): 120.

32. Ibid., 137, 124. Kidd notes that, by comparison, children's literature about the Holocaust emerged much more gradually and tends toward more narrative subtlety.

33. Writing about photographs of children from the Holocaust, Marianne Hirsch cautions, "The image of the child victim . . . facilitates an identification in which the viewer can too easily assume the position of a surrogate victim. Most important, the easy identification with children, their virtually universal availability for projection, risks the blurring of important areas of difference and alterity—context, specificity, responsibility, history"; Hirsch, "Projected Memory," in *The Generation of Postmemory: Writing and Visual Culture after the Holocaust* (New York: Columbia University Press: 2012), 167.

34. Lisa Stansbury (Library of Congress [LOC] GR 13), American Folklife Center.

35. My brief conversation with the curator revealed the drawings' ambiguous status. On the one hand, she intimated that the library was somewhat surprised by the request to archive them (and she seemed surprised, too, that I had asked to see them), but she also clarified that this is, to her knowledge, the only collection of its kind; Ann Hoog, conversation with the author, May 26, 2015.

36. Karen Sánchez-Eppler, "How Studying the Old Drawings and Writings of Kids Can Change Our View of History," *The Conversation*, January 25, 2016, http://theconversation.com/how-studying-the-old-drawings-and-writings-of-kids-can-change-our-view-of-history-46724.

37. Although psychiatry, as a field, has begun to accept the possibility that trauma can arise from a mediated encounter, the latest edition of the Diagnostic and Statistical Manual, the DSM-5, refuses the possibility of mediated trauma in children; Amit Pinchevski, "Screen Trauma: Visual Media and Post-Traumatic Stress Disorder," *Theory, Culture & Society* 33, no. 4 (2016): 64.

38. Kidd, "'A' is for Auschwitz," 141.

39. As Maira's ethnographic work demonstrates, September 11 and the War on Terror marked the beginning of political subject formation for a whole generation of Muslim youth in the United States; Maira, *9/11 Generation*, 16.

40. American Folklife Center, AFC 2001/015, September 11, 2001, Documentary Project Collection, Series III Graphic Images, LOC GR 01-15, box 37 of 48, accessed May 27, 2015. These drawings are by Deanna Sanders (LOC GR 12), Mollye O'Rourke (LOC GR 15A), Hannah Beach (LOC GR 15C), and Meagan Yoakley (LOC GR 15B), respectively.

41. Maira, *9/11 Generation*, 78ff.

42. See "9/11 Assignment Outrages Texas Elementary School Parents," *ABC News* video, 0:28, September 14, 2012, http://abcnews.go.com/US/video/911-assignment-outrages-texas-elementary-school-parents-17239839; Alex Alvarez, "Texas School Makes Children Draw Disturbing Pictures of 9/11," *Mediaite*, September 14, 2012, http://www.mediaite.com/tv/texas-school-makes-children-draw-disturbing-pictures-of-9/11; and "Texas Fourth-Graders Instructed to Draw Disturbing 9/11 Pictures," *Fox News*, September 13, 2012, http://www.foxnews.com/us/2012/09/13/parent-claims-11-class-assignment-encouraged-hate-racism.html.

43. Victoria Ford Smith, "Art Critics in the Cradle: Fin-de-Siècle Painting Books and the Move to Modernism," *Children's Literature* 43 (2015): 161–81. An example of the popular "Little Folks" painting books is available at https://archive.org/details/littlefolkspaintooweatiala.

44. Reeves Wiedeman, "A History of the Coloring Book to Damien Hirst," *New Yorker*, October 19, 2009, http://www.newyorker.com/books/page-turner/a-history-of-the-coloring-book-to-damien-hirst.

45. Donna Darling Kelly, *Uncovering the History of Children's Drawing and Art* (Westport, Conn.: Praeger, 2004), 2–3, 6.

46. Cathy Malchiodi, "What Art Therapy Learned from September 11th," *Psychology Today*, September 9, 2011, http://www.psychologytoday.com/blog/arts-and-health/201109/what-art-therapy-learned-september-11th. See also Jason Kane, "Then and Now: Children Draw to Cope with 9/11," *PBS News Hour*, September 10, 2011, http://www.pbs.org/newshour/rundown/then-now-children-draw-to-cope-with-911.

47. Robin F. Goodman, Ph.D., and Andrea Henderson Fahnestock, *The Day Our World Changed: Children's Art of 9/11* (New York: Harry N. Abrams, 2002). Though the book is often promoted as children's literature (the used copy I obtained came from the children's section of a public library), I am not sure this characterization is apt or complete, particularly because of the extensive textual component of essays by experts and politicians. Kidd describes this book as perhaps the least worst of all of the 9/11 literature for children ("'A' is for Auschwitz," 139).

48. Although her study focuses on pictures *of* children rather than pictures *by* children, Anne Higonnet's *Pictures of Innocence* is instructive. She identifies pictures of children as precarious documents, arguing that they "guard the cherished idea of childhood innocence, yet they contain within them the potential to undo that ideal. No subject seems cuter or more sentimental, and we take none more for granted, yet pictures of children have proved dangerously difficult to understand or control." She continues, "Childhood innocence was considered an attribute of the child's body, both because the child's body was supposed to be naturally innocent of adult sexuality, and because the child's mind was supposed to begin blank. Innocence therefore lent itself to visual representation," which also arguably instantiated a visual practice focused on reading children's bodies for the evidence (or absence) of that innocence. I would argue that we read their drawings in the same way; Higonnet, *Pictures of Innocence: The History and Crisis of Ideal Childhood* (London: Thames and Hudson, 1998), 7–8.

49. The larger discourse of childhood innocence requires children to be "psychically or sexually innocent to deserve protection," and innocence in this logic is descriptive and prescriptive at once; ibid., 12.

50. Bernstein, *Racial Innocence*, 6. She continues that this kind of innocence was associated with whiteness and that discourses of childhood innocence were essential to the development of racialized binaries of blackness and whiteness (8).

51. Stearns, "Childhood Emotions," 158.

52. Meyer, "Moral Rhetoric of Childhood," 102.

53. It barely even registered, as Nicholas Mirzoeff observes, as a factor in the 2004 presidential election; Mirzoeff, "Invisible Empire: Visual Culture, Embodied Spectacle, and Abu Ghraib," *Radical History Review* 95 (2006): 21–44.

54. Elizabeth Dauphinée, "The Politics of the Body in Pain: Reading the Ethics of Imagery," *Security Dialogue* 38, no. 2 (2007): 139.

55. Susan D. Moeller, *Packaging Terrorism: Co-Opting the News for Politics and Profit* (Malden, Mass.: Wiley Blackwell, 2009), 160. Frances Larsen reports, "One survey, conducted five months after Berg's death, found that between May and June, 30 million people, or 24 per cent of all adult Internet users in the United States, had seen images from the war in Iraq that were deemed too gruesome and graphic to be shown on television. . . . Americans were seeking these images out: 28 per cent of those who had seen graphic content online actively went looking for it. The survey found that half of those who had seen graphic content thought they had made a 'good decision' by watching"; Larson, "What a Beheading Feels Like: The Science, The Gruesome Spectacle—and Why We Can't Look Away," *Salon*, February 23, 2015, http://www.salon.com/2015/02/03/what_a_beheading_feels_like_the_science_the_gruesome_spectacle_and_why_we_cant_look_away/.

56. There is a confused element of this story that relates to the Berg video; the superintendent accused Brian Newark of also assigning the Berg video, which Newark denies, while one parent from the community argued that he should have also assigned it in tandem with the Abu Ghraib images to provide a more balanced view of wartime atrocity. For a detailed account, see Michael Kunzelman, "Teacher in Flap over Iraqi Prisoner Photos Files Lawsuit," *Milford Daily News*, August 25, 2004, http://www.milforddailynews.com/x349379805.

57. I develop a fuller account of the rights of the people captured in the Abu Ghraib images, including their right to have authority over how those photos are circulated, in Chapter 1 of Rebecca A. Adelman, *Beyond the Checkpoint: Visual Practices in America's Global War on Terror* (Boston: University of Massachusetts Press, 2014).

58. Quoted in Tal Barak, "Texas Teachers Are Suspended for Showing Video of Beheading," *Education Week* 23, no. 38 (May 26, 2004): 4.

59. "School Suspends Teachers for Video," *Washington Times*, May 20, 2004, http://www.washingtontimes.com/news/2004/may/20/20040520-115221-3528r/.

60. Donald E. Pease, *The New American Exceptionalism* (Minneapolis: University of Minnesota Press, 2009), 171.

61. Maira, *9/11 Generation*, 17. Bernstein describes adolescence as "an unstable and . . . permanently inadequate effigy for childhood"; Bernstein, *Racial Innocence*, 27.

62. For more on the complexities of adolescence as it relates to militarization, see Anna M. Agathangelou and Kyle D. Killian, "(Neo)zones of Violence: Reconstructing Empire on the Bodies of Militarized Youth," in *The Militarization of Childhood: Thinking Beyond the Global South*, ed. J. Marshall Beier (New York: Palgrave Macmillan, 2011), 17–42.

63. "Teacher Center—The Torture Question," *Frontline*, October 2005, http://www.pbs.org/wgbh/pages/frontline/teach/torture/hand2.html.

64. Jarrett Murphy, "Berg Video Shown in Classrooms," *CBS News*, May 11, 2004, http://www.cbsnews.com/news/berg-video-shown-in-classrooms-11-05-2004/.

65. Associated Press, "Teachers Disciplined for Showing Beheading," *New York Times*, May 24, 2004, http://www.nytimes.com/2004/05/24/us/teachers-disciplined-for-showing-beheading.html.

66. J. Marshall Beier, "Introduction: Everyday Zones of Militarization," in Beier, *Militarization of Childhood*, 6.

67. Maira, *Missing*, 15.

68. Lesley Copeland, "Mediated War: Imaginative Disembodiment and the Militarization of Childhood," in Beier, *Militarization of Childhood*, 133–52.

69. Karin Johnson, "School Investigates Report That Students Were Shown ISIS Execution Video," *WLWT5* (Cincinnati, Ohio), February 9, 2015, http://www.wlwt.com/article/school-investigates-report-that-students-were-shown-isis-execution-video/3551350.

70. Jessica Willey, "ISIS Homework Assignment Sparks Anger in Alvin," *KTRK* (Houston, Tex.), September 10, 2014, http://www.abc13.com/education/isis-homework-sparks-anger-in-alvin/303462. It is worth noting that this phenomenon is not, apparently, localized to the United States. For example, a priest in Poland drew reprimands for screening a video of ISIS beheading Egyptian Christians in a tenth-grade lesson about martyrdom; see the Associated Press, "Polish Priest Criticized for Showing ISIS Beheading Christians Video to High School Students," *CP 24* (Toronto), April 1, 2015, http://www.cp24.com/world/polish-priest-criticized-for-showing-isis-beheading-christians-video-to-high-school-students-1.2307494.

71. Sarah Larimer, "Utah School Realizes It Shouldn't Ask Students to Make Terrorist Propaganda Posters," *Washington Post*, November 20, 2015, https://www.washingtonpost.com/news/education/wp/2015/11/20/utah-school-realizes-it-shouldnt-ask-students-to-make-terrorist-propaganda-posters/.

72. Dora Scheidell, "Utah School Apologizes for Homework Assignment to Make Propaganda Poster for Terrorist Group," *Fox 13* (Salt Lake City, Utah), November 19, 2015, http://fox13now.com/2015/11/19/middle-school-in-utah-apologizes-after-homework-assignment-to-make-propaganda-poster-for-terrorist-group/.

73. Larimer, "Utah School Realizes."

2. AFFECTIVE PEDAGOGIES FOR MILITARY CHILDREN

1. *Sesame Street for Military Families* has a breathing app too, called "Breathe, Think, Do," meant to help all children, not just those connected to the military, use breathing as a behavioral reset. Breathing allows them to cope with frustration or sadness, think about better problem-solving strategies, and then enact them: feelings expressed wordlessly through the breath.

2. Thanks to Meghan Marx for alerting me to the politics of breathing.

3. John Protevi, *Political Affect: Connecting the Social and the Somatic* (Minneapolis: University of Minnesota Press, 2009), xiv.

4. Patricia Lester and Lt. Col. Eric Flake, "How Wartime Military Service Affects Children and Families," *The Future of Children* 23, no. 2 (Fall 2013): 125.

5. See, for example, the initiatives listed on the National Military Family Association, "Kids + Deployment," 2017, http://www.militaryfamily.org/kids-operation-purple/deployment.html, accessed August 9, 2017.

6. Spencer Ackerman, "Canada Frees Omar Khadr, Once Guantánamo Bay's Youngest Inmate," *Guardian*, May 7, 2015, http://www.theguardian.com/world/2015/may/07/canada-free-bail-omar-khadr-guantanamo-bay-youngest.

7. Khadr's case received extensive news coverage, particularly in Canada. See, for example, Michael Friscolanti, "The Secret Omar Khadr File," *Maclean's*, September 27, 2012, http://www.macleans.ca/news/canada/the-secret-khadr-file; Canadian Press, "Alta. Court Delays Ruling on Omar Khadr Appeal," *CBC News*, April 30, 2014, http://www.cbc.ca/news/canada/edmonton/alta-court-delays-ruling-on-omar-khadr-appeal-1.2627636; and Canadian Press, "Omar Khadr Bail Being Challenged, Canadian Government in Court Today," *CBC News*, May 5, 2015, http://www.cbc.ca/news/canada/edmonton/omar-khadr-bail-being-challenged-canadian-government-in-court-today-1.3061285. For courtroom sketches of Khadr at Guantánamo, see Spencer Ackerman, "10 Sketches from Inside Guantanamo Bay's Military Courtroom," *Wired*, March 22, 2013, http://www.wired.com/2013/03/guantanamo-sketches.

8. Most accounts suggest that his father, Ahmed Said Khadr, had ties to Al Qaeda. American officials maintained that he was a high-level operative

and financier; he always insisted that his work in Pakistan and Afghanistan was charitable; see "Profile: Omar Khadr," *BBC News*, September 29, 2012, http://www.bbc.com/news/world-us-canada-11610322.

9. Sherene H. Razack, "The Manufacture of Torture as Public Truth: The Case of Omar Khadr," in *At the Limits of Justice: Women of Colour on Terror*, ed. Suvendrini Perera and Sherene H. Razack (Toronto: University of Toronto Press, 2014), 60.

10. Ian Austen, "Blurry Peek at Questioning of a Guantánamo Inmate," *New York Times*, July 16, 2008, http://www.nytimes.com/2008/07/16/world/16khadr.html.

11. See "Guantanamo Detainee, 16 YO Omar Khadr, Interrogation Video Released," *LiveLeak* video, 9:54, July 15, 2008, https://www.liveleak.com/view?i=4c4_1216119003; and YouTube video, 9:54, posted by "a2zme," July 15, 2008, https://www.youtube.com/watch?v=yNCyrFV2G_o.

12. Prior to his military commission, Omar Khadr was interviewed by Michael Welner, a New York–based forensic psychiatrist who would later testify for the prosecution. During the interview, Khadr contended that his interrogators showed him a picture of his dead father and then mocked him for crying at the sight. The implacable calm of the interrogators, according to Joseph Pugliese, marks a "fault line" between their position and Khadr's, the "violent rupture between the irreconcilable levels of perceptual and experiential reality"; Pugliese, "Apostrophe of Empire: Guantánamo Bay, Disneyland," *Borderlands E-Journal* 8, no. 3 (2009): 14, http://www.borderlands.net.au/vol8no3_2009/pugliese_apostrophe.pdf.

13. Although the precise circumstances of his father's death remain unclear, he was likely killed in a firefight with Pakistani security forces in 2003; "Canadian al-Qaeda Suspect Dead: Pakistan," *CBC News*, January 24, 2004, http://www.cbc.ca/news/world/canadian-al-qaeda-suspect-dead-pakistan-1.470158.

14. Canadian Press, "Omar Khadr Explains War-Crimes Guilty Pleas in Court Filing," *CBC News*, December 13, 2013, http://www.cbc.ca/news/canada/omar-khadr-explains-war-crimes-guilty-pleas-in-court-filing-1.2463558.

15. Razack, "Manufacture of Torture," 57.

16. This presumed incongruity between children and militarization is reflected in a relative dearth of scholarship about their intersections. Sunaina Marr Maira's work, *Missing: Youth, Citizenship, and Empire After 9/11* (Durham: Duke University Press, 2009), is a notable exception to this. So are the collections edited by David Kieran and James Marten: Kieran, ed., *The War of My Generation: Youth Culture and the War on Terror* (New Brunswick, N.J.: Rutgers University Press, 2015); James Marten, ed.,

Children and War: A Historical Anthology (New York: New York University Press, 2002).

17. Marten, "Children and War," in *The Routledge History of Childhood in the Western World*, ed. Paula S. Fass (London: Routledge, 2013), 142.

18. Elizabeth McKee Williams argues that the war increased fear of death across all strata of society, including the very young; Williams, "Childhood, Memory, and the American Revolution," in Marten, *Children and War*, 19–20.

19. Steven Mintz, *Huck's Raft: A History of American Childhood* (Cambridge, Mass.: Belknap Press of Harvard University Press, 2004), 132.

20. Ibid., 127.

21. Ibid., 152.

22. Marten, "Children and War," 144.

23. For example, in Europe, the fairy-tale whimsicality of flight made the novelty of air war intelligible, even beguiling, to young audiences; Guillaume de Syon, "The Child in the Flying Machine: Childhood and Aviation in the First World War," in Marten, *Children and War*, 116–34.

24. Mintz, *Huck's Raft*, 255, 256, 258.

25. Lisa L. Ossian, "'Too Young for a Uniform': Children's War Work on the Iowa Farm Front, 1941–1945," in Marten, *Children and War*, 255, 257.

26. Chris O'Brien, "Mama, Are We Going to Die? America's Children Confront the Cuban Missile Crisis," in Marten, *Children and War*, 75.

27. Mintz, *Huck's Raft*, 282.

28. Mintz, in ibid., suggests that this enfranchisement was not as consequential as many observers might have hoped or feared (333).

29. Molly Clever and David R. Segal, "The Demographics of Military Children and Families," *Future of Children* 23, no. 2 (Fall 2013): 13–39.

30. Anneke Meyer, "The Moral Rhetoric of Childhood," *Childhood* 14, no. 1 (2007): 90–91.

31. Jennifer Sinor, "Inscribing Ordinary Trauma in the Diary of a Military Child," *Biography* 26, no. 3 (Summer 2003): 407–8.

32. Ibid., 408.

33. Lester and Flake, "Wartime Military Service," 123.

34. Lester and Flake elaborate: "Despite . . . challenges, living in a military family gives children a meaningful identity associated with strength, service, and sacrifices, which is a basic component of military culture not only for service members but also for their family members"; "Wartime Military Service," 123.

35. Cindy Dell Clark, "Summer, Soldiers, Flags, and Memorials: How U.S. Children Learn Nation-Linked Militarism from Holidays," in

The War of My Generation: Youth Culture and the War on Terror, ed. David Kieran (New Brunswick, N.J.: Rutgers University Press, 2015), 40, 42. She also observes that children's participation in militarized spectacles like parades and memorial ceremonies might actually help adults reckon with militarism by making it seem more playful and approachable (44).

36. Katrina Lee-Koo, "Horror and Hope: (Re)Presenting Militarised Children in Global North-South Relations," *Third World Quarterly* 32, no. 4 (2011): 726. For a historical account of the relationship between discourses of children's rights and militarization, see Dominique Marshall, "Humanitarian Sympathy for Children in Times of War and the History of Children's Rights: 1919–1959," in Marten, *Children and War*, 184–99.

37. Lee-Koo, "Horror and Hope," 727. Lorraine Macmillan analyzes the "many inconsistencies" about the militarization of childhood in Anglophone cultures. She specifically notes practices that allow students to be recruited and trained for future military service and encourage them to participate in nationalist projects. At the same time, she observes, there exists pervasive cultural concern about the deleterious influence of violence in children's lives (especially violence in mass media) more generally; Macmillan, "Militarized Children and Sovereign Power," in *The Militarization of Childhood: Thinking beyond the Global South*, ed. J. Marshall Beier (New York: Palgrave Macmillan, 2011), 75.

38. Lee-Koo, "Horror and Hope," 735.

39. James Schmidt, "Children and the State," in Fass, *Routledge History of Childhood*, 177.

40. Lester and Flake, "Wartime Military Service," 122.

41. Sinor, "Inscribing Ordinary Trauma," 416, 410. Sinor's formulation here might be a bit reductive. This characterization presumes that (previously innocent) military children are always and necessarily traumatized by their circumstances and so risks underestimating the extent to which they can manage them, while occluding the unsavory possibility that children might enjoy the experience of being militarized.

42. Anne Higonnet, *Pictures of Innocence: The History and Crisis of Ideal Childhood* (London: Thames and Hudson, 1998), 119.

43. Robin Bernstein, *Racial Innocence: Performing American Childhood from Slavery to Civil Rights* (New York: New York University Press, 2011), 33. She notes also that within these categories, representations were not monolithic.

44. Ibid., 34.

45. Ibid., 35.

46. Lester and Flake, "Wartime Military Service," 122. This amounts to 5 percent of American children overall.

47. Laura Browder, "How to Tell a True War Story . . . for Children: Children's Literature Addresses Deployment," in Kieran, *War of My Generation*, 86.

48. Ibid., 87–88.

49. Ibid., 92. Browder also observes that these books cast children as mini-soldiers who, by extension, must abide by the Uniform Code of Military Justice that governs the behavior of their parents. According to the UCMJ, military personnel cannot publicly question or undermine the commander in chief or, by extension, the mission that he overseas. Browder suggests that children's literature enacts a similar prohibition by emphasizing loyalty and obedience (102–3).

50. All of the content I describe in this section can be found fairly easily by navigating around the *SSMF* website. My analysis is based on research conducted between the fall of 2014 and the summer of 2017 and was current as of August 2017. I provide URLs for content that might be harder to find and access dates for content that moved or changed during that period.

51. The website was originally called *Families Near and Far* and rebranded to *Sesame Street for Military Families* in 2015. In the process of this change, the website eliminated the social networking component of *Families Near and Far*, which it described as "outdated"; "Military Families Resources for Young Children," *Sesame Street for Military Families*, http://www.sesamestreetformilitaryfamilies.org/faq/.

52. Ibid.

53. This faith in emotiveness is inherited from the Victorian era, with its new emphasis on the home as a predominantly emotional space.

54. Ibid.

55. "Military Families Resources for Young Children," *Sesame Street for Military Families*, http://www.sesamestreetformilitaryfamilies.org/partners/. Given these alliances, it seems that recent concerns about *Sesame Street*'s acquisition by premium-cable channel HBO are a bit misplaced. For commentary on *Sesame Street's* relocation, see Jonathan Gray, "*Sesame Street*'s New Landlord," *Antenna: Responses to Media, Arts, Culture*, August 15, 2015, blog.commarts.wisc.edu/2015/08/15/sesame-streets-new-landlord, and Josef Adalian, "Mitt Romney Won: In a Blow to the 47 Percent, *Sesame Street* Moves to HBO," *Vulture*, August 13, 2015, http://www.vulture.com/2015/08/sesame-street-hbo-mitt-romney-wins.html.

56. Lauren Berlant's analysis of "infantile citizenship" underscores the importance of seemingly "silly" pop cultural objects in shoring up this mode of national belonging; if, on the surface, *Sesame Street* Muppets are silly objects *par excellence*, they also have the weight of some decidedly un-silly organizations behind them.

57. Cynthia Enloe, *The Morning After: Sexual Politics at the End of the Cold War* (Berkeley: University of California Press, 1993), 64.

58. It also suggests that in the case of multiple deployments, the parents should not give the news about redeployment too soon.

59. "Self-Expression," *Sesame Street for Military Families*, accessed April 29, 2016, http://sesamestreetformilitaryfamilies.org/topic/self-expression/.

60. Lester and Flake, "Wartime Military Service," 127.

61. In subsequent edits of the site, the video was removed, but it is still available at https://www.youtube.com/watch?v=fIvyXn_MWA4.

62. Throughout the video, the children sing in concert with young adults who provide the emotional vocabulary and cues for acting out feelings and lead them in cheers and dances. In response to the calls "All the military kids say 'Yeah!'" and "All the military kids say 'Hey!'" the younger children shout, raise their arms, and begin to bop happily around. Clark's research emphasizes the role of sensorimotor involvement in children's militarization. Here, the embodied actions of cheering and dancing reinforce militarized identity, while the positioning of this cheer after the exercise in naming and expressing feelings implies that its outcome is a child happy with his or her place; Clark, "Summer Soldiers," 44.

63. "Grief—Communicating and Connecting," *Sesame Street for Military Families*, accessed May 20, 2017, http://sesamestreetformilitaryfamilies.org/topic/grief/.

64. Peter N. Stearns, "Consumerism and Childhood: New Targets for American Emotions," in *An Emotional History of the United States*, ed. Peter N. Stearns and Jan Lewis (New York: New York University Press, 1998), 409.

65. "Self-Expression," *Sesame Street for Military Families*, http://sesamestreetformilitaryfamilies.org/topic/self-expression/?ytid=1uioi2438Jw, accessed July 11, 2016. This content now appears on http://www.pbs.org/parents/cominghome/article-signsofstress.html, accessed July 9, 2017.

66. Stearns, "Consumerism and Childhood," 402.

67. Sianne Ngai, *Our Aesthetic Categories: Zany, Cute, Interesting* (Cambridge, Mass.: Harvard University Press, 2012), 15.

68. Ibid., 30.

69. Ibid., 4. Marita Sturken's analysis of the militarization of the teddy bear after terrorist attacks is also relevant here; Sturken, *Tourists of History: Memory, Kitsch, and Consumerism from Oklahoma City to Ground Zero* (Durham: Duke University Press, 2007).

70. The video is available at http://sesamestreetformilitaryfamilies.org/topic/injuries/?ytid=BfR5Mackh7w.

71. Ngai, *Ugly Feelings* (Cambridge, Mass.: Harvard University Press, 2005), 27.

72. Stearns, "Childhood Emotions," in Fass, *Routledge History of Childhood*, 170. More recently, this has taken the shape of a pervasive discourse of "resiliency," which many critics decry as a concession to the exigencies of neoliberalism.

73. In part, *SSMF* must be gentler because it is aimed at a younger demographic, but even in the sections of *MKC* where it targets younger children (ages 6–8), the approach is blunter, more matter-of-fact than that of *SSMF*.

74. For example, the warning prefacing its "Tough Topics" section suggests adult coviewing and thus places the responsibility on children to seek out adults to watch with them.

75. All of the content I describe in this section can be found fairly easily by navigating around the *MKC* website. My analysis is based on research conducted between the spring of 2015 and the summer of 2017 and was current as of August 2017. I provide URLs for content that might be harder to find and access dates for content that moved or changed during that period.

76. "Terms of Use Agreement," *Military Kids Connect*, last modified March 18, 2016, accessed August 9, 2017, http://militarykidsconnect.dcoe.mil/terms-of-use.

77. Ann S. Masten, "Afterword: What We Can Learn from Military Children and Families," *The Future of Children* 23, no. 2 (Fall 2013): 199.

78. "About Military Kids Connect®," *Military Kids Connect*, April 2, 2014, http://militarykidsconnect.dcoe.mil/about-us.

79. Yet *MKC* is also more candid in acknowledging the mixed possibilities inherent in the life of a military child, in which the experience of war is equally likely, in James Marten's terms, to be "catastrophic" or "transformative and even liberating," or perhaps both at once; Marten, "Children and War," 142.

80. National Center for Telehealth and Teletechnology, email communication, December 21, 2015, and "Military Kids Connect Website Wins Three More Awards," *National Center for Telehealth and Teletechnology*, June 21, 2015, http://t2health.dcoe.mil/news/military-kids-connect-website-wins-three-more-awards.

81. "Moving," *Military Kids Connect*, September 22, 2014, http://militarykidsconnect.dcoe.mil/tweens/connect/whats-on-your-mind. The thread has since been removed.

82. Though I have a range of hypotheses, I am not sure why this aspect of *MKC* is so underutilized. It may be that military kids have no need of an

additional social network beyond those that are freely available elsewhere or that something about this one is off-putting.

83. Cary L. Cooper and Philip Dewe, *Stress: A Brief History* (Malden, Mass.: Blackwell, 2004), 2.

84. Ibid., 9.

85. Mark Jackson, *The Age of Stress: Science and the Search for Stability* (Oxford: Oxford University Press, 2013), 34–35.

86. Ibid., 11.

87. Cooper and Dewe, *Stress*, 111.

88. Russell Viner, "Putting Stress in Life: Hans Selye and the Making of Stress Theory," *Social Studies of Science* 29, no. 3 (June 1999): 391–92.

89. Ibid., 399.

90. Kristian Pollock, "On the Nature of Social Stress: Production of a Modern Mythology," *Social Science & Medicine* 26, no. 3 (1988): 381.

91. Ibid., 387, 389, 390.

92. In his history of the "emotional body," Otniel E. Dror comments on the racialized and gendered history of the peptic ulcer in the nineteenth century, often perceived as a nervous symptom of civilizedness and susceptibility to the worries of modernity. There is, perhaps, a similar logic at work in the concern about military kids' digestive systems, as if they are straining under the weight of the U.S. military agenda; Dror, "Creating the Emotional Body: Confusion, Possibilities, and Knowledge," in Stearns and Lewis, *Emotional History*, 180.

93. This content is accompanied by a lengthy voiceover about where stress comes from, how it manifests, and what can be done about it. Parenthetically, this page is illustrated with a word cloud graphic that clearly was not meant for children, including stress-inducing phenomena like "neighbors," "bills," "new boss," . . . and "children."

94. National Center for Post-Traumatic Stress Disorder and Walter Reed Army Medical Center, "Iraq War Clinician Guide, 2nd edition," *U.S. Department of Veterans Affairs, PTSD: National Center for PTSD*, June 2004, https://www.ptsd.va.gov/professional/manuals/manual-pdf/iwcg/iraq_clinician_guide_v2.pdf, 9.

95. Derek Gregory, "'The Rush to the Intimate': Counterinsurgency and the Cultural Turn in Late Modern War," *Geographical Imaginations* (2012), Geographicalimaginations.files.wordpress.com/2012/07/gregory-rush-to-the-intimate-full.pdf.

96. Berlant, *Cruel Optimism* (Durham: Duke University Press, 2011), 109.

97. Berlant, "Thinking About Feeling Historical," in *Political Emotions: New Agendas in Communication*, ed. Janet Staiger, Ann Cvetkovich, and Ann Reynolds (New York: Routledge, 2010), 232.

3. RECOGNIZING MILITARY WIVES

1. Personal communication with author, March 2016.

2. See, for example, Military Spouse Team, "Spouse Stories: 35 Crazy Deployment Curses!," *MilitarySpouse*, 2014, accessed March 24, 2016, http://militaryspouse.com/spouse-101/spouse-stories-35-crazy-deployment-curses/.

The falling-apart house in the *Military Kids Connect* "My House" game strikes me as a cartoon version of this. It is available at "My House—Kids," *Military Kids Connect*, accessed April 10, 2016, http://militarykidsconnect.dcoe.mil/kids/games/my-house.

3. Shelley MacDermid Wadsworth and Kenona Southwell, "Military Families: Extreme Work and Extreme 'Work-Family,'" *Annals of the American Academy of Political and Social Science* 638 (2011): 165. Wadsworth and Southwell count upward of 1.9 million military children. The more recent and more expansive tally I cited in Chapter 1 is 4,000,000 "military-connected children"; See Patricia Lester and Lt. Col. Eric Flake, "How Wartime Military Service Affects Children and Families," *Future of Childhood* 23, no. 2 (Fall 2013): 122.

4. See David Kieran, *Signature Wounds: The Cultural Politics of Mental Health During the Iraq and Afghanistan Wars* (New York: New York University Press, forthcoming), especially Chapter 7.

5. United States Department of Defense, "Military Spouse Preference Program," last revised July 2014, https://www.cpms.osd.mil/Content/Documents/PPP-Program%20S.pdf. See also USAJOBS, "Military Spouses," https://www.usajobs.gov/Help/working-in-government/unique-hiring-paths/military-spouses/.

6. Erica Scott, "First Ladies Obama and Bush Urge Next President to Prioritize Veterans," *ABC News*, September 16, 2016, http://abcnews.go.com/Politics/ladies-obama-bush-urge-president-prioritize-veterans/story?id=42137937.

7. The White House, "Joining Forces," last updated January 19, 2017, https://obamawhitehouse.archives.gov/joiningforces/about.

8. Brooke Seipel, "White House Women Meet with Military Spouses," *The Hill*, August 2, 2017, http://thehill.com/homenews/administration/345102-white-house-women-meet-with-military-spouses.

9. Kenneth T. MacLeish, *Making War at Fort Hood: Life and Uncertainty in a Military Community* (Princeton: Princeton University Press, 2013), 151.

10. Wendy M. Christensen, "Technological Boundaries: Defining the Personal and the Political in Military Mothers' Online Support Forums," *WSQ: Women's Studies Quarterly* 37, nos. 1 and 2 (Spring/Summer 2009): 149.

11. Christine Sylvester, *War as Experience: Contributions from International Relations and Feminist Analysis* (New York: Routledge, 2013), 38.

12. Deborah Cohler, "Keeping the Home Front Burning: Renegotiating Gender and Sexuality in U.S. Mass Media after September 11," *Feminist Media Studies* 6, no. 3 (2006): 245.

13. Kelly Oliver, "Women: The Secret Weapon of Modern Warfare?," *Hypatia* 23, no. 2 (April–June 2008): 1.

14. Cohler, "Keeping the Home Front Burning," 250–54. Statistically, enlisted women are more likely to be married to military men than to civilians; Wadsworth and Southwell, "Military Families," 166.

15. Elsewhere, in nation-states with universal conscription, like Israel, governments often use images of female military personnel to make militarism seem both sexy and unthreatening; Eva Berger and Dorit Naaman, "Combat Cuties: Photographs of Israeli Women Soldiers in the Press Since the 2006 Lebanon War," *Media, War & Conflict* 4, no. 3 (2011): 269–86.

16. Oliver, "Women," 10.

17. For an extensive comparison of the media coverage of these three women, see Rebecca Wanzo, *The Suffering Will Not Be Televised: African American Women and Sentimental Political Storytelling* (Albany: State University of New York Press, 2009).

18. Mary Douglas Vavrus, "Lifetime's *Army Wives*, or I Married the Media-Military-Industrial Complex," *Women's Studies in Communication* 36 (2013): 93.

19. Jennifer Mittelstadt notes that in the late twentieth century, conservative activists in the military often contrasted the integrity of military families with the vision of the "putatively immoral, decaying civilian family"; Mittelstadt, *The Rise of the Military Welfare State* (Cambridge, Mass.: Harvard University Press, 2015), 167.

20. Catherine Lutz, *Homefront: A Military City and the American Twentieth Century* (Boston: Beacon Press, 2001), 236.

21. Christensen, "Technological Boundaries," 146–47.

22. Bree Kessler, "Recruiting Wombs: Surrogates as the New Security Moms," *WSQ: Women's Studies Quarterly* 37, nos. 1–2 (Spring/Summer 2009): 175.

23. Inderpal Grewal, "'Security Moms' in Twenty-First-Century U.S.A.: The Gender of Security in Neoliberalism," in *The Global and the Intimate: Feminism in Our Time*, ed. Geraldine Pratt and Victoria Rosner (New York: Columbia University Press, 2012), 195, 208.

24. Laurel Elder and Steven Greene, "The Myth of 'Security Moms' and 'Nascar Dads': Parenthood, Political Stereotypes, and the 2004 Election," *Social Science Quarterly* 88, no. 1 (March 2007): 5.

25. Lutz, *Homefront*, 187.

26. Mittelstadt's research reveals that these networks have operated insidiously, sometimes through military wives surveilling one another; *Rise of the Military Welfare State*, 177.

27. MacLeish, *Making War*, 102.

28. This affirms Judith Butler's contention that both the state and the family are organized around the propagation and protection of male citizens; Butler, *Antigone's Claim: Kinship between Life and Death* (New York: Columbia University Press, 2000), 11–12.

29. Molly Clever and David R. Segal, "The Demographics of Military Children and Families," *Future of Children* 23, no. 2 (Fall 2013): 15.

30. Lutz, *Homefront*, 187.

31. Ibid., 61.

32. Anne Yoder, "Military Classifications for Draftees," *Swarthmore*, last modified March 2014, accessed April 10, 2016, https://www.swarthmore.edu/library/peace/conscientiousobjection/MilitaryClassifications.htm.

33. "Postponements, Deferments, Exemptions," *Selective Service System*, accessed May 6, 2016, https://www.sss.gov/About/Return-to-the-Draft/Postponements-Deferments-Exemptions.

34. MacLeish, *Making War*, 151. Jesse Paul Crane-Seeber notes that in Western culture, the bodies of military men are objectified "in ways that are normally reserved only for women's bodies"; Crane-Seeber, "Sexy Warriors: The Politics and Pleasures of Submission to the State," *Critical Military Studies* 2, nos. 1–2 (2016): 51.

35. Lutz, *Homefront*, 56. Mittelstadt also notes that the army and its personnel have often been viewed as disreputable; *Rise of the Military Welfare State*, 75.

36. Margot Canaday, *The Straight State: Sexuality and Citizenship in Twentieth-Century America* (Princeton: Princeton University Press, 2011), 137–39.

37. Wadsworth and Southwell, "Military Families," 165.

38. Liz Montegary, "An Army of Debt: Financial Readiness and the Military Family," *Cultural Studies* 29, no. 5–6 (2015): 655. Over time, the military's status as a provider of benefits and financial stability has eroded.

39. Lutz, *Homefront*, 168.

40. Wadsworth and Southwell, "Military Families," 169.

41. Lutz, *Homefront*, 167.

42. Canaday, *Straight State*, 206–7.

43. Lutz, *Homefront*, 178.

44. Clever and Segal, "Demographics of Military Children," 20.

45. The 1979 film *The Great Santini*, directed by Lewis John Carlino, is an iconic representation of this phenomenon.

46. Lester and Flake, "Wartime Military Service," 128.

47. Clever and Segal, "Demographics of Military Children," 28.

48. Enloe frames the matter bluntly: "Insofar as any military cannot control the actions and the emotions of women married to soldiers, they think they cannot actually carry on effective foreign policy"; Cynthia Enloe and Elizabeth Miklya Legerski, "Taking Women Seriously Makes Us Smarter about the U.S. War in Iraq," *Social Thought & Research* 26 (2006): 10.

49. Montegary, "Army of Debt," 663.

50. Enloe and Legerski, "Taking Women Seriously," 7. See also Enloe, *Maneuvers: The International Politics of Militarizing Women's Lives* (Berkeley: University of California Press, 2000), 154–55, and Enloe, *Nimo's War, Emma's War: Making Feminist Sense of the Iraq War* (Berkeley: University of California Press, 2010), 176.

51. Ben Anderson, "Modulating the Excess of Affect: Morale in a State of 'Total War,'" in *The Affect Theory Reader*, ed. Melissa Gregg and Gregory J. Seigworth (Durham: Duke University Press, 2010), 163.

52. Montegary, "Army of Debt."

53. Kessler, for example, notes that a small but significant number of them turn to gestational surrogacy for other couples, labor that can garner more money than their husbands make in a year; Kessler, "Recruiting Wombs," 175.

54. Lutz, *Homefront*, 60.

55. Christensen, "Technological Boundaries," 149.

56. Tanya Biank, "The Home Fires Are Burning Out," *New York Times*, March 13, 2006, http://www.nytimes.com/2006/03/13/opinion/13biank.html.

57. Biank, "Army Wives: The Unwritten Code of Military Marriage," n.d., http://www.tanyabiank.com/books_armywives.html.

58. MacLeish, *Making War*, 167.

59. Enloe, *Nimo's War*, 185, 186.

60. David Kieran, "'We Combat Veterans Have a Responsibility to Ourselves and Our Families': Domesticity and the Politics of PTSD in Memoirs of the Iraq and Afghanistan Wars," *American Studies* 53, no. 2 (2014): 97.

61. Ibid.

62. Keith D. Renshaw, Camila S. Rodrigues, and David H. Jones, "Psychological Symptoms and Marital Satisfaction in Spouses of Operation Iraqi Freedom Veterans: Relationships with Spouses' Perceptions of Veterans' Experiences and Symptoms," *Journal of Family Psychology* 22, no. 3 (2008): 592.

63. Kieran, "Combat Veterans," 97.

64. Clever and Segal, "Demographics of Military Children," 21.

65. Wadsworth and Southwell, "Military Families," 172. They also mention that military personnel receive higher wages during deployment, a very partial financial offset for the increased risk.

66. Asma Abbas, "In Terror, In Love, Out of Time," in *At the Limits of Justice: Women of Colour on Terror*, ed. Suvendrini Perera and Sherene H. Razack (Toronto: University of Toronto Press, 2014), 503. MacLeish, *Making War*, makes a compatible point in his analysis of how love functions as a form of sovereignty and vector of power for army families (103).

67. Lauren Berlant's attention to the small things we all do to make existence more tolerable is instructive; she writes, "The ordinary [is] a zone of convergence of many histories, where people manage the incoherence of lives that proceed in the face of threats to the good life they imagine"; Berlant, *Cruel Optimism* (Durham: Duke University Press, 2011), 10.

68. MacLeish, *Making War*, writes that "for those outside the Army . . . the Army wife's love both domesticates and relays outward the shock and awe of war, making it *more* real, *more* comprehensible" (177–78).

69. Marta Zarzycka, "Outside the Frame: Reexamining Photographic Representations of Mourning," *Photography and Culture* 7, no. 1 (March 2014): 64.

70. MacLeish, *Making War*, 19.

71. Ibid., 153.

72. It is also worth noting the putative sexiness of the militarized male body; doubtless, fantasies about the military wife's access to that body are operative here. For more on the sexualization of military men, see Crane-Seeber, "Sexy Warriors."

73. Iris Marion Young, "The Logic of Masculinist Protection: Reflections on the Current Security State," *Signs: Journal of Women in Culture and Society* 29, no. 1 (2003): 4.

74. The White House, "President George W. Bush Commemorates Military Spouse Day," May 6, 2008, https://georgewbush-whitehouse.archives.gov/news/releases/2008/05/20080506-3.html.

75. Lutz, *Homefront*, 45.

76. Sue Campbell, *Interpreting the Personal: Expression and the Formation of Feelings* (Ithaca: Cornell University Press, 1997), 172.

77. Enloe and Legerski, "Taking Women Seriously," 11.

78. Ann S. Masten, "Afterward: What We Can Learn from Military Children and Families," *Future of Children* 23, no. 2 (Fall 2013): 204.

79. Deborah Gould, "On Affect and Protest," in *Political Emotions*, ed. Janet Staiger, Ann Cvetkovich, and Ann Reynolds (New York: Routledge, 2010), 33. As her ultimate interest is in how affects can be mobilized in protest against the state, she continues by noting that "[these] efforts do not always succeed," because affect operates unpredictably.

80. Senate Joint Resolution 115, "Designating the Last Sunday in September as 'Gold Star Mother's Day,' and for Other Purposes," Pub. Res., No. 123, 1895, June 23, 1936, https://www.loc.gov/law/help/statutes-at-large/74th-congress/c74.pdf.

81. Abraham Lincoln, "Second Inaugural Address," *Bartleby*, March 4, 1865, http://www.bartleby.com/124/pres32.html.

82. Mary L. Clark, "Keep Your Hands off My (Dead) Body: A Critique of the Ways in Which the State Disrupts the Personhood Interests of the Deceased and His or Her Kin in Disposing of the Dead and Assigning Identity in Death," *Rutgers Law Review* 58 (Fall 2005): 57.

83. This description is based on my search of archival records maintained by the University of California, Santa Barbara's American Presidency Project. Its collection of presidential proclamations from Washington onward can be accessed at http://www.presidency.ucsb.edu/proclamations.php.

84. Elisabeth R. Anker, *Orgies of Feeling: Melodrama and the Politics of Freedom* (Durham: Duke University Press, 2014), 8.

85. "Proclamation 5184—Military Spouse Day, 1984," April 17, 1984, available at http://www.presidency.ucsb.edu/ws/index.php?pid=39798.

86. Not everyone agrees that such initiatives benefit military families. One study notes that some "military policy-makers and service providers worry that providing too many entitlements encourages military families to believe that they are not able or expected to take action on their own behalf, ultimately making them feel *dis*empowered and less able to cope"; Wadsworth and Southwell, "Military Families," 178. This is, of course, a common criticism of entitlement programs in general.

87. Enloe, *Nimo's War*, 180.

88. Statistics on domestic violence in the military are notoriously hard to come by. For a summary of extant research, see "Intimate Partner Violence: Prevalence Among U.S. Military Veterans and Active Duty Servicemembers and a Review of Intervention Approaches," *Department of Veteran Affairs Health Services Research & Development Service*, August 2013, http://www.hsrd.research.va.gov/publications/esp/partner_violence-REPORT.pdf.

89. Wadsworth and Southwell, "Military Families," 175.

90. Clever and Segal, "Demographics of Military Children," 33.

91. Herman Gray, "Subject(ed) to Recognition," *American Quarterly* 65, no. 4 (2013): 778.

92. MacLeish, *Making War*, 95.

93. Lutz, *Homefront*, 189. Lutz also notes that Congressional action in the 1980s allowed women to retain some of their husband's military benefits after divorce.

94. J. G. Noll, "Trump Breaks with Military Spouse Appreciation Day Tradition," *MilitaryOneClick*, May 12, 2017, http://militaryoneclick.com/trump-military-spouse-appreciation-day-proclamation/.

95. See, for example, Ashley Fantz, "As ISIS Threats Online Persist, Military Families Rethink Online Lives," CNN, March 23, 2015, http://www.cnn.com/2015/03/23/us/online-threat-isis-us-troops/index.html; Catherine Herridge, "Feds Investigating ISIS 'Kill List,' Military Spouse Says Families Warned to be Vigilant," *Fox News*, March 23, 2015, http://www.foxnews.com/politics/2015/03/23/feds-investigating-isis-kill-list-military-spouse-says-families-told-to-be.html; "Military Spouses Hacked, Threatened by Alleged ISIS Sympathizer," *NBC News*, February 10, 2015, http://www.nbcnews.com/news/us-news/military-spouses-hacked-threatened-alleged-isis-sympathizer-n303971; Kristina Wong, "ISIS Hacker Targets Military Spouses," *The Hill*, February 10, 2015, http://thehill.com/policy/defense/232286-isis-hacker-targets-military-spouses.

96. Molly Blake, "After ISIS Threats, Some Military Moms Rethink their Facebook Lives," *Today*, April 17, 2015, https://www.today.com/money/after-isis-threats-military-families-safeguard-online-profiles-t14871.

97. Available at *National Archive Catalog*, https://research.archives.gov/id/513597.

98. Patricia Ticineto Clough and Craig Willse, "Gendered Security/National Security: Political Branding and Population Racism," *Social Text* 28, no. 4 (Winter 2010): 45–63.

99. Grewal, "Security Moms," 202.

100. Marita Sturken, *Tourists of History: Memory, Kitsch, and Consumerism from Oklahoma City to Ground Zero* (Durham: Duke University Press, 2007).

101. Christensen, "Technological Boundaries," 160.

102. Enloe, *Nimo's War*, 190.

103. Oliver, "Women," 1.

104. Brenton Malin notes that women who worked as telephone operators, so-called Hello Girls, during World War I served an important communicative function, facilitating connections between troops; Malin, *Feeling*

Mediated: A History of Media Technology and Emotion in America (New York: New York University Press, 2014), 60.

105. Although my primary focus here is on posters produced for circulation in the United States, some of my archive includes posters produced in the United Kingdom. I argue that given the U.S.-U.K. alliance and the fact that posters circulated between them, the iconographic similarities in their depictions of the threat women posed are more salient than differences in countries of origin.

106. See, for example, "Be Like Dad—Keep Mum," *Imperial War Museums*, http://www.iwm.org.uk/collections/item/object/22549; *Wikimedia Commons*, https://commons.wikimedia.org/wiki/File:INF3-231_Anti-rumour_and_careless_talk_Be_like_dad_-_keep_mum!_(set_of_twelve_human_figures_talking)_Artist_Grimes.jpg.

107. The corporeal wartime threat embodied by women was, of course, "venereal disease" (VD), also the subject of an intensive promotional campaign, targeting loose women rather than loose lips. A sample of the posters is available at Sandee LaMotte, "Meet the Shady Ladies of 'Penis Propaganda': Anti-VD Posters of World War II," CNN, August 26, 2015, www.cnn.com/2015/08/25/health/wwii-vd-posters-penis-propaganda.

108. MacLeish, *Making War*, 96. Malin's work on the emotional history of the telegraph suggests that such anxiety bound to the timely receipt of information precedes broadcast media. He notes that early telegraph users successfully sued telegraph companies for damages when important information—like the death of a parent—was delivered too slowly; Malin, *Feeling Mediated*, 46.

109. Lester and Flake, "Wartime Military Service," 125.

110. United States Special Operations Command, "OPSEC Way of Life: An OPSEC Family Guide," n.d., accessed March 10, 2016, http://www.socom.mil/ffrp/Documents/Family%20OPSEC%20trifold.pdf.

111. United States Africa Command, "OPSEC Hub," n.d., accessed March 14, 2016, http://www.africom.mil/staff-resources/opsec-hub.

112. Military OneSource, "How to Keep Your Family Safe through Operations Security," February 2016, accessed March 11, 2016, http://www.militaryonesource.mil/deployment?content_id=270487.

113. U.S. Army, "OPSEC and Safe Social Networking," n.d., accessed April 7, 2016, http://8tharmy.korea.army.mil/site/assets/doc/support/Social-Media-and-OPSEC.pdf.

114. United States Department of Defense, "OPSEC for Families," June 1, 2010, http://www.slideshare.net/DepartmentofDefense/opsec-for-families.

115. Lutz, *Homefront*, 209. Jennifer Mittelstadt observes that army wives often "operat[e] in a paradoxical relationship to autonomy and power." They are urged to put the needs of others—like spouses, children, and the military itself—first, while being expected to function independently, with minimal support from others; Mittelstadt, *Rise of the Military Welfare State*, 134.

116. Cohler, "Keeping the Home Front Burning," 249–50.

117. Ibid., 245. Zarzycka makes a compatible claim in her analysis of state appropriation of widows' grief in visual and rhetorical strategies that absorb it while utilizing it to deflect attention off its own militarism; Zarzycka, "Outside the Frame," 69–71.

118. Ruby C. Tapia, *American Pietàs: Visions of Race, Death, and the Maternal* (Minneapolis: University of Minnesota Press, 2011), 110.

119. Lutz, *Homefront*, 236.

120. Clark, "Off My (Dead) Body," 48.

121. Luc Capdevila and Danièle Voldman, *War Dead: Western Societies and the Casualties of War*, trans. Richard Veasey (Edinburgh: Edinburgh University Press, 2006).

122. Clark, "Off My (Dead) Body," 52.

123. Ibid., 60.

124. "American Widow Project on ABC World News," YouTube video, 3:59, posted by Taryn Davis, March 25, 2009, https://www.youtube.com/watch?v=xfs9BwELEag.

125. Butler, *Antigone's Claim*, 24.

126. Ibid., 57.

127. Abbas, "Voice Lessons: Suffering and the Liberal Sensorium," *Theory and Event* 13, no. 2 (2010).

128. Here, she is referring to the context of the postcolony, but I believe the analysis can be translated onto the circumstance of military widows.

129. Abbas, "In Terror, In Love," 511.

130. Widowers are welcome in the group but appear very infrequently in the AWP's documentation of its work.

131. Enloe, *Nimo's War*, 174.

132. Berlant, *Desire/Love* (Brooklyn, N.Y.: Punctum, 2012), 6.

133. MacLeish, *Making War*, 98.

134. Although the main AWP website includes WidowU among its list of initiatives, as of August 2017, the WidowU website (widowu.org) was unavailable. The text I quote here was current as of summer 2016.

135. Berlant, *Cruel Optimism*, 117.

136. The question of happiness, so elegantly untangled by Sara Ahmed in *The Promise of Happiness* (Durham: Duke University Press, 2010), is worthy of further consideration than space permits here.

137. See Butler, *Antigone's Claim*, and Butler, *Frames of War: When Is Life Grievable?* (London: Verso, 2010).

138. Jasbir K. Puar, *Terrorist Assemblages: Homonationalism in Queer Times* (Durham: Duke University Press, 2007), 3.

139. Tapia, *American Pietàs*, 120ff.

140. Elizabeth P. Van Winkle and Rachel N. Lipari, "The Impact of Multiple Deployments and Social Support on Stress Levels of Women Married to Active Duty Servicemen," *Armed Forces & Society* 41, no. 3 (2015): 396.

141. Ibid., 405.

142. Jennifer Terry notes that when women sustain these injuries, clinicians often take their reduced capacity to relate to spouses and children as an index of the severity of the damage; Terry, "Significant Injury: War, Medicine, and Empire in Claudia's Case," *WSQ: Women's Studies Quarterly* 37, nos. 1 and 2 (Spring/Summer 2009): 212. She specifically cites the case of a woman who, after her traumatic brain injury, could not remember that she was a mother.

143. Renshaw, Rodrigues, and Jones, "Psychological Symptoms," 592.

144. Renshaw and Sarah B. Campbell, "Combat Veterans' Symptoms of PTSD and Partners' Distress: The Role of Partners' Perceptions of Veterans' Deployment Experiences," *Journal of Family Psychology* 25, no. 6 (2011): 960.

145. Renshaw, Elizabeth S. Allen, Sarah P. Carter, Howard J. Markman, and Scott M. Stanley, "Partners' Attributions for Service Members' Symptoms of Combat-Related Posttraumatic Stress Disorder," *Behavior Therapy* 45, no. 2 (March 2014): 188.

146. Catherine M. Caska and Renshaw, "Perceived Burden in Spouses of National Guard/Reserve Service Members Deployed During Operations Enduring and Iraqi Freedom," *Journal of Anxiety Disorders* 25, no. 3 (April 2011): 350.

147. Renshaw, Allen, Galena K. Rhoades, Rebecca K. Blais, Markman, and Stanley, "Distress in Spouses of Service Members with Symptoms of Combat-Related PTSD: Secondary Traumatic Stress or General Psychological Distress?," *Journal of Family Psychology* 25, no. 4 (2011): 467. The authors note that "fewer than 20% of spouses reported that their symptoms were solely due to husbands' military experiences. Moreover, it is unclear how many spouses in this subset were attributing their symptoms specifically to their knowledge about their husbands' experiences, versus to their own experience of living with his reactions to those experiences (e.g., feeling jumpy because of their husbands' short temper)."

148. It is also possible that such a finding could be used to minimize the daily struggles that military wives face.

4. ECONOMIES OF POST-TRAUMATIC STRESS DISORDER AND TRAUMATIC BRAIN INJURY

1. "Wounded Warrior Project Collection," *Under Armour*, 2016, accessed June 29, 2016, https://www.underarmour.com/en-us/wwp-collection.

2. "Shop," *Wounded Warrior Project*, 2016, accessed June 29, 2016, https://www.woundedwarriorproject.org/shop.

3. See, for example, Thomas Gibbons-Neff, "Wounded Warrior Project Executives Fired Amid Controversy," *Washington Post*, March 10, 2016, https://www.washingtonpost.com/news/checkpoint/wp/2016/03/10/report-wounded-warrior-project-executives-fired-amid-controversy/.

4. Accessed August 18, 2017.

5. For more on how the fabricated "character" of "the warrior" "has eclipsed the soldier," see Andrew J. Bacevich, *Breach of Trust: How Americans Failed Their Soldiers and Their Country* (New York: Metropolitan, 2013), 185.

6. Zoë H. Wool, *After War: The Weight of Life at Walter Reed* (Durham: Duke University Press, 2015), 11.

7. Visually, this echoes the silhouette form of the POW/MIA flag. On the politics of the POW/MIA image, see Roger Stahl, "Why We 'Support the Troops': Rhetorical Evolutions," *Rhetoric & Public Affairs* 12, no. 4 (2009): 550.

8. Even if veterans with PTSD are newly visible in popular culture and political discourse, such representations can only show the exteriorized symptoms of trauma, not the veterans' interior experience of it; Christopher J. Gilbert and John Louis Lucaites, "Returning Soldiers and the In/Visibility of Combat Trauma," in *In/Visible War: The Culture of War in Twenty-First-Century America* (New Brunswick, N.J.: Rutgers University Press, 2017), 50.

9. On the gendered, racialized, and economic dimensions of amputation, see H. N. Lukes, "The Sovereignty of Subtraction: Hypo/Hyperhabilitation and the Cultural Politics of Amputation in America," *Social Text* 33, no. 2 (June 2015): 1–27.

10. On the politics of "signature injuries," Jennifer Terry writes, "The naming and treatment of the injury as a signature wound mar[k] a particular converging history of technology, geopolitics, and biopolitics [and] show the ways in which medical techniques and violent warfare function in a relationship of mutual provocation, provoking one another in a manner that indicates the close ties between hygienic and military logics in modern U.S. empire building"; Terry, "Significant Injury: War, Medicine, and Empire in Claudia's Case," *WSQ: Women's Studies Quarterly* 37, nos. 1 and 2 (Spring/

Summer 2009), 202. She also observes that patterns of injuries *inflicted* by U.S. forces do not receive such elevated recognition (206).

11. Wool, "On Movement: The Matter of U.S. Soldiers' Being after Combat," *Ethnos* 78, no. 3 (2013): 404.

12. David Kieran, "'We Combat Veterans Have a Responsibility to Ourselves and Our Families': Domesticity and the Politics of PTSD in Memoirs of the Iraq and Afghanistan Wars," *American Studies* 53, no. 2 (2014): 100.

This dynamic is not unique to the current conflict. For example, Joe Kember notes that in World War I–era depictions of veterans with severe facial disfigurements, their wives were tasked with returning the men to "normalcy" and "wholeness" at home, a process which included sexual intimacy; Kember, "Face Value: The Rhetoric of Facial Disfigurement in American Film and Popular Culture, 1917–1927," *Journal of War & Culture Studies* 10, no. 1 (February 2017), 56.

13. "DoD Worldwide Numbers for TBI," *Defense and Veterans Brain Injury Center*, accessed September 19, 2017, http://dvbic.dcoe.mil/dod -worldwide-numbers-tbi.

14. Nearly everyone concedes that even these numbers are likely too low, because of servicemember underreporting and misdiagnosis; "How Common is PTSD?," *U.S. Department of Veterans Affairs—National Center for PTSD*, last modified October 3, 2016, http://www.ptsd.va.gov/public/PTSD -overview/basics/how-common-is-ptsd.asp.

15. For a critique of the common ways that scholars integrate veterans' voices and experiences into their research, see Sarah Bulmer and David Jackson, "'You Do Not Live in My Skin': Embodiment, Voice, and the Veteran," *Critical Military Studies* 2, no. 1–2 (2016): 25–40.

16. John M. Kinder, *Paying with Their Bodies: American War and the Problem of the Disabled Veteran* (Chicago: University of Chicago Press, 2015), 21. Most historians agree that land offered very little benefit to its recipients, particularly those with severely disabling injuries.

17. Ibid., 25. The United States borrowed the practice of attributing financial value to injuries from a similar French system created in 1831; David A. Gerber, "Introduction: Finding Disabled Veterans in History," in *Disabled Veterans in History*, ed. David A. Gerber, enlarged and rev. ed. (Ann Arbor: University of Michigan Press, 2012), 17.

18. Kinder, *Paying with Their Bodies*, 31.

19. Daryl S. Paulson and Stanley Krippner, *Haunted by Combat: Understanding PTSD in War Veterans Including Women, Reservists, and Those Coming Back from Iraq* (Westport, Conn.: Praeger Security International, 2007), 9–10.

20. Diagnostic terms, as Tracey Loughran argues, are collections of ideas, not stable ontological facts; Loughran, "Shell Shock, Trauma, and the First World War: The Making of a Diagnosis and Its Histories," *Journal of the History of Medicine and Allied Sciences* 67, no. 1 (January 2012): 100. Diagnoses, as Alison Howell notes, are "invented, not discovered"; Howell, "The Demise of PTSD: From Governing through Trauma to Governing Resilience," *Alternatives: Global, Local, Political* 37, no. 3 (2012): 215.

21. Kinder, *Paying with Their Bodies*, 24.

22. However, the psychological symptoms of World War II veterans elicited less popular interest than the trauma of concentration-camp survivors; James K. Boehnlein and Devon E. Hinton, "From Shell Shock to PTSD and Traumatic Brain Injury: A Historical Perspective on Responses to Combat Trauma," in *Culture and PTSD: Trauma in Global and Historical Perspective*, ed. Devon E. Hinton and Byron J. Good (Philadelphia: University of Pennsylvania Press, 2016), 163.

23. Wool, "On Movement," 406.

24. Loughran notes that the construct of shell shock is central to our interpretations of both World War I and war in general; Loughran, "Shell Shock," 101. Kate MacDonald describes World War I itself as "crucial" for histories of disability; MacDonald, "The Woman's Body as Compensation for the Disabled First World War Soldier," *Journal of Literary & Cultural Disability Studies* 10, no. 1 (2016): 53.

25. Loughran, "Shell Shock," 105.

26. Ibid.

27. Ibid., 102. Some of these are similar to symptoms of PTSD and TBI, but Loughran argues that we should not regard these terms as three different names for the same thing (101–2).

28. Robert F. Worth, "What If PTSD Is More Physical than Psychological?," *New York Times Magazine*, June 10, 2016, http://www.nytimes.com/2016/06/12/magazine/what-if-ptsd-is-more-physical-than-psychological.html. Boehnlein and Hinton, "From Shell Shock to PTSD," note that prewar British psychiatrists relied heavily on the frameworks of hysteria and neurasthenia and that both of these informed their understandings of shell shock (160).

29. Annessa C. Stagner, "Making Broken Bodies Whole in a Shell-Shocked World," in *Body and Nation: The Global Realm of U.S. Body Politics in the Twentieth Century*, ed. Emily S. Rosenberg and Shannon Fitzpatrick (Durham: Duke University Press, 2014), 61. This, she argues, contrasted with more pessimism about the potential for recovery among European clinicians (63).

30. Ibid., 71. The soldier's body remains a potent contemporary metaphor for the nation-state. Zoë Wool argues that "the body of the injured soldier becomes a kind of avatar of the nation itself, both in its form and function and in the arrangements of life it entails"; Wool, "In-Durable Sociality: Precarious Life in Common and the Temporal Boundaries of the Social," *Social Text* 35, no. 1 (March 2017): 82.

31. As early as 2012, Alison Howell, citing the reinvigoration of a biomedical approach to trauma, surmised that PTSD might be approaching obsolescence as a diagnostic category; Howell, "Demise of PTSD," 214.

32. Boehnlein and Hinton, "From Shell Shock to PTSD," 167. While a full consideration of this repositioning is beyond the scope of my present analysis, I suggest that physiological explanations for emotional disorders are far more compatible with a masculine warrior ethos than psychological ones.

33. Howell, "Demise of PTSD," 222. Perhaps hedging its bets on this front, the U.S. military is also increasingly pursuing preventive measures for PTSD, which focus on positive psychology and training for resilience. On the discourse and practice of resilience in approaches to military mental health more generally, see David Kieran, *Signature Wounds: The Cultural Politics of Mental Health during the Iraq and Afghanistan Wars* (New York: New York University Press, forthcoming).

34. The questions of how, how much, and for how long a blast can injure the brain remain unanswered and controversial; Kieran, *Signature Wounds.*

TBI resulting from blast exposure is distinct from the Chronic Traumatic Encephalopathy (CTE) that often afflicts athletes like boxers and football players. In terms of the chemical changes that blast exposure induces in the brain, it may be more akin to the "heart concussions" suffered, for example, by baseball players who get hit in the chest; Jon Hamilton, "An Army Buddy's Call for Help Sends a Scientist on a Brain Injury Quest," *National Public Radio*, June 8, 2016, http://www.npr.org/sections/health-shots/2016/06/08/480608042/an-army-buddys-call-for-help-sends-a-scientist-on-brain-injury-quest.

35. Caroline Alexander, "Blast Force: The Invisible War on the Brain," Part 2, *National Geographic*, last modified March 28, 2015, http://www.nationalgeographic.com/healing-soldiers.

36. Worth, "What If PTSD Is More Physical?"

37. J. Martin Daughtry, "Thanatosonics: Ontologies of Acoustic Violence," *Social Text* 32, no. 2 (Summer 2014): 25. Daughtry theorizes that the sounds of combat, especially explosions, create surprising linkages, tenuous forms of acoustic community, between victims and perpetrators. On the sensory dimension of combat, see also John Hockey, "'Switch On': Sensory

Work in the Infantry," *Work, Employment & Society* 23, no. 3 (September 2009): 477–93.

38. Alexander, "Blast Force." Some experts use the language of magic to describe what happens, particularly in the earliest stages of the blast. A partial explanation for the lag in understanding TBI is that research on blast injuries long overlooked the brain and focused more on damage to other organs (Hamilton, "Army Buddy's Call for Help.").

39. E. Lanier Summerall, "Traumatic Brain Injury and PTSD," *U.S. Department of Veterans Affairs—National Center for PTSD*, last modified March 28, 2017, http://www.ptsd.va.gov/professional/co-occurring/traumatic-brain-injury-ptsd.asp.

40. "DoD Worldwide Numbers for TBI."

41. Summerall, "Traumatic Brain Injury."

42. Ibid., under "Mild TBI."

43. Kenneth T. MacLeish, *Making War at Fort Hood: Life and Uncertainty in a Military Community* (Princeton: Princeton University Press, 2013), 55.

44. Terri Tanielian and Lisa H. Jaycox, eds., *Invisible Wounds of War: Psychological and Cognitive Injuries, Their Consequences, and Services to Assist Recovery* (Santa Monica, Calif.: Rand Center for Military Health Policy Research: 2008), accessed June 29, 2016, http://www.rand.org/pubs/monographs/MG720.html,144.

45. Howell, "Demise of PTSD," 223.

46. John Protevi undertakes a long exegesis of Terri Schiavo's case, reflecting on whether or how she and her wishes might be recognized after "real organic damage permanently destroyed the neural bases of her subjectivity"; Protevi, *Political Affect: Connecting the Social and the Somatic* (Minneapolis: University of Minnesota, 2009), 115; see also 130. He positions the "trapped bare life" of the person in a persistent vegetative state who will not be allowed to die as a consequence of the same state of exception that licenses indefinite detention in places like Guantánamo (122).

47. Hamilton suggests that drug companies have been resistant to develop treatments specifically for TBI; "Army Buddy's Call for Help."

48. Terry, "Significant Injury," 202.

49. Bacevich, *Breach of Trust*, 107. Joseph Masco argues that the epidemic prevalence of PTSD and TBI are "but one immediate way the War on Terror loads new crises into the future, here in the form of vast medical and social problems, internal terrors of the most immediate kind"; Masco, *The Theater of Operations: National Security Affect from the Cold War to the War on Terror* (Durham: Duke University Press, 2014), 200.

50. Tanielian and Jaycox, eds., *Invisible Wounds*, 142.

51. Ibid., 143.

52. Boehnlein and Hinton, "From Shell Shock to PTSD," 159. Here, the authors are referring to railroad companies' deliberations about whether and how to compensate people afflicted with "railway spine."

53. Martin Crotty and Mark Edele note that veterans' entitlement claims have less authority in more elaborate or capacious welfare states; Crotty and Edele, "Total War and Entitlement: Towards a Global History of Veteran Privilege," *Australian Journal of Politics and History* 59, no. 1 (2013): 23–24. See also Loughran, "Shell Shock," 103.

54. Patrick G. Coy, Lynne M. Woehrle, and Gregory M. Money, "Discursive Legacies: The U.S. Peace Movement and 'Support the Troops,'" *Social Problems* 55, no. 2 (May 2008): 162. The authors note that even antiwar organizations base their platforms on the idea of the noble and heroic soldier (162). On the seeming unanimity of support for the Gulf War, see Stahl, "Why We 'Support the Troops,'" 534.

55. Stahl, "Why We 'Support the Troops,'" 534. Stahl suggests that this transformation occurs primarily through two rhetorical maneuvers: "deflection," which recasts the war as a fight to save U.S. soldiers, and "dissociation," which positions any dissent as an attack on the troops.

56. Ibid., 545.

57. It also precludes concern for other casualties of war. On the allocation of grievability, see, of course, Judith Butler, *Frames of War: When Is Life Grievable?* (London: Verso, 2010). Nigel Thrift's analysis of Western "victimology" as affect, aesthetic, and ethic is also instructive here; Thrift, "Immaculate Warfare?: The Spatial Politics of Extreme Violence," in *Violent Geographies: Fear, Terror, and Political Violence*, ed. Derek Gregory and Allan Pred (New York: Routledge, 2007), 281.

58. Grégoire Chamayou, *A Theory of the Drone*, trans. Janet Lloyd (New York: New Press, 2015), 77. This arrangement, he notes, "tends to compromise the traditional social division of danger, in which soldiers are at risk and civilians are protected." For more on the centrality of the body to warfare, see Kevin McSorley, "Toward an Embodied Sociology of War," *Sociological Review* 62, no. S2 (2014): 107–28, and Christine Sylvester, *War as Experience: Contributions from International Relations and Feminist Analysis* (London: Routledge, 2013).

59. MacLeish, "Armor and Anesthesia: Exposure, Feeling, and the Soldier's Body," *Medical Anthropology Quarterly* 26, no. 1 (2012): 49.

60. Gerber, "Creating Group Identity: Disabled Veterans and American Government," *OAH Magazine of History* (July 2009): 23.

61. Wool observes that such rhetoric conceptualizes sacrifice economically as an act that has been "*paid* rather than *offered* or *committed*"; Wool, *After War*, 104.

62. Gerber, "Introduction," 7.

63. For a canonical analysis of the challenges posed by extraordinary bodies to liberal ideals about democracy and individuality, see Rosemarie Garland Thomson, *Extraordinary Bodies: Figuring Physical Disability in American Culture and Literature* (New York: Columbia University Press, 1997).

64. David Serlin, "Crippling Masculinity: Queerness and Disability in U.S. Military Cultures, 1800–1945," *GLQ: A Journal of Lesbian and Gay Studies* 9, no. 1–2 (2003): 161–62.

65. In a fascinating study of the centrality of heterosexual coupling to ideas of veterans' recovery after World War I, Joanna Bourke describes how pity commingled, sometimes uncomfortably, with conjugal love. Referring to a somewhat short-lived movement to encourage women to marry disabled veterans, she explains how it was animated by a hope that this would be a way for women to "repay even one little bit of the debt" they owed to servicemen. Some men chafed at these arrangements, but no one seemed to doubt that civilian women had incurred such a duty; Bourke, "Love and Limblessness: Male Heterosexuality, Disability, and the Great War," *Journal of War & Culture Studies* 9, no. 1 (February 2016): 9.

66. Crotty and Edele, "Total War," 15–16. In Jennifer Mittelstadt's estimation, Ronald Reagan "arguably did more than anyone to advance the belief that the military constituted a special and elevated category"; Mittelstadt, *The Rise of the Military Welfare State* (Cambridge, Mass.: Harvard University Press, 2015), 9.

67. Mittelstadt, *Rise of the Military Welfare State*, 4.

68. Gerber, "Creating Group Identity," 24.

69. Kinder, *Paying with Their Bodies*, 86.

70. Ana Carden-Coyne, "Ungrateful Bodies: Rehabilitation, Resistance, and Disabled American Veterans of the First World War," *European Review of History/Revue Européenne d'Histoire* 14, no. 4 (December 2007): 546.

71. Wool, *After War*, 12.

72. On the coercive politics of gratitude more generally during the Trump administration, see Diana Butler Bass, "Thank Trump, Or You'll Be Sorry," *New York Times*, April 22, 2018, https://www.nytimes.com/2018/04/22/opinion/donald-trump-gratitude.html.

73. "Overcoming," as many disabilities studies scholars note, is central to Western narratives of disability; Carden-Coyne, "Ungrateful Bodies," provides an account of how the U.S. government and various charitable organizations colluded to promote this agenda to World War I veterans (544).

74. MacLeish, *Making War*, 110.

75. Beginning in the World War I era, this exercise of willpower was also taken as a sign of adult masculinity as opposed to petulant boyhood; Carden-Coyne, "Ungrateful Bodies," 557.

76. MacLeish, *Making War*, 12.

77. Wool, *After War*, 3.

78. Kinder, *Paying with Their Bodies*, 4.

79. Carden-Coyne, "Ungrateful Bodies," notes that other injured veterans were often enlisted to "cheer up" their sullen comrades but points out that forced optimism coexisted paradoxically alongside the widespread perception that their lives were hopeless (551).

80. Martin F. Norden, "Bitterness, Rage, and Redemption: Hollywood Constructs the Disabled Vietnam Veteran," in *Disabled Veterans in History*, ed. David A. Gerber, enlarged and rev. ed. (Ann Arbor: University of Michigan Press, 2012), 96–114.

81. Nancy Sherman, *Stoic Warriors: The Ancient Philosophy behind the Military Mind* (Oxford: Oxford University Press, 2005), 66.

82. Kinder, *Paying with Their Bodies*, 66.

83. Nate Bethea, "Sarah Palin, This Is What PTSD is Really Like," *New York Times*, January 22, 2016, http://www.nytimes.com/2016/01/23/opinion/sarah-palin-this-is-what-ptsd-is-really-like.html.

84. The miniseries is based on a book of the same title, by Evan Wright, the *Rolling Stone* journalist who embedded with a marine battalion during the 2003 invasion of Iraq. The clip is available at https://www.youtube.com/watch?v=cRg4-pCSMjM.

85. On the political consequences of the rhetorical turn to "troops," see Stahl, "Why We 'Support the Troops,'" 549.

86. Wesley Morgan, "The Bizarre, Unsatisfying Things Soldiers Receive in 'Care Packages,'" *Atlantic*, September 4, 2010, http://www.theatlantic.com/international/archive/2010/09/the-bizarre-unsatisfying-things-soldiers-receive-in-care-packages/62507/.

87. Bacevich, *Breach of Trust*, 30.

88. Parenthetically, the VA itself seems to believe in the magic of gratitude; its litany of tips for caregivers about how to help manage the stress and isolation of their roles includes the suggestion of keeping a gratitude journal; VA National Caregiver Training Program, "Module 1: Caregiver Self Care," 2015, accessed June 28, 2016, http://www.caregiver.va.gov/pdfs/Caregiver_Workbook_V3_Module_1.pdf, 41.

89. Mark E. Jonas, "Gratitude, *Ressentiment*, and Citizenship Education," *Studies in the Philosophy of Education* 31, no. 1 (2012): 30n1. Throughout this article, Jonas is building on the work of Patricia White; White, "Gratitude,

Citizenship and Education," *Studies in Philosophy and Education* 18, no. 1–2 (1999): 43–52.

Simon Keller, in a reflection on national loyalty, writes, "Gratitude is, in the first instance, a psychological state. It is a cluster of feelings and attitudes held towards someone (or something) that you take to have acted out of concern for you. It includes feelings of thankfulness and appreciation, and an understanding that something valuable has been done on your behalf; in being grateful for another's efforts, you take it not to be something trivial, but to matter, that they have been put towards to your interests, rather than towards something else"; Keller, *The Limits of Loyalty* (Cambridge: Cambridge University Press, 2007), 103.

90. The debt form of gratitude is deeply individualized, whereas the recognition form of gratitude can be generalized to a whole group or society, as a foundation for community or civility (Jonas, "Gratitude").

91. Bacevich describes thanking the troops as a key ritual in a "civic religion" observed with empty symbolic acts; *Breach of Trust*, 4.

92. Jonas contends that "simple thanks" are often preferable to grandly or excessively reciprocal gestures to convey gratitude and repay the initial gift; such displays, he suggests, reaffirm hierarchies and risk devolving into fawning or groveling, which prevents the initial gift from establishing a meaningful relationship between the parties ("Gratitude," 40).

93. Carden-Coyne, "Ungrateful Bodies," 556.

94. Wool, *After War*, 15.

95. Quil Lawrence, "Government Shutdown Will Add to VA's Backlog," *National Public Radio*, October 3, 2013, http://www.npr.org/2013/10/03/228733842/government-shutdown-will-add-to-vas-backlog; Jennifer Steinhauer, "Shutdown Denies Death and Burial Benefits to Families of 4 Dead Soldiers," *New York Times*, October 8, 2013, http://www.nytimes.com/2013/10/09/us/politics/shutdown-holds-up-death-benefits-for-military-families.html.

96. Brienne P. Gallagher, "Burdens of Proof: Veteran Frauds, PTSD Pussies, and the Spectre of the Welfare Queen," *Critical Military Studies* 2, no. 3 (2016): 139–54.

97. Elisabeth Anker, "Heroic Identifications: Or 'You Can Love Me Too—I Am So Like the State,'" *Theory & Event* 15, no. 1 (March 2012), under "The Desire for Freedom."

98. Gratitude for the soldiers is, by proxy, a form of gratitude to the state. This arrangement and felt obligation alarm political philosophers like Patricia White and Iris Marion Young. White, for example, surmises that gratitude toward the state might be more compatible with "benevolent dictatorship" than democracy; White, "Gratitude, Citizenship and Educa-

tion," 49. See also Young, "The Logic of Masculinist Protection: Reflections on the Current Security State," *Signs: Journal of Women in Culture and Society* 29, no. 1 (2003): 1–25.

99. For a thoughtful philosophical defense of the notion that we have a "sacred moral obligation to those who serve," see Sherman, *Afterwar: Healing the Moral Wounds of Our Soldiers* (Oxford: Oxford University Press, 2015), 3. On the other hand, when I presented a preliminary version of this argument at the American Studies Association conference in 2013, Meredith Lair replied with a provocative thought experiment, asking what would happen if we began from the premise that we did *not* owe a debt to our military personnel.

100. Trevor McCrisken, "Justifying Sacrifice: Barack Obama and the Selling and Ending of the War in Afghanistan," *International Affairs* 88, no. 5 (2012): 993–1007.

101. MacLeish, *Making War*, 186.

102. Wool, *After War*, 110. Wool argues that soldiers become "sacrificial victims" to the people who insist they owe a debt to the military (112).

103. Jacques Derrida, *Given Time*, vol. 1, *Counterfeit Money*, trans. Peggy Kamuf (Chicago: University of Chicago Press, 1992), 23.

104. Maurizio Lazzarato, *The Making of the Indebted Man: An Essay on the Neoliberal Condition*, trans. Joshua David Jordan (Los Angeles: Semiotext(e), 2012), 45.

105. United States Congress, "H.R. 258—Stolen Valor Act of 2013," *Congress.gov*, June 3, 2013, https://www.congress.gov/bill/113th-congress/house-bill/258.

106. Here again, Gallagher's analysis is instructive; "Burdens of Proof."

107. Kinder, *Paying with Their Bodies*, 100.

108. Ibid., 18, 19.

109. Wool, in her encounters with the overfull storerooms of VA residences, writes of "abundantly productive circuits of gratitude" that address the perceived needs of soldiers (for food, for hygiene products, for distraction, for comfort) but keep them from reclaiming ordinary lives; Wool, *After War*, 114. See also MacLeish, *Making War*, 200.

110. "Send a Letter," *A Million Thanks*, 2016, accessed July 2, 2016, http://www.amillionthanks.org/send_a_letter.php.

111. This is as of spring 2018.

112. In summer 2016, every wish featured on the site was from a veteran who self-identified as having at least one of these conditions.

113. On military indebtedness, see Liz Montegary, "An Army of Debt: Financial Readiness and the Military Family," *Cultural Studies* 29, no. 5–6 (2015): 652–68.

114. "Grant a Wish," *A Million Thanks*, http://amillionthanks.org/applyforawish.php, under "What kinds of wishes can I ask for?"

115. For a more sustained consideration of the notion of the aporia, see Nicholas Rescher, *Aporetics: Rational Deliberation in the Face of Inconsistency* (Pittsburgh: University of Pittsburgh Press, 2009).

116. Derrida, *Given Time*.

117. Derrida, *Deconstruction in a Nutshell: A Conversation with Jacques Derrida*, ed. John D. Caputo (New York: Fordham University Press, 1997), 18.

118. Lauren Berlant asks, "Which kinds of life engender ordinary anonymity and, in contrast, which unhistoric lives are exemplary only as waste, uncanny in trauma, and perfected in death?"; Berlant, "Uncle Sam Needs a Wife: Citizenship and Denegation," in *Visual Worlds*, ed. John R. Hall, Blake Stimson, and Lisa Tamiris Becker (London: Routledge, 2005), 35. This has broader implications for the question of recognition. Butler writes, "The differential distribution of grievability across populations has implications for why and when we feel politically consequential affective dispositions such as horror, guilt, righteous sadism, loss, and indifference"; Butler, *Frames of War*, 24.

119. Derrida again: "Justice and gift should go beyond calculation. This does not mean that we should not calculate. We have to calculate as rigorously as possible. But there is a point or limit beyond which calculation must fail, and we must recognize that"; Derrida, *Deconstruction in a Nutshell*, 19.

120. True gratitude cannot be commanded or required; Jonas, "Gratitude," 43.

121. Timothy V. Kaufman-Osborn, "'We Are All Torturers Now': Accountability After Abu Ghraib," *Theory & Event* 11, no. 2 (2008).

122. "Supporting the troops" while critiquing the state that sent them to war has become a favorite tightrope position of antiwar activists and politicians, but if support entails gratitude, that position may ultimately prove untenable.

123. Quoted in Levi Newman, "Purple Heart Group Says 'No' to Award for PTSD," *Veterans United Network*, July 18, 2012, accessed June 1, 2016, https://www.veteransunited.com/network/purple-heart-group-says-no-to-award-for-ptsd/.

124. Quoted in Barbara Salazar Torreon, "The Purple Heart: Background and Issues for Congress," *Congressional Research Service*, August 31, 2015, https://www.hsdl.org/?view&did=787816, 8.

125. Loughran, "Shell Shock," 106. Prior to diagnosis, the afflicted might also have been categorized as NYDN, or "Not Yet Diagnosed Nervous."

126. Kinder, *Paying with Their Bodies*, 161.

127. Wool, *After War*, 84.

128. Jim Garamone, "DoD Issues Purple Heart Standards for Brain Injury," *United States Army*, April 28, 2011, https://www.army.mil/article/60078/169586.

129. Torreon, "Purple Heart," under "Summary."

130. A lengthy addition to the United States Code after September 11 specifies the parameters of what counts as an attack by a foreign terrorist organization.

131. "Frequently Asked Questions on the Purple Heart and Mild Traumatic Brain Injury MTBI," *United States Army Human Resources Command*, last modified March 21, 2017, accessed June 14, 2016, https://www.hrc.army.mil/tagd/purple%20heart.

132. As of 2011, the marines' standards no longer required a loss of consciousness, provided that the person with TBI was unable to perform his duties for at least forty-eight hours; Capt. Patrick Boyce, "Marines with Concussion, Mild Brain Injury May Qualify for Purple Heart," *United States Marine Corps*, December 21, 2011, http://www.hqmc.marines.mil/News/News-Article-Display/Article/553006/marines-with-concussion-mild-brain-injury-may-qualify-for-purple-heart/.

133. Gene Beresin, "Why Are We Denying Purple Hearts to Veterans with PTSD?," *Huffington Post*, May 3, 2015, http://www.huffingtonpost.com/gene-beresin/why-are-we-denying-purple_b_6786318.html.

134. "Parity for Patriots: The Mental Health Needs of Military Personnel, Veterans, and their Families," *National Alliance on Mental Illness*, June 2012, accessed June 20, 2016, https://www.nami.org/getattachment/About-NAMI/Publications/Reports/ParityforPatriots.pdf.

135. This also has implications for military personnel who receive other-than-honorable discharges, a status that can severely restrict the veteran's access to benefits; Dave Philipps, "Wounded Troops Discharged for Misconduct Often Had PTSD or T.B.I.," *New York Times*, May 16, 2017, https://www.nytimes.com/2017/05/16/us/military-misconduct-ptsd.html.

136. Purple Heart awardees are eligible for additional medical benefits, but unlike recipients of the Medal of Honor, they do not receive additional pension funds.

137. Veteran suicide has, increasingly, drawn sympathetic official recognition, most notably in President Obama's signing of the Clay Hunt Act—named for a twenty-eight-year-old marine, diagnosed with depression and PTSD—meant to improve veterans' access to mental health services. The politics of such posthumous recognition are worthy of further consideration; Jenna Brayton, "The Clay Hunt Act: What the President Just Signed," *The White House—President Barack Obama*, February 12, 2015, accessed May 29,

2016, https://www.whitehouse.gov/blog/2015/02/12/clay-hunt-act-what-president-just-signed.

138. The book also inspired a 2017 movie by the same title.

139. David Finkel, *Thank You for Your Service* (New York: Sarah Crichton/Farrar, Straus and Giroux, 2013), 127. Sherman raises the possibility, surely unfathomable to the charities that promote civilian shows of thanks, that military resentment of civilians is common (*Afterwar*, 26).

140. For an overview of the implications of PTSD and TBI for spouses and children, see Allison K. Holmes, Paula K. Rauch, and Colonel Stephen J. Cozza (U.S. Army, Retired), "When a Parent Is Injured or Killed in Combat," *Future of Children* 23, no. 2 (Fall 2013): 143–62. The authors raise the possibility that TBI might be "uniquely traumatic for children" relative to other parental medical conditions (148).

141. Wool, *After War*, 3.

142. Beth Linker and Whitney Laemmli, "Half a Man: The Symbolism and Science of Paraplegic Impotence in World War II America," *OSIRIS* 30 (2015): 229.

143. Finkel, *Thank You*, 11.

144. Ibid., 7–9.

145. Gerber, "Creating Group Identity," 26. Gerber describes how injured veterans' status as "ward[s] of the state" grates against their own visions of themselves as capable and independent.

146. Kinder, *Paying with Their Bodies*, 34.

147. MacDonald, "Woman's Body," 54.

148. Kieran, "'We Combat Veterans,'" 96. When a woman is injured in this way, gendered expectations about maternal nurture and affection get rerouted through medical discourses to index the severity of the injury; Terry, "Significant Injury," 212.

149. Hera McLeod, a children's rights activist, argues that the government should treat the spouses and children who are abused by veterans as additional casualties of his brain injury; McLeod, "The Other Wounded Warriors," *New York Times*, August 3, 2017, https://www.nytimes.com/2017/08/03/opinion/military-veterans-abuse-violence.html.

150. Finkel, *Thank You*, 7–9, 39, 10, 193–208, 92, and 101–2.

151. Gallagher, "Burdens of Proof," 151.

152. In keeping with the trend across most media accounts of people with disabilities, the main protagonists in *Thank You for Your Service* are young white men; Kinder, *Paying with Their Bodies*, 8, 9. Although there are men of color in *Thank You for Your Service*, the narrative focuses more on the struggles of the white characters.

153. Finkel, *Thank You for Your Service*, 165.

154. Ibid., 39.
155. Ibid., 57.
156. Ibid., 159–69.
157. Ibid., 13.
158. Ibid., 175.
159. Ibid., 191.
160. Ibid., 221.
161. MacDonald argues that differential valuation of women's and men's bodies is a staple of wartime cultural production. In the accounts that she studies, "a woman's body was shown to be worth less than a man's body, in some stories only worth as much as his missing limb or damaged sight" ("Woman's Body," 66).
162. Terry, "Significant Injury," 202. Gerber raises the possibility that the prevalence of PTSD (presumably in the absence of TBI) among women in the military might be twice that among men, citing the compounding factor of sexual assault; Gerber, "Preface to the Enlarged and Revised Edition: The Continuing Relevance of the Study of Disabled Veterans," in *Disabled Veterans in History*, enlarged and rev. ed., ed. David A. Gerber (Ann Arbor: University of Michigan Press, 2012), xix.
163. Finkel, *Thank You*, 197.
164. Worth, "What If PTSD Is More Physical than Psychological?" These scar-related lesions occur as part of the brain's repair process.
165. Family members might not know whether the decedent had blast exposure or have only sketchy details about their combat history.
166. Summerall, "Traumatic Brain Injury."
167. As military, governmental, and public attention shifts toward TBI as arguably the most vexing health crisis among military personnel, the repository garners regular media attention in outlets like the *New York Times*, *National Geographic*, and *Al Jazeera*; http://www.researchbraininjury.org/news-items.
168. Sharon Baughman Shively, Iren Horkayne-Szakaly, Robert V. Jones, James P. Kelly, Regina C. Armstrong, and Daniel P. Perl, "Characterisation of Interface Astroglial Scarring in the Human Brain after Blast Exposure: A Post-Mortem Case Series," *Lancet Neurology*, June 9, 2016 (corrected proofs). "Astroglial scarring" describes a compensatory increase in astroglia brain cells in response to damage.
169. Ibid., 9.
170. Alexander, "'Shell Shock': The 100-Year-Old Mystery May Now Be Solved," *National Geographic*, June 9, 2016, http://news.nationalgeographic.com/2016/06/blast-shock-tbi-ptsd-ied-shell-shock-world-war-one/.

In its January 2017 issue, the *Lancet* published correspondence related to the initial study. One such letter critiques news stories like this one as "sensationalised"; Charles W. Hoge, Jonathan Wolf, and David Williamson, "Astroglial Scarring after Blast Exposure: Unproven Causality," *Lancet Neurology* 16, no. 1 (January 2017): 26n4.

171. This is precisely the sort of recategorization that Howell describes ("Demise of PTSD," 222). In an earlier version of the analysis, I described this as a reassertion of "biomedical authority" over trauma, but Dr. Perl and his staff suggested that I had overstated their position.

172. Alexander, "'Shell Shock.'"

173. In their review of this section, CNRM staff noted that they had not encountered such thoughts.

174. Greg Moorlock, Jonathan Ives, and Heather Draper, "Altruism in Organ Donation: An Unnecessary Requirement?," *Journal of Medical Ethics*, March 28, 2013, doi:10.1136/ medethics-2012-100528, 2. For a consideration of the ethical complexities of "altruism" in organ donation, see Ben Saunders, "Altruism or Solidarity? The Motives for Organ Donation and Two Proposals," *Bioethics* 26, no. 7 (September 2012): 376–81.

175. Personal correspondence, October 2016.

176. This includes their rigorous protections of donor confidentiality. Toward that end, no one but CNRM-authorized personnel are permitted inside the repository.

177. Specimens are preserved so that they can be stored or analyzed as long as necessary, to be safely discarded only if the research ends or a donor's family asks to have their tissue removed. CNRM staff notes that, to date, no family has requested removal of their donated specimen, but they do have a protocol in place for this eventuality; personal correspondence.

178. Personal correspondence.

179. Alexander, "Behind the Mask: Revealing the Trauma of War," Part 1, *National Geographic*, last modified March 28, 2015, www.nationalgeographic .com/healing-soldiers.

5. LIBERAL IMAGINARIES OF GUANTÁNAMO

1. *CNN Larry King Live*, "Interview with Donald Rumsfeld" (Transcript), CNN, May 25, 2006, http://transcripts.cnn.com/TRANSCRIPTS /0605/25/lkl.01.html.

2. The same is true of their counterparts at places like Bagram Airfield, about whom we know even less. Bagram has been exempted from even the relatively meager protections that the Obama administration has afforded to Guantánamo detainees. The relative dearth of scholarly and activist atten-

tion to the occupants of Bagram mirrors administrative efforts to secure it at as failsafe for Guantánamo, should the U.S. government feel the need to operate with even less oversight; Rebecca A. Adelman, *Beyond the Checkpoint: Visual Practices in America's Global War on Terror* (Boston: University of Massachusetts Press, 2014), 195.

3. Elizabeth A. Povinelli, *Economies of Abandonment: Social Belonging and Endurance in Late Liberalism* (Durham: Duke University Press, 2011), 11. Later, Povinelli observes that "empathetic deliberation" often persists even as its objects die awaiting a resolution (179).

4. Banu Bargu, "Sovereignty as Erasure: Rethinking Enforced Disappearance," *Qui Parle: Critical Humanities and Social Sciences* 23, no. 1 (Fall/Winter 2014): 63.

5. Joseph Masco, *The Theater of Operations: National Security Affect from the Cold War to the War on Terror* (Durham: Duke University Press, 2014), 16.

6. Michael Barkun encapsulates predominant imaginaries of terrorists as evocative of the "features of the stereotypical fictional villain: functional invisibility, allowing him to move across borders undetected and disappear at will; a sophisticated hiding place that is undetectable and/or impenetrable; and access to weapons of mass destruction concocted by the rogue scientists at his disposal"; Barkun, *Chasing Phantoms: Reality, Imagination, and Homeland Security Since 9/11* (Chapel Hill: University of North Carolina Press, 2011), 81.

7. Liz Philipose, "The Politics of Pain and the Uses of Torture," *Signs: Journal of Women in Culture and Society* 32, no. 4 (Summer 2007): 1055.

8. Furthermore, Philipose argues, the mere fact of detention indelibly marks detainees, even when wrongfully held, as terrorists; ibid., 1048. She describes this as a process of racialization.

9. Terri Tomsky, "The Guantánamo Lawyers: Life Writing for the 'Courts of Public Opinion,'" *Biography* 38, no. 1 (Winter 2015): 27.

10. Sunaina Marr Maira, "'Good' and 'Bad' Muslim Citizens: Feminists, Terrorists, and U.S. Orientalisms," *Feminist Studies* 35, no. 3 (Fall 2009): 631–56.

11. Povinelli, *Economies of Abandonment*, 80.

12. Sasha Torres, "Televising Guantánamo: Transmission of Feeling during the Bush Years," in *Political Emotions: New Agendas in Communication*, ed. Janet Staiger, Ann Cvetkovich, and Ann Reynolds (New York: Routledge, 2010), 45. Torres writes that "the forms of affect torture produces may be connected in unpredictable ways, and that exploring those connections could serve as the beginning of a durable and broadly based political mobilization against torture" (48).

13. Elspeth Van Veeren, "Guantánamo Does Not Exist: Simulation and the Production of 'The Real' Global War on Terror," *Journal of War and Culture Studies* 4, no. 2 (September 2011): 201n4.

14. Lauren Berlant, *Cruel Optimism* (Durham: Duke University Press, 2011), 182.

15. While the notion of "speaking for" the less privileged has been widely criticized in feminist and antiracist activism, I have not seen that critique developed in theorizations of indefinite detention.

16. Wendy Brown, *States of Injury: Power and Freedom in Late Modernity* (Princeton: Princeton University Press, 1995), xii.

17. Adelman, "'Safe, Humane, Legal, Transparent': State Visions of Guantánamo Bay," *Reconstruction: Studies in Contemporary Culture* 12, no. 4 (2013), http://reconstruction.eserver.org/Issues/124/Adelman_Rebecca.shtml.

18. Rebecca Wanzo, *The Suffering Will Not Be Televised: African American Women and Sentimental Political Storytelling* (Albany: SUNY Press, 2009), 32.

19. Van Veeren, "Captured by the Camera's Eye: Guantánamo and the Shifting Frame of the Global War on Terror," *Review of International Studies* 37 (2011): 1728, 1739.

20. "JTF-GTMO Photo Gallery," *Joint Task Force-Guantanamo (JTF-GTMO)*, 2016, accessed June 20, 2016, http://www.jtfgtmo.southcom.mil /xWEBSITE/photos/photos.html.

21. Susan L. Carruthers, "Why Can't We See Insurgents? Enmity, Invisibility, and Counterinsurgency in Iraq and Afghanistan," *Photography & Culture* 8, no. 2 (July 2015): 195.

22. Claire Birchall, "Radical Transparency?," *Cultural Studies–Critical Methodologies* 14, no. 1 (2014): 85n5.

23. Adelman, "'Safe, Humane, Legal, Transparent.'"

24. As I was making final revisions on the book manuscript in the summer and fall of 2017, the implications of the Trump presidency on the future of Guantánamo remained unknown.

25. Spencer Ackerman, "Guantánamo Detainee Says His 'Comfort Items' Were Taken to Force Interrogations," *Guardian*, July 29, 2015, http://www.theguardian.com/us-news/2015/jul/29/guantanamo-memoir -comfort-items-interrogations.

26. Bargu, "Sovereignty as Erasure," 63.

27. Kelly Oliver, *Witnessing: Beyond Recognition* (Minneapolis: University of Minnesota Press, 2001), 2.

28. Carruthers, "Why Can't We See Insurgents?," 193. Marita Sturken queries the lack of widespread outrage on the part of the American public over the torture at Abu Ghraib, the continued detention at Guantánamo, and Obama's failure to put a decisive end to these practices. That line of

questioning is worth considering, but beyond the scope of my argument here; see Sturken, "Comfort, Irony, and Trivialization: The Mediation of Torture," *International Journal of Cultural Studies* 14, no. 4 (2011): 423–40.

29. This epistemic impossibility generates gaps in academic knowledge production as well. Jasbir K. Puar observes that most scholarship on Guantánamo focuses on the legal status of detainees, while few academics undertake the more logistically and ideologically daunting project of analyzing "the impact of indefinite detention on the detainees and people connected to them"; Puar, *Terrorist Assemblages: Homonationalism in Queer Times* (Durham: Duke University Press, 2007), 143.

30. Carruthers, "Why Can't We See Insurgents?," 192. Anjali Nath, "Seeing Guantánamo, Blown Up: Banksy's Installation in Disneyland," *American Quarterly* 65, no. 1 (March 2013): 187.

31. Talal Asad, *On Suicide Bombing* (New York: Columbia University Press, 2007), and Ghassan Hage, "'Comes a Time We Are All Enthusiasm': Understanding Palestinian Suicide Bombers in Times of Exighophobia," *Public Culture* 15, no. 1 (Winter 2003): 65–89. Asad notes, for example, that we can only learn about the motivations of suicide bombers from those who fail (45).

32. My own interest in detainee anger is inspired by Sianne Ngai's work on "ugly feelings," her analysis of a cultural tendency to suppress or devalue unpleasant, noncathartic emotions. Given that detainees will either be held indefinitely or released for reasons unrelated to their emotions, the detainees' feelings are profoundly noncathartic. Consequently, the people who serve as conduits between them and the outside world must manage multiple layers of emotional "ugliness"; Ngai, *Ugly Feelings* (Cambridge, Mass.: Harvard University Press, 2005).

33. Neel Ahuja, "Abu Zubaydah and the Caterpillar," *Social Text* 29, no. 1 (Spring 2011): 127–49.

34. Laleh Khalili, *Time in the Shadows: Confinement in Counterinsurgencies* (Stanford, Calif.: Stanford University Press, 2013), 45. On the intimacies that such tactics generate, see Derek Gregory, "'The Rush to the Intimate': Counterinsurgency and the Cultural Turn in Late Modern War," *Geographical Imaginations*, 2012, geographicalimaginations.files.wordpress.com/2012/07/gregory-rush-to-the-intimate-full.pdf.

35. Bargu, "Sovereignty as Erasure," 43.

36. Adam Ewing, "In/Visibility: Solitary Confinement, Race, and the Politics of Risk Management," *Transition* 119 (January 2016): 112, 113.

37. Hugh Gusterson notes that in most media representations, the insurgent is "fundamentally unknowable and must be so"; Gusterson, "Can the Insurgent Speak?," in *Orientalism and War*, ed. Tarak Barkawi and Keith

Stanski (London: Hurst, 2012), 85. He further notes that they appear only indirectly, as through the casualty counts of their actions (87).

38. Gusterson argues that "to make insurgents' voices heard, and to invite audiences to move into the insurgent subject-position, is deeply threatening to the ideological status quo"; ibid., 96.

39. For a history of the politics and performance of multiculturalism and inclusivity in Silicon Valley and the Bay Area, see Maira, *The 9/11 Generation: Youth, Rights, and Solidarity in the War on Terror* (New York: New York University Press, 2016), 37.

40. Pioneer Valley No More Guantánamos, "Article 14: Information Supporting 'Resolution to Assist in the Safe Resettlement of Guantánamo Detainees,'" n.d., accessed May 5, 2016, https://www.amherstma.gov/DocumentCenter/Home/View/2744.

41. City of Berkeley Peace and Justice Commission, "The Safe Resettlement of Cleared Guantánamo Detainees," February 15, 2011, http://www.ci.berkeley.ca.us/uploadedFiles/Clerk/Level_3_-_City_Council/2011/02Feb/2011-02-15_Item_18a_The_Safe_Resettlement_of_Cleared_Guantinamo_Detainees-PJC.pdf.

42. The press release requests that a country in Latin America volunteer to resettle him.

43. With their emphasis on the docility of the detainees as evidence of their fitness for integration into Western society, these depictions are reminiscent of the government-commissioned photos of interned Japanese Americans during World War II. In her analysis of the photos, Wendy Kozol notes the lack of racist imagery and cheerful focus on the internees' enjoyment of familiar domestic pursuits. While this might seem a counterintuitive visual strategy, Kozol argues that the photos were commissioned and shot with an eye toward the postwar reintegration of their subjects. By identifying specific detainees, the resolutions follow a similar logic; Kozol, "Relocating Citizenship in Photographs of Japanese Americans During World War II," in *Haunting Violations: Feminist Criticism and the Crisis of the "Real,"* ed. Wendy S. Hesford and Wendy Kozol (Urbana: University of Illinois Press, 2001), 217–50.

44. On the perils of aspiring to the "good life," see Berlant, *Cruel Optimism*.

45. Deborah Becker, "Amherst Welcomes Cleared Guantánamo Detainees," WBUR, November 6, 2009, www.wbur.org/2009/11/06/amherst-gitmo. One of the orchestrators of the resolution, Rita Hook, mentions that she protested outside the Supreme Court in the previous year, wearing an orange jumpsuit.

46. "Mission Statement," *Move America Forward*, n.d., accessed June 8, 2016, http://www.moveamericaforward.org/mission-statement/.

47. "Berkeley Votes against Welcoming Gitmo Detainees," *CBS News*, February 15, 2011, http://www.cbsnews.com/news/berkeley-votes-against-welcoming-gitmo-detainees/.

48. Doug Oakley, "Berkeley Council Rejects Proposal to Invite Guantánamo Detainees to Live in City," *Mercury News*, February 16, 2011, http://www.mercurynews.com/bay-area-news/ci_17401616?source=rss.

49. City of Berkeley Office of the City Manager, "Response to the Peace and Justice Commission's Report entitled 'The Safe Resettlement of Cleared Guantánamo Detainees,'" February 15, 2011, http://www.ci.berkeley.ca.us/uploadedFiles/Clerk/Level_3_-_City_Council/2011/02Feb/2011-02-15_Item_18b_The_Safe_Resettlement of Cleared Guantanamo_Detainees-CM.pdf.

50. Elisabeth R. Anker, *Orgies of Feeling: Melodrama and the Politics of Freedom* (Durham: Duke University Press, 2014), 181.

51. City of Berkeley City Clerk, "Guidelines for Public Comment, Written Communications, and Council Meeting Order," n.d., accessed April 9, 2016, https://www.cityofberkeley.info/Clerk/City_Council/City_Council__General_Information.aspx. During his request, the mayor asks for one language change, so that the resolution indicates that, upon lifting of the Congressional ban, detainees "would be" (rather than the previous "will be") invited to resettle in Berkeley.

52. A full video of the meeting is available at http://berkeley.granicus.com/MediaPlayer.php?publish_id=817. The discussion of the resolution begins roughly forty-four minutes in.

53. Quoted in Becker, "Amherst Welcomes Cleared Guantánamo Detainees."

54. Charlie Savage, "U.S. Frees Last of the Chinese Uighur Detainees from Guantánamo Bay," *New York Times*, December 31, 2013, http://www.nytimes.com/2014/01/01/us/us-frees-last-of-uighur-detainees-from-guantanamo.html; Savage, "Guantánamo Detainee Refuses Offer of Release After 14 Years in Prison," *New York Times*, January 21, 2016, http://www.nytimes.com/2016/01/22/us/politics/guantanamo-detainee-refuses-release-offer.html; Ackerman, "Two Guantánamo Detainees Transferred as Third Refuses Resettlement Offer," *Guardian*, January 21, 2016, http://www.theguardian.com/us-news/2016/jan/21/two-guantanamo-bay-detainees-transferred.

55. Puar, *Terrorist Assemblages*, 145.

56. Anker, *Orgies of Feeling*, 34.

57. Anker writes, "Melodrama legitimates state power not only because citizens demand it or because it offers the truth of the nation's own virtue. The feeling of legitimacy proves the virtue of each feeling citizen, and thus captures why melodramatic political discourse becomes so potent at multiple affective registers. Felt legitimacy creates a way for citizens who experience

the feeling of victimization to index that this feeling proves their virtue. Felt legitimacy then doubles to demonstrate one's qualification for membership in a national identity defined by goodness"; ibid., 136. In this instance, the citizens feel legitimate in their desire to act in the state's stead, on its reparative behalf.

58. Quoted in Common Dreams, "Leverett (MA) Town Meeting Approves Resolution to Welcome Cleared Guantánamo Detainees," *Common Dreams*, April 26, 2010, http://www.commondreams.org/newswire/2010/04/26/leverett-ma-town-meeting-approves-resolution-welcome-cleared-guantanamo.

59. Timothy V. Kaufman-Osborn, "'We Are All Torturers Now': Accountability After Abu Ghraib," *Theory & Event* 11, no. 2 (2008).

60. Observing the ease with which liberal societies condone imperial violence, Talal Asad highlights the "ingenuity of liberal discourse in rendering inhuman acts humane"; Asad, *On Suicide Bombing*, 38.

61. City of Berkeley Peace and Justice Commission, "The Safe Resettlement of Cleared Guantánamo Detainees," February 15, 2011, http://www.ci.berkeley.ca.us/uploadedFiles/Clerk/Level_3_-_City_Council/2011/02Feb/2011-02-15_Item_18a_The_Safe_Resettlement_of_Cleared_Guantinamo_Detainees-PJC.pdf, 12–15.

62. Simultaneously, counterinsurgent propaganda images of happy insurgents provide the "alibi the counterinsurgent needs to persuade the people at home that his imperial intents . . . are ultimately humanitarian"; Khalili, "The Uses of Happiness in Counterinsurgencies," *Social Text* 32, no. 1 (Spring 2014): 25, 38, 39.

63. When I first started talking about these resolutions, someone observed that they could be for the premise for a horrible sitcom. Maybe. As I was researching, I found myself trying to imagine what would happen if, improbably, one of these communities eventually got its wish.

I imagine that just after sunrise on the appointed day (no more middle-of-the-night renditions), things would start to happen. His last Guantánamo dawn prayers. His last Guantánamo breakfast. There would be paperwork, small bureaucratic delays, even perhaps a moment when some last-minute glitch would threaten to foil the whole thing. There would be an instruction to pack his few possessions, which wouldn't take long. There would be goodbyes, quick clutching embraces and whispered benedictions, from the others who were not so lucky (if that is the right word, if there is any luck about it), as he. Some of the guards would seem happy for him. Some of the others would seem sorry, whether because of fondness or malice or both commingled, to see him go. Some would mutter about liberals. Others would wish for an unseen group of do-gooders to arrange an escape for them.

Stateside, at his destination, an anticipatory flurry of activity. The reporters, having arrived a few days prior, would be gathering b-roll, interviewing passersby. Protestors would be assembling and supporters arranging themselves for counter-demonstrations. And the people who had orchestrated it all, underslept and anxious and about to see this unbelievable thing made real, would be making final preparations. Prearranged, there would be English lessons, an apartment, mental health screenings, doctor visits, the rudiments of a social network, a job. Leaders at the local mosque, who had not really been consulted in the planning but could not refuse and risk alienating their few allies, would agree to help their newest neighbor get acclimated.

Presumably, there would be a private security briefing. Some officials would worry that someone somewhere along the line had miscalculated. Or that the financial support and good will of the private benefactors would run out. Or that the detainee had successfully concealed his hatred of America in a ploy to finagle his release. Something would go wrong, possibly very, and inevitably city government would get the blame. Outwardly, speaking to journalists and chary constituents, officials would offer reassurances.

At the airport, the designated committee of greeters, along with a few citizens whose curiosity had gotten the better of them, would be scanning the horizon for sight of the plane, smiling and brandishing their hand-lettered welcome signs even though he wasn't even on the ground yet.

And then he would arrive.

Waiting for him, a new life ready, fully constructed, bespoke to their vision of him, like a uniform waiting to pulled on.

64. "Ahmed Belbacha," *Reprieve*, 2014, https://www.reprieve.org.uk/case-study/ahmed-belbacha/.

65. The Guantánamo Docket, "Ravil Minzagov," *New York Times*, January 2017, https://www.nytimes.com/interactive/projects/guantanamo/detainees/702-ravil-mingazov.

66. Center for Constitutional Rights, "Djamel Ameziane," April 10, 2015, https://ccrjustice.org/djamel-ameziane.

67. "The Project," *Witness to Guantánamo*, 2016, accessed May 2, 2016, http://witnesstoguantanamo.com/about/.

68. Lilie Chouliaraki, *The Ironic Spectator: Solidarity in the Age of Post-Humanitarianism* (Cambridge: Polity, 2013), 17.

69. "Student Reactions," *Witness to Guantánamo*, 2016, accessed June 10, 2016, http://witnesstoguantanamo.com/react/student-reactions/.

70. An earlier iteration of the site characterized these men as "former detainees," but as of summer 2017, the site describes them as "detainees," a mislabeling that I find curious.

71. Gusterson cautions against such romanticization in the case of insurgents' voices; Gusterson, "Can the Insurgent Speak?," 100.

72. Asma Abbas, "Voice Lessons: Suffering and the Liberal Sensorium," *Theory & Event* 13, no. 2 (2010): n. 3.

73. Oliver, *Witnessing*, 99.

74. "Comments from Participants," *Witness to Guantánamo*, 2016, accessed May 4, 2016, http://witnesstoguantanamo.com/comments-from-participants/.

75. Puar and Amit S. Rai, "Monster, Terrorist, Fag: The War on Terrorism and the Production of Docile Patriots," *Social Text* 20, no. 3 (Fall 2002): 137.

76. Chad Shomura, "'These Are Bad People': Enemy Combatants and the Homopolitics of the 'War on Terror,'" *Theory & Event* 13, no. 1 (2010).

77. Philipose, "Politics of Pain," 1055.

78. Tomsky, "Guantánamo Lawyers," 29.

79. Lauren Wilcox, "Dying Is Not Permitted: Sovereignty, Biopower, and Force-Feeding at Guantánamo Bay," in *Torture: Power, Democracy, and the Human Body*, ed. Shampa Biswas and Zahi Zalloua (Seattle: University of Washington Press, 2011), 115.

80. Asad, *On Suicide Bombing*, 91.

81. Sami al-Hajj, "The Good That Came Out of It," *Witness to Guantánamo*, January 2013, http://witnesstoguantanamo.com/wp-content/uploads/2013/01/SamiAlHajj_TheGoodThatCameOutofIt.pdf.

82. Moazzam Begg, "Human Contact," *Witness to Guantánamo*, January 2013, http://witnesstoguantanamo.com/wp-content/uploads/2013/01/MoazzamBegg_HumanContact.pdf.

83. Mosa Zemmouri, "Hunger Strike," *Witness to Guantánamo*, November 2014, http://witnesstoguantanamo.com/wp-content/uploads/2014/11/MosaZemmouri_HungerStrike.pdf.

84. Murat Kurnaz, "No Rules," *Witness to Guantánamo*, January 2013, http://witnesstoguantanamo.com/wp-content/uploads/2013/01/MuratKurnaz_NoRules.pdf.

85. Bisher al Rawi, "Unnecessary Punishment," *Witness to Guantánamo*, June 2012, http://witnesstoguantanamo.com/wp-content/uploads/2012/06/BisheralRawi_UnnecessaryPunishment.pdf.

86. Khalil Mamut, "Dentist Visit," *Witness to Guantánamo*, 2015, http://witnesstoguantanamo.com/interviews/detainees/khalil-mamut-uyghur/.

87. Brahim Yadel, "Interrogation During Surgery, *Witness to Guantánamo*, 2016, http://witnesstoguantanamo.com/interviews/detainees/brahim-yadel-french/.

88. Feroz Ali Abbasi, "Beyond the Veil," *Witness to Guantánamo*, January 2013, http://witnesstoguantanamo.com/wp-content/uploads/2013

/01/FerozAliAbbasi_BeyondTheVeil.pdf; Abbasi, "Enduring Guantánamo," *Witnessing to Guantánamo*, January 2013, http://witnesstoguantanamo.com/wp-content/uploads/2013/01/FerozAliAbbasi_EnduringGuantanamo.pdf.

89. Abdul Rahim Janko, "Not a Bad Person," *Witness to Guantánamo*, 2016, http://witnesstoguantanamo.com/interviews/detainees/abdul-rahim-janko-syrian-kurd/. Relatedly, Terri Tomsky, "Guantánamo Lawyers," notes that lawyers for their detainees often plead their clients' cases with appeals to "notions of *American* legitimacy, openness, integrity, and fairness" (36).

90. Donald E. Pease, *The New American Exceptionalism* (Minneapolis: University of Minnesota Press, 2009), 20.

91. Ibid., 173.

92. Gusterson, "Can the Insurgent Speak?," 85.

93. By an alphabetic coincidence that nearly edged him, and his recalcitrance, off the screen, Zemmouri's collection of videos was last on the WTG page when I was researching in summer 2016. The site was subsequently redesigned and reorganized; Zemmouri, "Hunger Strike," *Witness to Guantánamo*, November 2014, http://witnesstoguantanamo.com/wp-content/uploads/2014/11/MosaZemmouri_HungerStrike.pdf; Zemmouri, "Isolation," *Witness to Guantánamo*, November 2014, http://witnesstoguantanamo.com/wp-content/uploads/2014/11/MosaZemmouri_Isolation.pdf; Zemmouri, "Worshippers of America," *Witness to Guantánamo*, November 2014, http://witnesstoguantanamo.com/wp-content/uploads/2014/11/MosaZemmouri_WorshippersofAmerica.pdf.

94. Al Rawi, "Getting Back to Normal," *Witnessing to Guantánamo*, http://witnesstoguantanamo.com/wp-content/uploads/2012/06/BisheralRawi_GettingBacktoNormal.pdf. These ellipses are in the original transcript and correspond to places where al Rawi trails off.

95. Allen Feldman, *Archives of the Insensible: Of War, Photopolitics, and Dead Memory* (Chicago: University of Chicago Press, 2015), 52.

96. Amira Jarmakani, *An Imperialist Love Story: Desert Romances and the War on Terror* (New York: New York University Press, 2015).

97. Elisabeth Weber, "Guantánamo Poems: Guantánamo, Amas, Amat," *Journal of Literature and Trauma Studies* 2, no. 1–2 (Spring/Fall 2013): 159. She notes that there are thousands of censored poems held in a "secure facility" in Virginia, their confinement mirroring that of their authors; Ahuja argues that the existence of the poems "suggest secondary affect economies that take shape in the space of confinement, materializing bodies in geographically and historically specific ways unexpected by the logic of torture" ("Abu Zubaydah," 139).

98. Weber, "Guantánamo Poems," 160.

99. Erin Trapp, "The Enemy Combatant as Poet: The Politics of Writing in *Poems from Guantánamo*," *Postmodern Culture* 21, no. 3 (May 2011).

100. Judith Butler, *Frames of War: When Is Life Grievable?* (London: Verso, 2009), 58, 59.

101. Ariel Dorfman, "Where the Buried Flame Burns," in *Poems from Guantánamo: The Detainees Speak*, ed. Marc Falkoff (Iowa City: University of Iowa Press, 2007), 72.

102. Michael Richardson suggests that of all the writings produced by the detainees, the poems were unique because they were initially created with no intended purpose beyond Guantánamo; Richardson, *Gestures of Testimony: Torture, Trauma, and Affect in Literature* (London: Bloomsbury, 2016), 66.

103. Weber, "Guantánamo Poems," 169.

104. Ahuja, "Abu Zubaydah," 148n46.

105. In the introductory essay, Flagg Miller argues otherwise: "Alert to the many investments at stake in representing and controlling their identities, the poets struggle intelligently, with what resources they have, to engage the sympathy and responses of the broadest possible audience" (15). But then again, Dorfman's concluding essay, "Where the Buried Flame Burns," says they didn't intend the poems for publication, didn't expect that "anyone other than their God would listen or care" (70). However, on the very first page of the book, the matter is settled with a claim of the ultimate significance of the work for its readers: "But now that the poems have been declassified and collected, they offer the world a unique opportunity to hear directly from the detainees themselves about their time in America's notorious prison camp" (1). This is followed by a long list of the degradations that the detainees endured; Miller, "Forms of Suffering in Muslim Prison Poetry," in *Poems from Guantánamo*; Falkoff, "Notes on Guantánamo," in *Poems from Guantánamo*.

106. Falkoff, "Notes on Guantánamo," 5. The press's website for the book includes MP3 recordings of others reading the poems, which seems to be unique to this book.

107. For a methodological reflection on how we might sift meaning from heavily redacted documents, see Nath, "Beyond the Public Eye: On FOIA Documents and the Visual Politics of Redaction," *Cultural Studies–Critical Methodologies* 14, no. 1 (2013): 21–28.

108. In the summer of 2016, Larry Siems began publicly sharing his cautious optimism that Slahi would be released; Siems, "Will the Author of *Guantánamo Diary* Finally Be Set Free?," *Slate*, June 8, 2016, http://www.slate.com/articles/news_and_politics/foreigners/2016/06/mohamedou_ould_slahi_author_of_guantanamo_diary_might_finally_be_freed.html.

109. Mohamedou Ould Slahi, *Guantánamo Diary*, ed. Larry Siems (New York: Little, Brown, and Company, 2015), xi.

110. Ibid., xliv.

111. Ibid., xi, xlvii, xlviii.

112. Ibid.

113. American Civil Liberties Union, "'Guantánamo Diary' Author to Rejoin Family in Mauritania after U.S. Review Board Cleared Way for Release," October 17, 2016, https://www.aclu.org/news/mohamedou-slahi-released-guantanamo-after-14-years-without-charge-or-trial.

114. Khalili, *Time in the Shadows*, 143.

115. Jessica Winegar, "The Humanity Game: Art, Islam, and the War on Terror," *Anthropological Quarterly* 81, no. 3 (Summer 2008): 652. In practice, however, this regard was extended almost exclusively to art that appeared to be both secular and critical of social arrangements in the Middle East, particularly those related to gender (660).

116. Sturken, "Comfort, Irony, and Trivialization," 424.

117. "Still Life with Enemy Combatant: An Exclusive Look at the Artwork of Guantanamo Bay Prisoners," *Slate*, September 2, 2010, http://www.slate.com/articles/news_and_politics/politics/2010/09/still_life_with_enemy_combatant.html. I do not know whether the timing is significant.

118. In addition to the article in *Slate*, see also "Artwork by Inmates from the Guantanamo Bay Detention Center (PHOTOS)," *Huffington Post*, July 13, 2011, http://www.huffingtonpost.com/2011/07/13/artwork-by-inmates-from-t_n_892641.html; Robert Johnson, "The Hauntingly Beautiful Artwork of Guantanamo Detainees," *Business Insider*, March 25, 2013, http://www.businessinsider.com/guantanamo-bay-detainee-artwork-2013-3; "In Pictures: The Art of Guantanamo's Inmates," BBC, June 29, 2011, http://www.bbc.com/news/world-us-canada-13967341. The *Business Insider* article has the most curatorial approach and includes glosses on the emotional and thematic content of the paintings. Captions in the BBC article, by contrast, usually bear no connection to the paintings.

119. Andreja Zevnik, "Life at the Limit: Body, Eroticism, and the Excess," *Theory & Event* 16, no. 4 (2013).

120. Department of Islamic Art, "Figural Representation in Islamic Art," in *Heilbrunn Timeline of Art History* (New York: Metropolitan Museum of Art, 2000–), October 2001, http://www.metmuseum.org/toah/hd/figs/hd_figs.htm; Winegar, "Humanity Game," argues that there is a "politics of knowledge" to this supposed fact about Islamic art, which is often deployed as evidence of Muslim "backwardness" (653). This is an important counter to conventional wisdom; nonetheless, it is true that none of the detainees' drawings or paintings in the corpus I consider here depict living beings.

121. Following Lisa Marie Cacho's critique of the practice of ascribing radical politics to all marginalized subjects, I cannot presume resistance on the part of the detainees or their creative praxis; Cacho, "Conclusion: Racialized Hauntings of the Devalued Dead," in *Social Death: Racialized Rightlessness and the Criminalization of the Unprotected* (New York: New York University Press, 2012).

122. In an apposite analysis of the role of beauty in smoothing over the rough edges of liberalism, Berlant writes, "Even if conscience links you to others by virtue of your recognition of their pain, to the degree that it remains a story of their pain and your compassion for it, the prosthesis of the image keeps you and your world safe from risky transformation. This is to say that right feeling turns the repulsive into the beautiful. Such homeopathy through consumption of the wounded image is central to the liberal aesthetics of political experience, in which atrocity and the therapy culture that now subsumes politics redemptively draw out lessons from lesions, weaving gold out of straw"; Berlant, "Uncle Sam Needs a Wife: Citizenship and Denegation," in *Visual Worlds*, ed. John R. Hall, Blake Stimson, and Lisa Tamiris Baker (London: Routledge, 2005), 31.

123. Berlant, "Uncle Sam Needs a Wife," 16.

124. *Guantanamo Bay Museum of Art and History*, http://www.guantanamo baymuseum.org/?url=welcome.

125. M. Rebekah Otto, "The Guantanamo Bay Museum of Art and History," *Art Practical*, October 27, 2013, http://www.artpractical.com /review/the-guantanamo-bay-museum-of-art-and-history/.

126. Alexis C. Madrigal, "The Imaginary Art Museum at Gitmo," *Atlantic*, August 29, 2012, http://www.theatlantic.com/technology/archive /2012/08/the-imaginary-art-museum-at-gitmo/261740/.

127. "Presenting the Jumah al-Dossari Center for Critical Studies, *Guantanamo Bay Museum of Art and History*, http://www.guantanamo baymuseum.org/?url=critstudies.

128. "Introducing the Tipton Three Exhibition Space," *Guantanamo Bay Museum of Art and History*, http://www.guantanamobaymuseum.org/?url =exhibitions.

129. Carling McManus and Jen Susman, "Arrows to Mecca," *Guantanamo Bay Museum of Art and History*, http://www.guantanamobaymuseum.org/?url =mcmanussusmanwork2.

6. Feeling for Dogs in the War on Terror

1. Ellen Brait, "Canine PTSD: How the U.S. Military's Use of Dogs Affects Their Mental Wellbeing," *Guardian*, November 11, 2015, http:// www.theguardian.com/society/2015/nov/11/canine-ptsd-us-military-working

-dogs; Ryan Cavarero, "Four-Legged Warriors Show Signs of PTSD," *National Public Radio*, March 11, 2013, http://www.npr.org/2013/03/11/173812785/four-legged-warriors-show-signs-of-ptsd.

2. Lisa Gye argues that the relatively low quality of cell phone camera images and videos means that most users employ them to record events that are deemed insignificant, ephemeral, or banal; Gye, "Picture This: The Impact of Mobile Camera Phones on Personal Photographic Practices," *Continuum: Journal of Media & Cultural Studies* 21, no. 2 (June 2007): 283–84.

3. "Video Appears to Show Marine Abusing Puppy," CNN, March 4, 2008, http://www.cnn.com/2008/US/03/04/puppy.marine/.

4. Adriana Cavarero, *Horrorism: Naming Contemporary Violence*, trans. William McCuaig (New York: Columbia University Press, 2011), 76.

5. Jacques Derrida, *The Beast & the Sovereign*, ed. Michel Lisse, Marie-Louise Mallet, and Ginette Michaud, trans. Geoffrey Bennington (Chicago: University of Chicago Press, 2009), 1:138.

6. Battlefield images of dead animals, particularly horses, would have been far more common during conflicts like the American Civil War.

7. One compilation was available at http://youtube.com/watch?v=yQ1mZBfoZXw (the video has since been removed). Reaction videos are often filmed with cameras trained on spectators' faces to document their apparently immediate, reflexive responses to whatever they are seeing.

8. Quoted in "Video Appears to Show Marine Abusing Puppy."

9. Colleen Glenney Boggs, *Animalia Americana: Animal Representations and Biopolitical Subjectivity* (New York: Columbia University Press, 2013), 74.

10. Derrida, *The Animal That Therefore I Am*, ed. Marie-Louise Mallet, trans. David Wills, 3rd ed. (New York: Fordham University Press, 2008), 11.

11. Jean Baudrillard, "The Animals: Territory and Metamorphoses," in *Simulacra and Simulation*, trans. Sheila Faria Glaser (Ann Arbor: University of Michigan Press, 1994), 134.

12. Alice A. Kuzniar, *Melancholia's Dog: Reflections on Our Animal Kinship* (Chicago: University of Chicago Press, 2006), 3. Kelly Oliver has also written affirmatively about the overwhelming emotion provoked when we see ourselves through the eyes of animals; Oliver, *Animal Lessons: How they Teach Us to Be Human* (New York: Columbia University Press, 2009), 125.

13. Donna J. Haraway has argued that an ethical relationship between human and canine species requires meeting dogs as "strangers first," rather than through a screen of "assumptions and stories" about what the animals are; Haraway, *When Species Meet* (Minneapolis: University of Minnesota Press, 2008), 232.

14. Anat Pick, *Creaturely Poetics: Animality and Vulnerability in Literature and Film* (New York: Columbia University Press, 2011), 15.

15. Anne McClintock, "Paranoid Empire: Specters from Guantánamo and Abu Ghraib," in *States of Emergency: The Object of American Studies*, ed. Russ Castronovo and Susan Gillman (Chapel Hill: University of North Carolina Press, 2009), 89.

16. This is not to imply that every American loves or even cares about the suffering of dogs. Surely there are many people who find them uninteresting, frightening, or even disgusting. In this instance, I am speaking not as much of individuals as of broader historical and cultural trends.

17. In her analysis of the phenomenon of "cruel optimism," Lauren Berlant argues that "cruelty is the 'hard' in a hard loss. It is apprehensible as an affective event in the form of a beat or a shift in the air that transmits the complexity and threat of relinquishing what is different about the world." She writes further about the need to provide an account of the "process of knotty tethering to objects, scenes, and modes of life that generate so much overwhelming yet sustaining negation"; Berlant, *Cruel Optimism* (Durham: Duke University Press, 2011), 52.

18. Berlant, "Compassion (and Withholding)," in *Compassion: The Culture and Politics of an Emotion*, ed. Lauren Berlant (New York: Routledge, 2004), 10.

19. Rebecca Wanzo, *The Suffering Will Not Be Televised: African American Women and Sentimental Political Storytelling* (Albany: SUNY Press, 2009), 3.

20. The earliest humane societies focused on protecting abused animals and neglected children.

21. Marjorie Garber, *Dog Love* (New York: Simon & Schuster, 1996), 233.

22. Katherine C. Grier, *Pets in America: A History* (Chapel Hill: University of North Carolina Press, 2006), 153. The anti-cruelty movement in England also had its own class dynamics and generally emphasized criminalizing abuse of animals by the poor, but not more aristocratic pursuits like hunting for sport; Kathleen Kete, "Animals and Ideology: The Politics of Animal Protection in Europe," in *Representing Animals*, ed. Nigel Rothfels (Bloomington: Indiana University Press, 2002), 26–27.

23. Grier, *Pets in America*, 117.

24. Ibid., 131.

25. Ibid., 130–31.

26. Ibid., 174.

27. Judith Butler has written, in a different context, of the "differential distribution of grievability" among various kind of humans; Butler, *Frames of War: When Is Life Grievable?* (London: Verso, 2009), 24. Certainly, human existence has long been attended by animal deaths. In the nineteenth century, the significance of these deaths began to change in contradictory ways. Comparing modern forms of death-dealing to animals with premodern rituals, Jean Baudrillard claims that with the wane of animal sacrifice,

"we have made of [animals] a racially inferior world, no longer even worthy of our justice, but only of our affection and social charity, no longer worthy of punishment and of death, but only of experimentation and extermination like meat from the butchery"; Baudrillard, "Animals," 135.

28. "History of the ASPCA," *ASPCA*, last modified March 23, 2016, https://www.aspca.org/about-us/history-of-the-aspca.

29. Susan J. Pearson, *The Rights of the Defenseless: Protecting Animals and Children in Gilded Age America* (Chicago: University of Chicago Press, 2011).

30. Raymond Madden has described this relationship as "co-constitutive" of both species at once; Madden, "Imagining the Greyhound: 'Racing' and 'Rescue' Narratives in a Human and Dog Relationship," *Continuum: Journal of Media & Cultural Studies* 24, no. 4 (August 2010): 503. Of course, this sympathy for animals, dogs in particular, is not evenly allocated, and increasingly, certain breeds like pit bulls are criminalized by local and state governments. These measures coexist, as Colin Dayan points out, with newly intensified affective investments in those creatures deemed acceptably loveable; Dayan, *The Law Is a White Dog: How Legal Rituals Make and Unmake Persons* (Princeton: Princeton University Press, 2011).

31. Maneesha Deckha, "Welfarist *and* Imperial: The Contributions of Anticruelty Laws to Civilizational Discourse," *American Quarterly* 65, no. 3 (2013): 515. Deckha's work points to a troubling impasse wherein the anti-cruelty practices that afforded a measure of protection to animals also perpetuated an imperialist agenda against non-Western humans.

32. Janet M. Davis, "Cockfight Nationalism: Blood Sport and the Moral Politics of American Empire and Nation Building," *American Quarterly* 65, no. 3 (September 2013): 549–74.

33. Deckha, "Welfarist *and* Imperial," 522.

34. Ibid., 516.

35. On the use of police dogs to terrorize African American communities, see Tyler Wall, "'For the Very Existence of Civilization': The Police Dog and Racial Terror," *American Quarterly* 68, no. 4 (December 2016): 861–82.

36. Oliver, "Ambivalence Toward Animals and The Moral Community," in *Invited Symposium: Feminists Encountering Animals, Hypatia* 27, no. 3 (July 2012): 493–98. Oliver takes this as evidence that African Americans are excluded from the moral community of their white counterparts.

37. Megan H. Glick, "Animal Instincts: Race, Criminality, and the Reversal of the 'Human,'" *American Quarterly* 65, no. 3 (September 2013): 640. Glick moves toward the surprising conclusion that these very discourses contributed to the "lenience" of the sentence Vick ultimately received. This, she suggests, reflects the "discourses of banality of criminality, of the banality of monstrosity, that shape popular imaginings of black masculinity" (656).

38. Vick's relatively seamless resumption of his football career after his release from federal prison reveals that other prized attributes—like fame and athletic prowess—can ultimately override public animosity toward perpetrators of animal cruelty.

39. This is different than the case of World War II, for example, when civilians were called upon to do specific things—from rationing to recycling to buying war bonds—that directly contributed to the war effort.

40. Chris Hedges, *War Is a Force That Gives Us Meaning* (New York: Random House, 2003), 167.

41. "Seeing my dog the day I got back from Afghanistan," YouTube video, 0:38, posted by Itschmidt02, April 11, 2008, http://www.youtube.com/watch?v=ysKAVyXioJ4. In March 2017, Gracie's beloved owner posted an update to let her fans know that she was still doing well at the age of fourteen and that he'd be "retiring from the Guard after twenty years to spend more time with my dog"; accessed August 1, 2017.

42. One of the many places it appeared is Anahad O'Connor, "A Dog at the Funeral Captured on Video," *Well* (blog), *New York Times*, August 26, 2011, http://well.blogs.nytimes.com/2011/08/26/a-dog-at-the-funeral-captured-on-video/.

43. There is also a subgenre of war memoirs from Iraq and Afghanistan about these relationships. See, for example, Jay Kopelman, *From Baghdad to America: Lessons from a Dog Named Lava* (New York: Skyhorse, 2008); Mike Dowling, *Sergeant Rex: The Unbreakable Bond between a Marine and His Military Working Dog* (New York: Atria, 2011); Mike Ritland, *Trident K9 Warriors: My Tale from the Training Ground to the Battle Field with Elite Navy SEAL Canines*, with Gary Brozek (New York: St. Martin's, 2013). Reflecting on the possibility that sensitive dogs might also be undone by the horrors of wartime, Kopelman diagnoses his dog Lava with PTSD (110), seeing echoes of his own trauma in the dog as he notes "the way [Lava] so often just loses it for no apparent reason" (*From Baghdad to America*, 101).

44. Purnima Bose, "The Canine-Rescue Narrative, Civilian Casualties, and the Long Gulf War," in *In/Visible War: The Culture of War in Twenty-First-Century America*, ed. Jon Simons and John Louis Lucaites (New Brunswick, N.J.: Rutgers University Press, 2017), 193–210. Bose argues that the dogs "displace Iraqi civilians as affective objects of identification in need of rescue by American soldiers" (202).

45. The movie adaptation of *Thank You for Your Service* includes an added subplot wherein one of the returned soldiers rescues a badly injured pit bull from an illegal dogfight.

46. Stories about these programs are myriad and far too numerous to list exhaustively. Some examples include Chris Colin, "How Dogs Can Help

Veterans Overcome PTSD," *Smithsonian* (July 2012), http://www.smithsonianmag.com/science-nature/How-Dogs-Can-Help-Veterans-Overcome-PTSD-160281185.html; Karen Jones, "Veterans Helped by Healing Paws," *New York Times*, November 10, 2008, http://www.nytimes.com/2008/11/11/giving/11DOGS.html; "Soldier's Best Friend," *Maryland Morning*, March 19, 2012, http://mdmorn.wordpress.com/2012/03/19/319124/; Mary Murray, "Healing Soldiers, One Dog at a Time," *NBC News*, January 30, 2012, http://dailynightly.nbcnews.com/_news/2012/01/30/10271218-healing-soldiers-one-dog-at-a-time?lite.

47. This logic also underpins programs that use dog-training to rehabilitate prisoners.

48. Arwa Damon, "Stray Dogs Being Killed in Baghdad," CNN, February 12, 2009, http://www.cnn.com/2009/WORLD/meast/02/12/baghdad.dogs/.

49. Pearson, *Rights of the Defenseless*, 64.

50. The dog was working with the British Special Forces, which had been engaged in a "fatal firefight" on December 23, 2013. It is unclear how the dog got separated from his humans; Ernesto Londoño, "Military Dog Captured by Taliban Fighters, Who Post Video of Their Captive," *Washington Post*, February 6, 2014, http://www.washingtonpost.com/world/national-security/military-dog-captured-by-taliban-fighters-who-post-video-of-their-captive/2014/02/06/c8d0f8f0-8f44-11e3-84e1-27626c5ef5fb_story.html. See also "Taliban Claims to Capture U.S. Military Dog," *Military Video Center*, 4:03, posted by vlogger, February 6, 2014, http://www.military.com/video/operations-and-strategy/afghanistan-conflict/taliban-claims-to-capture-military-dog/3162004694001.

51. Garber, writing of the melodramatic trope of the homeless dog in film and literature, contends that these lost creatures are poignant because we tend to associate dogs closely and essentially with home and imagine them bereft if they do not have one. Here, that association is overlaid with national concern; Garber, *Dog Love*, 40.

52. Londoño, "Military Dog," para. 1, 2.

53. Ibid., para. 18.

54. In a phenomenological consideration of the challenges inherent in conceptualizing animal perception, Eric Luft outlines the difficulty of knowing how animals perceive and the necessity of trying to imagine their sensations and feelings. I suggest that, especially in the case of companion animals, we do this kind of imaginative work all the time, with little resistance, even though we cannot know whether or not our imaginings are accurate; Eric v.d. Luft, "Bullough, Pepper, Merleau-Ponty and the Phenomenology of Perceiving Animals," *Evental Aesthetics* 2, no. 2 (2013): 111–23.

55. In one episode of the HBO series *Generation Kill*, there is an intimation that some marines, desperate for supplies that never seem to arrive, shot and ate a stray dog. The show emphasizes the extreme lack that constitutes their daily existence; they are underfed, underslept, and consistently deprived. Still, within the diegesis, this behavior is marked as aberrant, one of the few kinds of conduct that other marines recognize as crossing a line.

56. "Military Animals," *ASPCA*, 2013, http://www.aspca.org/about-us/aspca-policy-and-position-statements/military-animals.

57. Wendy Kozol describes the "affective tug" of sentimental fantasies about U.S. military personnel; Kozol, "Battlefield Souvenirs and Ethical Spectatorship," text of paper presented at Feeling Photography Conference, University of Toronto, Toronto, Ont., October 17, 2009.

58. Glick, "Animal Instincts," 647.

59. Robert G. W. Kirk's history of ideas about the subjectivity of mine-detecting dogs in World War II is both meticulous and fascinating; Kirk, "In Dogs We Trust? Intersubjectivity, Response-Able Relations, and the Making of Mine Detector Dogs," *Journal of the History of the Behavioral Sciences* 50, no. 1 (Winter 2014): 1–36.

60. Boggs, *Animalia Americana*, 66.

61. Ibid., 69–75. Boggs argues that this bestialization engenders a representational crisis by blurring distinctions between subjects and "nonsubjects," while the attacking dog straddles this line.

62. Neel Ahuja, "Abu Zubaydah and the Caterpillar," *Social Text* 29, no. 1 (Spring 2011): 127–49.

63. Oliver continues, "Either way, animals serve to shore up the boundaries of what we consider the proper—and properly human—moral community. Either way, animals (and humans associated with them) are outside of moral and civil law. Either way, they are not considered moral agents; or they are liminal moral agents who can be disposed of outside of any civil or moral codes"; Oliver, "Ambivalence," 495.

64. There are also indications that he commanded the dog to lick peanut butter off the bodies of his fellow soldiers.

65. Elisabeth de Fontenay, *Without Offending Humans: A Critique of Animal Rights*, trans. Will Bishop (Minneapolis: University of Minnesota Press, 2012), 68.

66. This impression is heightened by the way that many of the photos are framed, with the human handlers either partially or entirely excluded from view.

67. Boggs, "American Bestiality: Sex, Animals, and the Construction of Subjectivity," *Cultural Critique* 76 (Fall 2010): 115.

68. The full text of the bill, "Canine Members of the Armed Forces Act," S. Res. 2134, 112th Cong. (2012), is available at *Govtrack.us*, 2012, https://www

.govtrack.us/congress/bills/112/s2134. For the FY 2013 defense authorization, see "An Act," H.R. Res. 4310, 112th Cong. 2nd Session. (2013), http://www.gpo.gov/fdsys/pkg/BILLS-112hr4310eh/pdf/BILLS-112hr4310eh.pdf. For information on advocacy groups' support for the bill, see "Military Dogs: Groups Fight to Elevate Military Dogs' Status from 'Equipment' to 'Members of Armed Forces,'" *Huffington Post*, June 7, 2012, http://www.huffingtonpost.com/2012/06/07/military-dogs-adoption_n_1576426.html.

69. H.R. 2810, "National Defense Authorization Act for Fiscal Year 2018," June 25, 2017, accessed August 28, 2017, https://www.congress.gov/bill/115th-congress/house-bill/2810/text#toc-H1E00E26E5B154426893BE4FE4F2B1D75.

70. For the copies of the General Orders, see United States Central Command, "General Order Number 1A (GO-1A)," *Cdn.Factcheck*, December 19, 2000, http://cdn.factcheck.org/UploadedFiles/2013/08/GeneralOrderGO-1A.pdf; United States Central Command, "General Order Number 1B (GO-1B)," *Cdn.Factcheck*, March 13, 2006, http://cdn.factcheck.org/UploadedFiles/2013/08/General-Order-1B1.pdf; and United States Central Command, "General Order Number 1C (GO-1C)," *FedBiz Opps*, May 21, 2013, https://www.fbo.gov/utils/view?id=aa41a22810e650c7886be968c2b2845f, respectively.

71. Of course, the military is not the only American institution that holds contradictory views about animals. Indeed, there often exists a gap between professed care for animals and our actual treatment of them. For example, many animal-rights activists and scholars identify hypocrisy in the twinned phenomena of lavish treatment of pets and the killing and consumption of livestock for food. Even among privileged species like cats and dogs, we often allocate care, love, and protection differentially. Yet the operative discourses are durable and expansive enough to abide this kind of inconsistency.

72. General Order 1-C provides an additional rationale for this as a measure against the spread of zoonotic disease.

73. In distinction to the seeming heartlessness of the General Orders, Jay Kopelman's wildly popular *From Baghdad to America: Life Lessons from a Dog Named Lava* is replete with stories from military personnel forced to leave their newly beloved animals behind when returning home and his own struggle to care for, rescue, and repatriate Lava to the United States. Stories like these about the ties between soldiers and animals situate the state, or the institution of the military, as the entity that interferes in the formation or maintenance of these bonds. In turn, this enables further idealization of the beings involved and the relationships between them as a touching act of rebellion. In turn, this portrayal intensifies the fantasy.

74. In the spring of 2017, an army veteran, Marinna Rollins, and her boyfriend shot a PTSD service dog and posted the video on Facebook. A few weeks later, she was found dead in an apparent suicide. The story did not receive as much coverage as Motari, Encarnacion, and the puppy, but a preliminary review of the comments posted in response to news stories about the incident reveals a similar dynamic at play. The nexus of gender, cruelty, mediation, and trauma is worthy of far more extended consideration than space permits here; CBS/AP, "Veteran Who Killed Her Service Dog on Camera Found Dead," *CBS News*, May 8, 2017, https://www.cbsnews.com/news/marinna-rollins-veteran-who-killed-her-service-dog-found-dead/.

75. Various versions of alleged apologies, some more contrite than others, from David Motari and his mother have been circulating online since shortly after the video became public. Sourcing these statements satisfactorily has been impossible, and to the best of my knowledge, no major news organizations have quoted or corroborated then. In the absence of definitive proof of their authorship, we must assume that at least one is a forgery. The existence of such fake apologies reveals a wish that Motari would have expressed remorse.

One such letter is rather defensive; "David Motari's Sorry Letter!," *Newgrounds*, March 8, 2008, http://www.newgrounds.com/bbs/topic/873384, accessed August 1, 2014). In this piece, Motari avers that no one can know what it is like to be deployed and that he killed the dog because it was sick and military personnel are not allowed to have dogs. A different, more contrite document offers an unequivocal and abject apology and a plea that the blame be directed at him alone, not his family or the Marine Corps (David Motari, "A Letter to Everyone," *Care2*, March 10, 2008, http://www.care2.com/c2c/share/detail/666896). This document restores, at least partially, the fantasy of the noble servicemember. Its authorship is apocryphal, and if Motari did not write it, the fictionalizing of his remorse reveals the intensity of our attachment to that fantasy of contrition and, hence, the upstanding morality of American military personnel.

76. On the legal discourse of wantonness in general, see Dayan, *Law Is a White Dog.*

77. Asma Abbas, "Voice Lessons: Suffering and the Liberal Sensorium," *Theory & Event* 13, no. 2 (2010): para. 1.

78. Berlant, "Uncle Sam Needs a Wife: Citizenship and Denegation," in *Visual Worlds*, ed. John R. Hall, Blake Stimson, and Lisa Tamiris Becker (London: Routledge, 2005), 32.

79. Brett Mills, "Television Wildlife Documentaries and Animals' Right to Privacy," *Continuum: Journal of Media & Cultural Studies* 24, no. 2 (April 2010): 193–202.

80. Justin Arnold and Yoshiaki Nohara, "Monroe Marine Kicked Out of Corps for Videotaped Puppy Tossing Incident," *HeraldNET*, July 11, 2008, http://www.heraldnet.com/news/monroe-marine-kicked-out-of-corps-for-videotaped-puppy-tossing-incident.

81. Sianne Ngai, *Ugly Feelings* (Cambridge, Mass.: Harvard University Press, 2005), 1.

82. Cavarero, *Horrorism*, 8. She writes, "Gripped by revulsion in face of a form of violence that appears more inadmissible than death, the body reacts as if nailed to the spot, hairs standing on end." Reflecting on the human encounter with the animal form, Marcus Bullock observes that "the muteness of the animal also imposes a moment of muteness on us"; Bullock, "Watching Eyes, Seeing Dreams, Knowing Lives," in *Representing Animals*, ed. Nigel Rothfels (Bloomington: Indiana University Press, 2002), 99.

83. Deckha, "Welfarist *and* Imperial," 519.

84. Wanzo, *The Suffering Will Not Be Televised*, 16.

85. Boggs has argued that humans like the Abu Ghraib detainees and animals like dogs occupy structurally similar and vulnerable positions in American culture; Boggs, "American Bestiality," 120.

86. Carrie A. Rentschler, "Witnessing: U.S. Citizenship and the Vicarious Experience of Suffering," *Media, Culture, & Society* 26, no. 2 (2004): 299. Rentschler suggests that this kind of mediated witnessing is integral to the experience of American citizenship.

87. Marie-José Mondzain, "Can Images Kill?," trans. Sally Shafto, *Critical Inquiry* 36 (Autumn 2009): 25.

88. Already, as Ann Pellegrini and Jasbir Puar note, we are learning that affect might be something beyond language or cognition; Pellegrini and Puar, "Affect," *Social Text* 27, no. 3 (Fall 2009): 37.

89. The question of animal utterances in the face of suffering points to a possible way around often intractable debates about whether traumatic events can ever be fully communicated. Pick reframes those debates with an important reminder that "trauma's language of unspeakability is the sole preserve of the human because only those who speak may be said to experience the unspeakable"; Pick, *Creaturely Poetics*, 143.

90. Berlant, "Uncle Sam Needs a Wife," 32.

91. For a critique of the limits of animal rights discourse, see Dominick LaCapra, *History and Its Limits: Human, Animal Violence* (Ithaca: Cornell University Press, 2009), 152.

92. Baudrillard writes that the speechlessness of animals can strengthen our intimacy with them; Baudrillard, "Animals," 137.

93. John Berger, "Why Look at Animals?," in *About Looking* (New York: Pantheon, 1980), 5.

94. Tricia Bishop, "Mistrial in Case against Twins Accused of Burning Pit Bull," *Baltimore Sun*, February 7, 2011, http://articles.baltimoresun.com/2011-02-07/news/bs-md-ci-phoenix-mistrial-20110207_1_travers-and-tremayne-johnson-caroline-griffin-jennifer-rallo.

95. Belatedly, the widespread enmity toward the brothers seemed vindicated when, in the spring of 2012, it was reported that both of them had found their ways back into police custody for different cases, Tremayne on drug charges and Travers for attempted murder (to which he pled guilty) and burglary.

96. See, for example, LaCapra, *History and Its Limits*, 152–53. LaCapra detects a flaw in many animal "rights" discourses, arguing that they may actually reinscribe a human-animal hierarchy. He also argues that when we reinforce a separation between humans and animals, we reduce them either to "infra-ethical" considerations or transform them into "supra-ethical" beings (153), both of which are inimical to the kind of serious ethical deliberation that might truly transform intraspecies relationships.

97. Even if it had survived, there would be no satisfactory way to make amends. Our culture does not have established practices for apologizing to or asking forgiveness from animals. Boggs, in "American Bestiality," observes that we "reserve . . . apologies exclusively for human beings, because only they participate as subjects in the structural and representational schemes that make up the symbolic order" (98–99).

Dayan's *The Law Is a White Dog* is a haunting description of the place of dogs in the law. She writes, "Bearing the brunt of the most extreme denigration and at the same time enjoying the greatest affection, dogs are in a precarious space—so ignoble that they are the least cared for and so good that they are the most loved. In the law, this seeming paradox becomes the working definition of dog: so empty of substance that it can accrue to itself all kinds of projections. The excesses of sentiment are as dangerous to the lives of dogs as their cruel treatment by humans" (232).

98. De Fontenay argues for a reconfiguration of the ethical and political framework that "opposes subjective, powerless, and particular pity with objective, universal, and efficient law, and the animal question must once again become a social question"; de Fontenay, *Without Offending Humans*, 131. This is not an uncomplicated or unproblematic proposition, but it does highlight the emotional insufficiency of laws meant to prohibit and punish cruelty to animals.

99. By all accounts, this advertising campaign has been wildly successful; see Stephanie Strom, "Ad Featuring Singer Proves Bonanza for the A.S.P.C.A," *New York Times*, December 25, 2008, http://www.nytimes.com/2008/12/26/us/26charity.html.

100. Deborah Rose implores us to attend to the "implications of the howling of living beings in a time of death"; Rose, "What If the Angel of History Were a Dog?," *Cultural Studies Review* 12, no. 1 (March 2006): 67. Kalpana Rahita Seshadri sets out a provocative alternative position in defense of silence, both human and animal: the possibility that it might function as a creative "space for opposition." She further contends that the remedy for animalization—whether of animals that are human or not—is not simply a matter of endowing them with speech via "resemanticization"; Seshadri, *HumAnimal: Race, Law, Language* (Minneapolis: University of Minnesota Press, 2012), 21, 33.

Of course, there are instances when animals need human interlocutors for practical reasons; see, for example, Rick Rojas, "Abused Dogs and Cats Now Have a (Human) Voice in Connecticut Courts," *New York Times*, August 27, 2017, https://www.nytimes.com/2017/08/27/nyregion/animal-abuse-connecticut-court-advocates.html.

101. Boggs, *Animalia Americana*, 24.

102. Hélène Cixous dramatizes the impossibility of such movement in "Stigmata, or Job the Dog," trans. Eric Prenowitz, in *Stigmata: Escaping Texts* (London: Routledge, 2005), 149–58.

103. On the value of dwelling in unpleasant feelings, see Ann Cvetkovich, *Depression: A Public Feeling* (Durham: Duke University Press, 2012).

CONCLUSION: A RADICAL AND UNSENTIMENTAL ATTENTION

1. Judith Butler, *Senses of the Subject* (New York: Fordham University Press, 2015), 12. Relatedly, Danielle Celermajer writes, "Insofar as we are attuned to the suffering or injustice that appears before us, not as a performance of an ethical precept, but as a response to the phenomenon itself, our future orientation towards injustice will be nourished"; Celermajer, "Unsettling Memories and the Irredeemable," *Theory & Event* 19, no. 2 (April 2016), under "The Messianic Rubble."

2. This resonates with Dominick LaCapra's ideal of "empathic unsettlement," which he posits as an alternative to problematic forms of sympathy that rely on false identification with the other or appropriation of their suffering. Empathic unsettlement, as I understand it, is a way of keeping otherness intact within a framework of deep and profound compassion; LaCapra, *History and Its Limits: Human, Animal, Violence* (Ithaca: Cornell University Press, 2009), 65–66.

3. My thinking here is inspired, in part, by Sara Ahmed's endorsement of the "feminist killjoy"; Ahmed, *The Promise of Happiness* (Durham: Duke University Press, 2010).

4. On failure as a strategy, see Jack Halberstam, *The Queer Art of Failure* (Durham: Duke University Press, 2011).

5. The thousands of books published on "anger management" testify to this.

6. Butler, *Senses of the Subject*, 16. This exposure is also central to the process of subject formation, a paradoxical process that Butler describes as follows: "I am not affected just by this one other or set of others, but by a world in which humans, institutions, and organic and inorganic processes all impress themselves upon this me who is, at the outset, susceptible in ways that are radically involuntary" (7).

INDEX